VOLUME ONE HUNDRED AND SIX

CURRENT TOPICS IN DEVELOPMENTAL BIOLOGY

Endocrine Gland Development and Disease

CURRENT TOPICS IN DEVELOPMENTAL BIOLOGY

"A meeting-ground for critical review and discussion of developmental processes"
A.A. Moscona and Alberto Monroy (Volume 1, 1966)

VOLUME ONE HUNDRED AND SIX

CURRENT TOPICS IN DEVELOPMENTAL BIOLOGY

Endocrine Gland Development and Disease

Edited by

PAUL THOMAS
School of Molecular and Biomedical Science
University of Adelaide, Adelaide
South Australia, Australia

ELSEVIER

AMSTERDAM • BOSTON • HEIDELBERG • LONDON
NEW YORK • OXFORD • PARIS • SAN DIEGO
SAN FRANCISCO • SINGAPORE • SYDNEY • TOKYO
Academic Press is an imprint of Elsevier

Academic Press is an imprint of Elsevier
525 B Street, Suite 1800, San Diego, CA 92101-4495, USA
225 Wyman Street, Waltham, MA 02451, USA
The Boulevard, Langford Lane, Kidlington, Oxford, OX5 1GB, UK
32 Jamestown Road, London NW1 7BY, UK
Radarweg 29, PO Box 211, 1000 AE Amsterdam, The Netherlands

First edition 2013

ISBN: 978-0-12-416021-7
ISSN: 0070-2153

CONTENTS

CONTRIBUTORS

Fabio Avolio
Univ. Nice Sophia Antipolis, iBV, UMR 7277; Inserm, iBV, U1091; and CNRS, iBV, UMR 7277, Nice, France

Nouha Ben-Othman
Univ. Nice Sophia Antipolis, iBV, UMR 7277; Inserm, iBV, U1091; and CNRS, iBV, UMR 7277, Nice, France

Michelle L. Brinkmeier
Department of Human Genetics, University of Michigan Medical School, Ann Arbor, Michigan, USA

Sally A. Camper
Department of Human Genetics, University of Michigan Medical School, Ann Arbor, Michigan, USA

Elise Cau
Université de Toulouse, UPS, Centre de Biologie du Développement (CBD), and CNRS, CBD UMR 5547, Toulouse, France

Dominic Cavlan
Centre for Endocrinology, William Harvey Research Institute, Barts and the London School of Medicine and Dentistry, Queen Mary, University of London, London, United Kingdom

Katy Cogger
Centre for Endocrinology, William Harvey Research Institute, Barts and the London School of Medicine and Dentistry, Queen Mary, University of London, London, United Kingdom

Patrick Collombat
Univ. Nice Sophia Antipolis, iBV, UMR 7277; Inserm, iBV, U1091; and CNRS, iBV, UMR 7277, Nice, France

Monica Courtney
Univ. Nice Sophia Antipolis, iBV, UMR 7277; Inserm, iBV, U1091; and CNRS, iBV, UMR 7277, Nice, France

Shannon W. Davis
Department of Biological Sciences, University of South Carolina, Columbia, South Carolina, USA

Noémie Druelle
Univ. Nice Sophia Antipolis, iBV, UMR 7277; Inserm, iBV, U1091; and CNRS, iBV, UMR 7277, Nice, France

Buffy S. Ellsworth
Department of Physiology, Southern Illinois University, Carbondale, Illinois, USA

Henrik Fagman
Department of Pathology, Sahlgrenska University Hospital, Göteborg, Sweden

Biljana Faurite
Univ. Nice Sophia Antipolis, iBV, UMR 7277; Inserm, iBV, U1091; and CNRS, iBV, UMR 7277, Nice, France

Nastaran Foyouzi
Department of Obstetrics and Gynecology, University of Michigan Medical School, Ann Arbor, Michigan, USA

Peter Gergics
Department of Human Genetics, University of Michigan Medical School, Ann Arbor, Michigan, USA

Elisabet Gjernes
Univ. Nice Sophia Antipolis, iBV, UMR 7277; Inserm, iBV, U1091; and CNRS, iBV, UMR 7277, Nice, France

Leonardo Guasti
Centre for Endocrinology, William Harvey Research Institute, Barts and the London School of Medicine and Dentistry, Queen Mary, University of London, London, United Kingdom

Claire Hughes
Centre for Endocrinology, William Harvey Research Institute, Barts and the London School of Medicine and Dentistry, Queen Mary, University of London, London, United Kingdom

Harshini Katugampola
Centre for Endocrinology, William Harvey Research Institute, Barts and the London School of Medicine and Dentistry, Queen Mary, University of London, London, United Kingdom

Peter King
Centre for Endocrinology, William Harvey Research Institute, Barts and the London School of Medicine and Dentistry, Queen Mary, University of London, London, United Kingdom

Eirini Meimaridou
Centre for Endocrinology, William Harvey Research Institute, Barts and the London School of Medicine and Dentistry, Queen Mary, University of London, London, United Kingdom

Louise Metherell
Centre for Endocrinology, William Harvey Research Institute, Barts and the London School of Medicine and Dentistry, Queen Mary, University of London, London, United Kingdom

Amanda H. Mortensen
Department of Human Genetics, University of Michigan Medical School, Ann Arbor, Michigan, USA

Mikael Nilsson
Sahlgrenska Cancer Center, Institute of Biomedicine, University of Gothenburg, Göteborg, Sweden

Caroline Alayne Pearson
Department of Neurobiology and Broad Center for Regenerative Medicine and Stem Cell Research, David Geffen School of Medicine at UCLA, Los Angeles, California, USA

María Inés Peréz Millan
Department of Human Genetics, University of Michigan Medical School, Ann Arbor, Michigan, USA

Anja Pfeifer
Univ. Nice Sophia Antipolis, iBV, UMR 7277; Inserm, iBV, U1091; and CNRS, iBV, UMR 7277, Nice, France

Marysia Placzek
MRC Centre for Developmental and Biomedical Genetics, and Department of Biomedical Science, University of Sheffield, Sheffield, United Kingdom

Dora Sapède
Université de Toulouse, UPS, Centre de Biologie du Développement (CBD), and CNRS, CBD UMR 5547, Toulouse, France

Vanessa Schade
Department of Human Genetics, University of Michigan Medical School, Ann Arbor, Michigan, USA

Paul Thomas
School of Molecular and Biomedical Science, University of Adelaide, Adelaide, South Australia, Australia

Andhira Vieira
Univ. Nice Sophia Antipolis, iBV, UMR 7277; Inserm, iBV, U1091; and CNRS, iBV, UMR 7277, Nice, France

Dagmar Wilhelm
Department of Anatomy and Developmental Biology, Monash University, Clayton, Victoria, Australia

Jennifer X. Yang
Department of Anatomy and Developmental Biology, Monash University, Clayton, Victoria, Australia

Rebecca Yates
Barts and the London School of Medicine and Dentistry, Queen Mary, University of London, London, United Kingdom

PREFACE

Endocrine organs are a diverse bunch. Linked, it appears, only by their ability to exert physiological control of target organs through hormonal secretions, they arise from all three germ layers during development and are located at many sites throughout the body, often at some distance from their targets. While their function in adults is generally well characterized, understanding of the genes and pathways that control their development has for the most part lagged behind. This is partly due to the difficultly in studying endocrine organ development, as they are small and relatively inaccessible and therefore lack the advantages of classic developmental systems such as the limb where developmental defects can be readily characterized by visual inspection. Nevertheless, in recent years, there has been considerable progress in our understanding of endocrine organ development and disease, which has been driven by at least two factors. The first stems from research into the genetic basis of congenital endocrinopathies in human, many of which are relatively simple to diagnose due to the absence or depletion of the relevant hormones in the bloodstream. Advances in sequencing technology have greatly facilitated the identification of causative genes for such disorders, thereby providing valuable insight into endocrine organ development in humans and in model species such as the mouse. The second key factor relates to the enormous interest in the use of stem cells for therapy. Endocrine organs are attractive targets for regenerative medicine, as they are composed of relatively few highly specialized cell types each of which secretes a cognate hormone. Furthermore, diseases resulting from chronic hormonal imbalances in some cases are highly prevalent (e.g., diabetes). While efficient directed differentiation of induced pluripotent cells and embryonic stem cells into terminally differentiated hormone-secreting cells remains a challenging task, this can be greatly facilitated by a more detailed knowledge of the signaling factors and transcriptional networks that coordinate endocrine organ development *in vivo*.

Thus, the invitation to compile a series of review articles for *Current Topics in Developmental Biology* on *Endocrine Organ Development and Disease* was both timely and welcome. My approach to the task was to seek reviews from prominent developmental biologists with an interest in each of the major endocrine organs, asking them to provide their perspective on recent developments in their field. I hope that you agree that all of the authors have provided insightful and comprehensive reviews and I would like to thank each of them for their contribution to this volume. In Chapter 1, Davis et al. provide an overview of the pituitary, a particularly interesting endocrine organ due to its ability to control multiple target organs through regulated hormonal secretion from five terminally differentiated "trophic" cell types. The authors show how our understanding of pituitary development has advanced considerably in recent years, particularly with respect to induction of Rathke's pouch (the primordium of the anterior pituitary) by a ventral diencephalic organizer, transcriptional control of progenitor cell differentiation, and identification of stem/progenitor cells in the adult gland. Although the pituitary was traditionally considered to be the "master endocrine gland," it is now known that it has its own master—the hypothalamus—which regulates pituitary function via secretion of regulatory hormones into the portal vasculature system that connects these structures. As explained by Pearson and Placzek in Chapter 2, our understanding of hypothalamic development has generally lagged behind other CNS regions due to its complexity, small size, and lack of cell type markers. Nevertheless, recent analysis of hypothalamic development in several species including mouse, chick, and zebrafish has revealed the key signaling pathways that induce and pattern the hypothalamus and provided a basic understanding of the transcription factor code that specifies neuronal subtypes.

Chapter 3 focuses on one of the most intriguing questions related to endocrine organ development and disease: how can a single primordium, the urogenital ridge, respond to sex-specific signals to generate different endocrine organs in males (testis) versus females (ovary)? Of course, the simple answer is that, at least in mammals, only males express the Y-linked testis determining transcription factor *Sry*. However, although *Sry* was identified over 20 years ago, it is only in recent years that detailed molecular genetic understanding of testis and ovary organogenesis has emerged. As outlined by Wilhelm et al. in Chapter 3, it is now clear that *Sry* has a surprisingly tenuous hold over male development and that its direct target *Sox9* is the true workhorse of testis differentiation. It is also now apparent that ovarian specification and development is an active process and that cross-repression pathways canalize gonad differentiation during embryogenesis and, surprisingly, remain active throughout life.

In Chapter 4, Nilsson and Fagman provide an update on thyroid organogenesis and disease. As is the case for all reviews in this volume, greater understanding of progenitor cell specification and differentiation has emerged through detailed analysis of constitutive and conditional mouse mutants. A particularly interesting recent study outlined in this review is the demonstration that transient overexpression of two transcription factors that are critical for thyroid development *in vivo* (NKX2.1 and PAX8) in murine embryonic stem cells can induce differentiation into functional thyroid "organoids." This discovery provides a very useful *in vitro* system for thyroid development and is an exciting platform for the development of cell therapy for hypothyroidism.

Chapter 5 focuses on the pineal gland which is one of the less well-characterized endocrine organs. This small neuroendocrine organ is primarily known for its important role in controlling the sleep–wake cycle through secretion of melatonin. Unlike most other endocrine organs, pineal development and function has been studied most extensively in lower vertebrates where it is directly photosensitive. However, as nicely outlined by Sapède and Cau, melatonin is implicated is several human disorders including cancer, autism, and depression, underlining the importance of pineal function in health and disease.

In contrast to the pineal gland, the development of the mammalian pancreas has been intensively studied, not least because of the clinical importance of diabetes and the quest for new therapeutic tools. In Chapter 6, Avolio et al. describe our current understanding of the pancreatic differentiation program. In addition to a detailed description of the signaling factors and transcriptional hierarchy that coordinate pancreatic development, the authors outline recent studies which collectively show that terminally differentiated pancreatic cells, under certain conditions, can undergo transdifferentiation into insulin-secreting B-cells, which is obviously an exciting step toward regenerative medicine.

The final chapter by Yates et al. provides one of the most comprehensive descriptions of adrenal development and disease that has been published in recent years. Similar to the pancreas, endocrinopathies related to adrenal dysfunction have spurred interest in adrenal development, resulting in a detailed understanding of the gene and signaling pathways that control these developments of this important endocrine organ. Yates et al. also provide a very useful overview of adrenal cancer.

I am pleased to have played a small role in advancing this field through compilation of this review volume. Thanks again to all of my colleagues for their commitment and effort—I have no doubt that it has been more than worthwhile.

Paul Thomas

CHAPTER ONE

Pituitary Gland Development and Disease: From Stem Cell to Hormone Production

Shannon W. Davis[*], Buffy S. Ellsworth[†], María Inés Peréz Millan[‡], Peter Gergics[‡], Vanessa Schade[‡], Nastaran Foyouzi[§], Michelle L. Brinkmeier[‡], Amanda H. Mortensen[‡], Sally A. Camper[‡,1]

[*]Department of Biological Sciences, University of South Carolina, Columbia, South Carolina, USA
[†]Department of Physiology, Southern Illinois University, Carbondale, Illinois, USA
[‡]Department of Human Genetics, University of Michigan Medical School, Ann Arbor, Michigan, USA
[§]Department of Obstetrics and Gynecology, University of Michigan Medical School, Ann Arbor, Michigan, USA
[1]Corresponding author: e-mail address: scamper@umich.edu

Contents

Abstract

Many aspects of pituitary development have become better understood in the past two decades. The signaling pathways regulating pituitary growth and shape have emerged, and the balancing interactions between the pathways are now appreciated. Markers for multipotent progenitor cells are being identified, and signature transcription factors have been discovered for most hormone-producing cell types. We now realize that pulsatile hormone secretion involves a 3D integration of cellular networks. About a dozen genes are known to cause pituitary hypoplasia when mutated due to their essential roles in pituitary development. Similarly, a few genes are known that predispose to familial endocrine neoplasia, and several genes mutated in sporadic pituitary adenomas are documented. In the next decade, we anticipate gleaning a deeper appreciation of these processes at the molecular level, insight into the development of the hypophyseal

Current Topics in Developmental Biology, Volume 106
ISSN 0070-2153
http://dx.doi.org/10.1016/B978-0-12-416021-7.00001-8

portal blood system, and evolution of better therapeutics for congenital and acquired hormone deficiencies and for common craniopharyngiomas and pituitary adenomas.

1. INTRODUCTION

The pituitary gland is known as the "master gland" of the body, acting as central endocrine regulator of growth, reproduction, metabolism, and response to stress. To exert its function, unique cell types in the anterior pituitary gland, including lactotrophs, somatotrophs, thyrotrophs, corticotrophs, and gonadotrophs, secrete polypeptide hormones: prolactin (PRL), growth hormone (GH), thyroid-stimulating hormone (TSH), adrenocorticotropic hormone (ACTH), and the gonadotropins-luteinizing (LH) and follicle-stimulating (FSH) hormones, respectively.

The anatomical steps of pituitary development have been described in many species, beginning a century or more ago (Rathke, 1839; Schwind, 1928). Because the basic aspects of pituitary development and functions of the pituitary hormones are fairly well conserved across all vertebrates, lessons learned in bird (chick-quail), amphibian (bullfrog), fish (zebrafish), and mammal (mouse and rat) provided important contributions to our current understanding of pituitary development and disease. Transplant studies laid the foundation for understanding the signaling that influences pituitary hormone production (Couly & Le Douarin, 1985; Gleiberman, Fedtsova, & Rosenfeld, 1999; Hermesz, Mackem, & Mahon, 1996; Takuma, Sheng, Furuta, et al., 1998). Electron microscopy ushered in the ability to distinguish hormone-producing cell types based on their size, shape, and secretory granules, and this was superseded by the availability of antibodies specific to individual hormones, permitting the emergence of differentiating cells to be tracked during embryogenesis (Japon, Rubinstein, & Low, 1994; Mikami, 1992; Simmons, Voss, Ingraham, et al., 1990). The molecular biology era brought the discovery of signature transcription factors that are important for cell specification and lineage determination, and the discovery of the signaling molecules that were predicted in early transplant experiments (reviewed in Bancalari, Gregory, McCabe, & Dattani, 2012; Zhu, Wang, Ju, & Rosenfeld, 2007; Schlosser, 2006).

A recent area of active investigation is aimed at understanding the nature of pituitary progenitors, including cells with stem-like characteristics during embryogenesis and in the adult organ (reviewed in Castinetti, Regis,

Dufour, & Brue, 2010; Drouin, Bilodeau, & Roussel-Gervais, 2010). Much still needs to be learned about the recruitment of progenitors, differentiated cell hypertrophy and hyperplasia during puberty, pregnancy, wound healing, and cases of unusual physiological demand. For a deeper basic understanding of pituitary development, we need to know how the hypophyseal portal system develops, which is necessary for hypothalamic releasing hormones to reach the pituitary gland and for transporting hormones to their target tissues (Szabo & Csanyi, 1982). The mechanisms that regulate the formation of pituitary cell networks and the role of these networks in hormone secretion are also under investigation (reviewed in Mollard, Hodson, Lafont, Rizzoti, & Drouin, 2012).

Studies of pituitary development have given us the ability to carry out molecular diagnoses for many of the rare familial pituitary adenomas and congenital pituitary hormone deficiency disorders. This is important for predicting risk, disease progression, and assessing treatments. Despite this progress, at least half of the congenital disorders still do not have diagnoses, and the common pituitary adenomas are mostly still mysterious and can be extremely difficult to treat (Kelberman, Rizzoti, Lovell-Badge, Robinson, & Dattani, 2009; Melmed, 2011; Vandeva, Vasilev, Vroonen, et al., 2010). In this review, we intend to focus on the areas where future investigation is needed and refer to recent reviews that cover the aspects of pituitary development and disease that are fairly well understood. We hope that future basic science studies will usher the way for improved detection, treatment, and prevention.

2. REGULATING THE PITUITARY ORGANIZER AND THE GROWTH AND SHAPE OF RATHKE'S POUCH

The pituitary gland is primarily derived from two ectodermal structures, the neural ectoderm, which gives rise to the posterior lobe, and the surface ectoderm, which produces Rathke's pouch, the precursor to the anterior and intermediate lobes (adenohypophysis or pars distalis and pars intermedia, respectively). The posterior lobe, or neurohypophysis, forms from the ventral diencephalon, and its formation and patterning is detailed in Chapter 2. The patterning of the ventral diencephalon is critical not only for establishing the pituitary posterior lobe but also for producing an organizing center that establishes the proper size and shape of Rathke's pouch (Fig. 1.1). Analysis of genetically modified mice has greatly advanced our understanding of the roles of various signaling pathways in pituitary

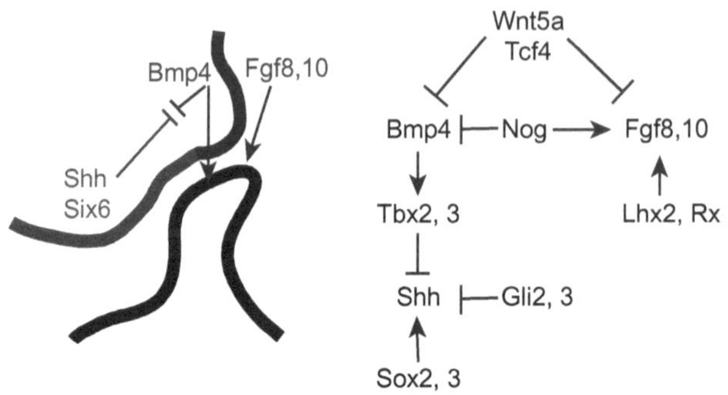

Figure 1.1 Signaling pathways initiating in the organizing center regulate anterior pituitary gland growth and shape. The oral ectoderm invaginates at the roof of the mouth to produce Rathke's pouch (black). The overlying neural ectoderm is defined molecularly by expression domain of *Shh* and *Six6* and a more dorsal region expressing *Bmp* and *Fgf*. These regions exert inhibitory effects on each other, creating a balance that induces normal patterning and growth of Rathke's pouch. The interaction between stimulatory and inhibitory transcription factors and signaling molecules is diagrammed (right). (See color plate.)

development (Table 1.1). The organizing center consists of an overlapping expression domain of bone morphogenetic protein and fibroblast growth factors (BMP4, FGF8, and FGF10) in the ventral diencephalon where it evaginates to form an infundibulum (Ericson, Norlin, Jessell, & Edlund, 1998; Treier, Gleiberman, O'Connell, et al., 1998). Rostral to the organizing center and the infundibulum is a domain of sonic hedgehog (SHH) expression (Treier et al., 1998). BMP4 is an essential inductive signal for Rathke's pouch formation because *Bmp4*$^{-/-}$ mice fail to form the pituitary placode or Rathke's pouch at e9.5 (Takuma et al., 1998). After placode formation and pouch induction, FGF signaling is necessary for cell proliferation in Rathke's pouch. The ligand and receptor mutants, *Fgf10*$^{-/-}$ and *Fgfr2IIIb*$^{-/-}$, form Rathke's pouch, but it fails to expand and is lost through apoptosis (De Moerlooze et al., 2000; Ohuchi, Hori, Yamasaki, et al., 2000). *Nkx2.1*$^{-/-}$ mice do not express *Fgf8* in the ventral diencephalon and Rathke's pouch is hypoplastic, which phenocopies the *Fgf10*$^{-/-}$ pituitary (Takuma et al., 1998). Mice with a hypomorphic mutation in *Fgf8* have a variable phenotype, including reduction in the size of the pituitary anterior lobe (adenohypophysis), loss of the pituitary posterior lobe, and neural ectoderm midline defects, including holoprosencephaly (McCabe, Gaston-Massuet, Tziaferi, et al., 2011). Thus, both BMP and FGF are critical at early stages of pouch induction and growth, and there is evidence for dosage sensitivity.

Table 1.1 Effects of signaling pathways demonstrated in genetically engineered mice

Gene	Disruption	Phenotype	References
BMP related			
Bmp4	Mouse knockout	Failure to induce Rathke's pouch	Treier et al. (1998)
Bmpr1a	Conditional knockout, *Cga-cre*	Hypoplastic Rathke's pouch, loss of *Isl1* expression	Davis and Camper (2007)
Noggin	Mouse knockout	Expanded Rathke's pouch, selective expansion of organizing center	Davis and Camper (2007)
Cga-Bmp4	Transgenic expression	Prevent terminal differentiation, get intermediate markers	Treier et al. (1998)
Pitx1-Noggin	Transgenic expression	Hypoplastic Rathke's pouch	Treier et al. (1998)
Cga-ΔBmprII	Transgenic expression	Loss of *Pou1f1* lineage, expanded ACTH, LH still present	Treier et al. (1998)
FGF related			
Fgf8	*Nkx2.1* knockout, reduced *Fgf8* expression	Hypoplastic Rathke's pouch, increased apoptosis	Takuma et al. (1998)
Fgf8	*Fgf8* hypomorph from Neo insertion	Variable, loss of anterior lobe to normal morphology with loss of LH	McCabe et al. (2011)
Cga-Fgf8	αGSU-FGF8 transgenic expression	Increased proliferation, loss of all but ACTH, maintain progenitor state	Treier et al. (1998)
Fgf10	Mouse knockout	Hypoplastic Rathke's pouch, increased apoptosis	Ohuchi et al. (2000)

Continued

Table 1.1 Effects of signaling pathways demonstrated in genetically engineered mice—cont'd

Gene	Disruption	Phenotype	References
Fgfr2	Mouse knockout	Hypoplastic Rathke's pouch, increased apoptosis	De Moerlooze et al. (2000)
WNT signaling			
Wnt5a	Mouse knockout	Expanded Rathke's pouch, expanded organizing center	Cha et al. (2004)
Wnt4	Mouse knockout	Reduction in GH, TSHβ, and αGSU (*Cga*)	Treier et al. (1998)
Wnt4	Mouse knockout	Slight reduction in GH and TSHβ	Treier et al. (1998)
Wnt6	Mouse knockout	No affect	Potok et al. (2008)
Tcf7l2 (Tcf4)	Mouse knockout	Rathke's pouch hyperplasia, expanded organizing center	Potok et al. (2008)
Ctnnb1 (β-catenin)	*Pitx1-cre* conditional knockout	Loss of POUF1 lineage	Potok et al. (2008)
Ctnnb1	*Pou1f1-cre* conditional knockout	No affect	Brinkmeier et al. (2007)
Lef1	Mouse knockout	Increased POUF1 lineage	Olson et al. (2006)
Ctnnb1	*Pitx1-cre* conditional activation of *Ctnnb1*	Hypoplastic Rathke's pouch	Olson et al. (2006)
Ctnnb1	*Hesx1-cre* conditional activation of *Ctnnb1*	Craniopharyngioma; increase in stem cell population	Olson et al. (2006)

Pou1f1-Ctnnb1	*Pou1f1*-cre conditional activation of *Ctnnb1*	No affect	Olson et al. (2006)
Gh-Ctnnb1	*Gh-cre* conditional activation of *Ctnnb1*	No affect	Gaston-Massuet et al. (2011)
Prl-Ctnnb1	*Prl-cre* conditional activation of *Ctnnb1*	No affect	Gaston-Massuet et al. (2011)
Aes	Mouse knockout	Dysmorphic intermediate lobe	Gaston-Massuet et al. (2011)
Shh signaling			
Hip	*Pitx1-Hip* transgenic	Hypoplastic pituitary, loss of BMP2	Treier et al. (2001)
Shh	*Cga-Shh* transgenics	Increased BMP2, expanded thyrotrophs and gonadotrophs	Treier et al. (2001)
Gli2	Mouse knockout	Loss of BMP4, reduction of FGF8 in organizing center, reduced proliferation in anterior lobe	Wang et al. (2010)
Gli2, *Gli3*	Mouse knockout	No pituitary	Wang et al. (2010)
Shh	SBE2 (SHH brain enhancer) conditional knockout	Expansion of BMP4 and FGF10 in organizing center	Zhao et al. (2012)
Notch signaling			
Rbp-j	*Pitx1-cre* conditional knockout	Expanded corticotrophs, loss of POU1F1 lineage	Zhu et al. (2006)

Continued

Table 1.1 Effects of signaling pathways demonstrated in genetically engineered mice—cont'd

Gene	Disruption	Phenotype	References
NICD	*Pit1-NICD* transgenics (Notch intracellular domain)	Prevention of terminal differentiation	Zhu et al. (2006)
Hes1	Mouse knockout	Conversion of melanotrophs to somatotrophs	Raetzman et al. (2007)
NICD	POMC-cre conditional activation of NICD	Inhibition of differentiation for corticotrophs and melanotrophs	Goldberg et al. (2011)
Dll3	Mouse knockout	No affect	Raetzman et al. (2004)
Dlk1	Mouse knockout	Reduction in somatotroph number	Puertas-Avendano et al. (2011)
Dlk1	Mouse knockout	Reduction in all anterior lobe cell types, somatrophs more significantly reduced	Cheung et al. (2013)
Pituitary organizer affects			
Sox2, *Sox3*	Double heterozygotes	Expansion of FGF10 in organizing center	Zhao et al. (2012)
Sox3	Mouse knockout	Expansion of BMP4 and FGF8 in organizing center	Rizzoti et al. (2004)
Rx	Mouse knockout	Reduction in FGF10 in organizing center	Medina-Martinez et al. (2009)
Tbx3	Mouse knockout	Expansion of SHH reduction in BMP4 and FGF in organizing center	Trowe et al. (2013)
Lhx2	Mouse knockout	Expansion of FGF8 in organizing center	Zhao et al. (2010)

The FGF family is large, and many of the genes are expressed in the pituitary gland. Unique roles of FGF8 and FGF10 in the pituitary-organizing center are not completely clear. FGF8 plays a central role in establishing the ectoderm midline, while *Fgf10*$^{-/-}$ mice do not display midline defects (McCabe et al., 2011; Meyers, Lewandoski, & Martin, 1998; Ohuchi et al., 2000). We hypothesize that FGF8 is more broadly required for patterning the neuroectoderm-derived pituitary-organizing center. FGF8 and 10 may work in concert for mouse infundibulum development in a manner similar to FGF3 and FGF10 in chick infundibular development. It would be useful to analyze the expression pattern of *Fgf10* in the *Nkx2.1*$^{-/-}$ and *Fgf8* hypomorphic mutants, and *Fgf8* expression in *Fgf10*$^{-/-}$ mice, to understand whether there are compensatory changes in expression. *Fgf18* is expressed in the organizing center (Maruoka et al., 1998; Ohbayashi et al., 1998), and FGFs 13, 14, and 17 are detected in the embryonic pituitary transcriptome (Brinkmeier, Davis, Carninci, et al., 2009). Thus, the potential for functional redundancy among the FGF family is great.

There is an intricate interplay between the signaling pathways in the ventral diencephalon. Single gene disruptions in one pathway influence expression of genes in a different signaling pathway. Noggin is expressed in the pituitary organizer and inhibits BMP4 activity. *Nog*$^{-/-}$ mice have an expanded domain of BMP4, and a reduction in FGF10 expression, revealing interaction between these signaling pathways (Davis & Camper, 2007). A larger domain of surface ectoderm is induced to become Rathke's pouch, and the pituitaries have highly variable dysmorphologies. The Wnt signaling pathway also affects BMP and FGF expression. *Wnt5a*$^{-/-}$ mice and *Tcf7l2*$^{-/-}$ (TCF4) mice have expanded expression domains of both BMP4 and FGF10 and enlargement of the gland. The effects are consistent, with *Wnt5a* mutants exhibiting a modest dysmorphology that resolves by birth and *Tcf7l2* mutants having a greatly enlarged pituitary gland that protrudes through the cartilage plate (Brinkmeier, Potok, Cha, et al., 2003; Brinkmeier, Potok, Davis, & Camper, 2007; Cha et al., 2004; Potok, Cha, Hunt, et al., 2008). WNT5A is likely to act through the noncanonical Wnt signaling pathway, and little or no stabilized β-catenin is detectable in the nuclei of Rathke's Pouch at that stage. Thus, the overgrowth characteristic of *Tcf7l2* mutants is likely due to a loss of transcriptional repression.

The WNT gene family is large, and many members are expressed in the developing pituitary and surrounding areas. *Wnt11* and *Wnt16* are expressed in the ventral diencephalon (Potok et al., 2008) and generally act as noncanonical and canonical WNTs, respectively. Thus, *Wnt16* is one candidate for regulating the pituitary organizer through TCF7L2. Recently, the ROR1 and ROR2 receptors have been identified as mediating noncanonical WNT5A signaling. The features of *Ror1*$^{-/-}$; *Ror2*$^{-/-}$ mice such as limb truncation phenocopy many aspects of the *Wnt5a*$^{-/-}$ mice, although the pituitary phenotype was not reported (Ho, Susman, Bikoff, et al., 2012). More work is needed to define the important players and to understand the roles of canonical and noncanonical WNT signaling in development of the pituitary organizer.

The SHH signaling pathway is important for regulating pituitary growth, and transcription factors from the SOX, T-BOX, and GLI families are involved in SHH expression and activity. SHH is expressed in both the oral ectoderm and the pituitary organizer within the ventral diencephalon. The pituitary placode arises from a patch of oral ectoderm that is negative for SHH expression, which is similar to the emergence of the pancreatic bud from the gut tube in an SHH negative zone. SHH expression in the ventral diencephalon is necessary to restrict the growth of the pituitary gland. Conditional loss of function of *Shh* in the ventral diencephalon is associated with expansion of the organizing center and pituitary enlargement (Zhao et al., 2012). *Shh* transcription in the ventral diencephalon is regulated by SOX2 and SOX3, which bind the *Shh* enhancer, SBE2, and activate expression. A dose-dependent reduction in *Sox2* and *Sox3* leads to reduced *Shh* expression in the ventral diencephalon and an expansion of the pituitary organizer (Zhao et al., 2012). The action of SOX2 and SOX3 are blocked by the T-BOX transcription factors TBX2 and TBX3. They bind SOX2 and SOX3, preventing the activation of *Shh* expression through the SBE2 enhancer. *Tbx3*$^{-/-}$ mice exhibit expanded SHH expression and reduced expression of BMP4 and FGF10, leading to a hypomorphic pituitary (Trowe et al., 2013). SHH signals are transduced through the Gli transcription factors, GLI2 and GLI3. GLI2 primarily activates and GLI3 primarily represses SHH transcriptional targets (Hui & Angers, 2011). *Gli2*$^{-/-}$ embryos have reduced *Bmp4* and *Fgf8* expression in the pituitary organizer and hypomorphic pituitaries, while *Gli2*$^{-/-}$; *Gli3*$^{-/-}$ embryos have no pituitary at all (Wang, Martin, & Bai, 2010). Given the active and repressive roles of GLI proteins, it is difficult to determine if early, active SHH signaling is necessary to induce the expression of *Bmp4* and *Fgf8* in the pituitary organizer, or if the repressive activity of GLI2 and GLI3 are necessary in the pituitary organizer to ensure the expression of *Bmp4* and *Fgf8*.

The homeobox transcription factors LHX2 and RX, which contain LIM and paired type homeodomains, respectively, are important regulators of the pituitary-organizing center. Both *Lhx2*$^{-/-}$ and *Rx*$^{-/-}$ embryos have enlarged pituitaries that are typical of genetic mutants that have expanded expression of both BMP4 and FGF8 and reduced SHH expression in the ventral diencephalon. *Lhx2* mutants have expanded FGF8 expression, but BMP4 expression appeared unchanged (Zhao, Mailloux, Hermesz, Palkovits, & Westphal, 2010). The *Rx*$^{-/-}$ mice have reduced FGF10 expression, and while neither BMP4 nor FGF8 expression was examined, we predict that their expression domains are expanded (Medina-Martinez, Amaya-Manzanares, Liu, et al., 2009). Additional characterization of these mouse models may reveal additional transcriptional regulation of the pituitary organizer and the subsequent induction of Rathke's pouch.

3. ACTIVITIES OF SIGNALING PATHWAYS INTRINSIC TO RATHKE'S POUCH

BMPs and FGFs are expressed in and around Rathke's pouch, and the roles of these signaling factors in cell specification within the anterior lobe are controversial. Loss-of-function models support the role of these signaling molecules in growth and shape, but not cell specification, while gain-of-function models suggest excess signaling can influence cell specification and/or affect the size of specific cell populations. At e10.5 of mouse development, FGF8 and FGF10 are expressed in the infundibulum, dorsal to Rathke's pouch, and BMP2 is expressed on the ventral side and adjacent mesoderm (Ericson et al., 1998; Treier et al., 1998). Counteracting gradients of FGF and BMP signaling have been proposed to regulate specification of anterior lobe cell types depending on where the progenitor cells are located relative to the gradient (Ericson et al., 1998; Treier et al., 1998). Gonadotrophs are enriched on the ventral side of the developing anterior lobe, and somatotrophs are initially located more dorsally, which could mean that progenitor cells closer to the source of BMP2 become gonadotrophs, while progenitor cells closer to the source of FGF become somatotrophs. No experiments, such as diI and genetic labeling, have been performed to follow progenitor cells and their descendants from a specific starting position near the lumen of Rathke's pouch to a final location and specific cell type in the anterior lobe. The discovery of cell type-specific networks within the anterior lobe and the movement of hormone secreting cells to form those networks suggest that the final position of cells in the anterior lobe cannot be directly correlated with a starting position in Rathke's pouch (Le Tissier et al., 2012). In fact, a birth dating study showed that progenitor cells that exit the cell cycle concurrently are scattered throughout the anterior lobe, implying the active movement of cells throughout the anterior lobe (Davis, Mortensen, & Camper, 2011). Therefore, we do not know if all progenitor cells near the lumen are equivalent or if the progenitor cells are patterned dependent on the location within Rathke's pouch prior to cell cycle exit.

Anterior lobe cell types begin to exit the cell cycle and start to differentiate between e11.5 and e13.5 (Davis et al., 2011; Seuntjens & Denef, 1999). As progenitor cells exit the cell cycle, they enter a noncycling, undifferentiated state that is characterized by the expression of p57Kip2 (*Cdkn1c*) (Bilodeau, Roussel-Gervais, & Drouin, 2009). These cells are visible on the ventral side of the lumen as they leave the epithelia of the lumenal area and enter the anterior lobe. The timing of cell cycle exit beginning at e11.5 correlates with a period when Rathke's pouch explants become refractory to exogenous signals including FGF and BMP (Ericson et al., 1998; Treier et al., 1998), suggesting that signals intrinsic to Rathke's pouch are likely to drive cell specification. Altering the expression of BMP and FGF in the pituitary organizer does not significantly alter anterior lobe cell specification (Brinkmeier et al., 2007; Cha et al., 2004; Davis & Camper, 2007; Potok et al., 2008; Zhao et al., 2010). Embryos homozygous for an FGF8 hypomorphic mutation, however, have fewer gonadotroph cells, indicating that extrinsic FGF may influence anterior lobe cell specification and/or population size at birth (McCabe et al., 2011; Zhao et al., 2010). The model employing counteracting gradients of BMP and FGF at e10.5 predicts that reduced FGF signaling should result in the loss of dorsal cell types, such as somatotrophs, and not a ventral cell type, like the gonadotrophs. Additional experiments are therefore necessary to determine the contribution and timing of extrinsic FGF signaling to cell specification.

More work is also necessary to elucidate the intrinsic roles of BMP and WNT within Rathke's pouch. BMP2, WNT4, WNT6, WNT11, and WNT16 are expressed in the pouch and could have roles (Davis & Camper, 2007; Potok et al., 2008; Treier et al., 1998). Expression of a dominant negative *Bmpr2* receptor in Rathke's pouch effectively reduces BMP signaling, and the consequences are loss of the POU1F1 (*Pit1*) lineage, which comprises thyrotrophs, somatotrophs, and lactotrophs, and a concomitant expansion of corticotrophs (Treier et al., 1998). Stimulating BMP signaling by driving BMP4 expression in Rathke's pouch promotes the differentiation of intermediate cell types, especially those expressing *Gata2* and *Isl1*, but it prevents the terminal differentiation of all hormone cell types except corticotrophs (Treier et al., 1998). It is difficult to be certain whether the consequences of nonphysiological gain-of-function experiments are truly reflective of intrinsic signaling pathway functions.

The expression of multiple WNTs in Rathke's pouch raises the possibility of functional redundancy and the use of both canonical and noncanonical pathways. *Wnt4* deficiency leads to a reduction in somatotrophs and thyrotrophs, whereas a loss of *Wnt6* has no obvious effect on pituitary cell specification (Potok et al., 2008; Treier et al., 1998). Canonical Wnt signaling appears critical for cell specification because the conditional inactivation of *β-Catenin* in the early pouch leads to a loss or reduction in all cell types except corticotrophs (Olson, Tollkuhn, Scafoglio, et al., 2006). Expression of an activated form of *β-Catenin* in the early Rathke's pouch causes variable phenotypes depending on the cre driver used to activate *β-Catenin*. With *Pitx1-cre*, the pituitary is arrested early in organogenesis (Olson et al., 2006). With *Hesx1-cre*, an increase in pituitary stem cells is observed leading to the formation of craniopharyngiomas and cell specification is altered, reducing all cell types, except corticotrophs (Gaston-Massuet, Andoniadou, Signore, et al., 2011). Despite the presence of a gene encoding a degradation resistant form of β-Catenin in all anterior lobe cells, nuclear-localized β-Catenin is not observed in the cells outside the stem cell niche that have begun to differentiate. The anterior pituitary appears to have mechanisms to suppress activation of *β-Catenin*. While both *Pitx1* and *Hesx1* are expressed very early in Rathke's pouch formation, they are also expressed in other anterior structures prior to pouch formation (Hermesz et al., 1996). Spatial and temporal differences in the expression of these two cre drivers likely contribute to the disparate phenotypes that are observed. In addition, the dosage sensitivity demonstrated for HESX1 and other pituitary transcription factors could be contributors in cases where cre expression occurs at the expense of an endogenous allele (Dattani, Martinez-Barbera, Thomas, et al., 1998; Gage, Suh, & Camper, 1999; Sheng, Moriyama, Yamashita, et al., 1997).

The role of SHH signaling within the pouch is suggested by the results of gain- and loss-of-function experiments in mice. *Shh* is initially expressed throughout the oral ectoderm, and it is excluded from the placode that forms Rathke's pouch. Despite this, Rathke's pouch cells are apparently receiving SHH signals because the downstream target gene patched (*Ptc1*) is expressed (Treier, O'Connell, Gleiberman, et al., 2001). Receipt of these signals must be important because blocking SHH signaling by driving expression of the *Shh* inhibitor, *Hip*, in the pouch reduces the proliferation of progenitor cells; *Bmp2* and *Lhx3* are not expressed; and the pituitary is very hypoplastic (Treier et al., 2001). Similarly, overexpression of *Shh* in Rathke's pouch causes an increase in *Bmp2* expression and an increase in thyrotrophs and gonadotrophs (Treier et al., 2001). Embryos with a conditional inactivation of *Gli2* in the pituitary, however, exhibit a reduction in progenitor proliferation, but the pituitary is well formed and, other than a reduction in corticotrophs, the hormone-producing cells are unaffected (Wang et al., 2010). Compensatory changes in gene expression may provide a partial rescue in loss-of-function mutants, but gain of function may exceed the ability make adjustments. Activating *Shh* signaling in the pituitary with ectopic expression of *SmoM2* increases proliferation without altering cell specification (Wang et al., 2010). The differences between the *HIP* transgenic and conditional *Gli2* loss-of-function studies may be indicative of a broader range of action for the secreted inhibitor, HIP, such as inhibiting SHH signaling in the ventral diencephalon as well as Rathke's pouch, or the differences may implicate non-canonical SHH signaling in the pituitary, such as the activation of RAC1 or RHO in a Gli-independent manner (Brennan, Chen, Cheng, Mahoney, & Riobo, 2012). The differing results for the ectopic stimulation of the SHH pathway in Rathke's pouch may be explained by noncanonical SHH signaling because *Ptc1* has *Smo*- and *Gli*-independent functions (Brennan et al., 2012).

The Notch signaling pathway has proved to be a prime candidate for driving anterior lobe cell specification. *Hes1* is a Notch-responsive transcription factor, and *Hes1*$^{-/-}$ embryos have a cell fate switch from melanotrophs to somatotrophs (Raetzman, Cai, & Camper, 2007). The conditional loss of an intracellular mediator of Notch signaling, *Rbpjk*, in Rathke's pouch promotes the differentiation of corticotrophs and the loss of the POU1F1 lineage (Zhu, Zhang, Tollkuhn, et al., 2006). Stimulation of Notch signaling in the corticotrophs and melanotrophs prevents their differentiation (Goldberg, Aujla, & Raetzman, 2011), while ectopic Notch signaling in the POU1F1 lineage prevents terminal differentiation (Zhu et al., 2006). The Notch ligand *Dll3* is expressed in corticotrophs; although it is not required for corticotroph differentiation (Raetzman et al., 2004). The *Dlk1* Notch ligand is expressed in all hormonal cell types, and loss of *Dlk1* leads to a decrease in all cell types, with the most significant reduction occurring in somatotrophs (Cheung, Rizzoti, Lovell-Badge, & Le Tissier, 2013; Puertas-Avendano, Gonzalez-Gomez, Ruvira, et al., 2011). More studies are necessary to define the role of NOTCH family receptors, ligands, and target genes in regulating pituitary progenitor transitions to differentiation and cell specification.

In sum, BMP, FGF, WNT, and NOTCH each have important roles in and around the pituitary gland. The multiplicity of ligands and receptors and interactions between pathways confer a degree of complexity that is difficult to unravel completely. Future research is needed to clarify the roles of individual signaling pathways in cell specification, and to understand the compensatory changes that are possible to ensure the proper distribution of cell types within the pituitary anterior lobe.

4. THE ROLE OF SIGNATURE TRANSCRIPTION FACTORS IN CELL SPECIFICATION

A collection of transcription factors have been identified that play important roles in the specification and/or expansion of pituitary hormone-producing cells, and many are relevant in human disease (Table 1.2). The first of these was POU1F1, which was identified based on its role in *trans*-activating the GH and PRL genes (Bodner et al., 1988; Ingraham, Chen, Mangalam, et al., 1988; Karin, Theill, Castrillo, McCormick, & Brady, 1990; Mangalam, Albert, Ingraham, et al., 1989). Mice and humans with inactivating mutations in this gene generally have recessive hypopituitarism, characterized by a congenital lack of GH, PRL, and TSH. POU1F1 is the signature transcription factor for the lineage that gives rise to the somatotrophs, lactotrophs, and thyrotrophs (Li et al., 1990). These cells fail to develop in POU1F1 mutants. The approach of identifying the *trans*-acting factors that bind critical *cis*-acting sites necessary for hormone gene expression was successful in identifying other critical transcription factors like PITX1, the orphan nuclear hormone receptor, NR5A1 or steroidogenic factor 1, and the helix-loop-helix factor NEUROD1, and the T-BOX factor TPIT (Table 1.2) (Lala, Rice, & Parker, 1992; Lamolet et al., 2004; Lamolet, Pulichino, Lamonerie, et al., 2001; Lamonerie et al., 1996; Pulichino, Lamolet, Vallette-Kasic, et al., 2004). In some cases, the transcription factor deficiency does not result in the complete absence of the cell type, but the differentiation is incomplete. For example, NR5A1-deficient mice do not produce gonadotropins, but hyperstimulation with GnRH is an effective inducer, suggesting that gonadotroph differentiation does not require NR5A1 (Ingraham, Lala, Ikeda, et al., 1994). Similarly, corticotroph development does not depend on either NEUROD1 or TPIT, but POMC expression is delayed and/or reduced if they are deficient (Lamolet et al., 2004; Pulichino et al., 2004). The failure to promote differentiation along one particular path can be permissive for alternative pathways. In the absence of TPIT, intermediate lobe cell types differentiate into gonadotrophs and POU1F1-independent thyrotrophs, implying an important role of TPIT in repressing anterior lobe cell fates, in part by antagonizing NR5A1

Table 1.2 Pituitary transcription factors

Family, gene	Human disease	Pituitary function	References[a]
Paired homeo			
PITX1	Congenital club foot, polydactyly, Liebenberg syndrome (homeotic arm to leg transformation)	Modest, overlaps with PITX2	Charles et al. (2005), Gurnett, Alaee, Kruse, et al. (2008), Spielmann, Brancati, Krawitz, et al. (2012), Szeto, Rodriguez-Esteban, Ryan, et al. (1999)
PITX2	Rieger syndrome, eyes, teeth, umbilicus	Rathke's pouch expansion	Gage et al. (1999), Semina, Reiter, Leysens, et al. (1996)
HESX1	Septo-optic dysplasia, mild to severe hypopituitarism	Affects midline and pituitary growth	Dattani et al. (1998)
PROP1	Evolving hypopituitarism	Silencing HESX1, OTX2 and activating NOTCH2 and POU1F1	Dattani et al. (1998), Gage et al. (1996), Raetzman et al. (2004)
PAX6	Various eye and optic nerve anomalies	Increased growth, expansion of TSH cells at expense of GH and LH	Bentley et al. (1999), Kioussi, O'Connell, St-Onge, et al. (1999), Solomon, Pineda-Alvarez, Balog, et al. (2009)
PAX7	Rhabdomyosarcoma 2, alveolar	Chromatin remodeling for selecting melanotroph fate	Budry et al. (2012), Davis, D'Cruz, Lovell, Biegel, and Barr (1994)
Pou homeo			
POU1F1 (Pit1)	Hypopituitarism	Signature factor for somatotrophs, lactotrophs, and thyrotrophs	Li et al. (1990), Tatsumi, Miyai, Notomi, et al. (1992)

POU5F1 (Oct4)	Sarcoma if fused with EWS (KO mice die at gastrulation)	Pituitary stem cell marker	Garcia-Lavandeira et al. (2009), Yamaguchi et al. (2005)
OTX1	Not known (critical role in head development)	Expressed in pituitary postnatally. Delayed growth and puberty Functional overlap with OTX1 and EMX1, two in early head development	Acampora et al. (1998)
OTX2	Variable, anophthalmia, microophthalmia, hypopituitarism	Expressed in pituitary organizer, neural ectoderm and transiently in Rathke's pouch	Dateki et al. (2008), Diaczok, Romero, Zunich, Marshall, and Radovick (2008), Mortensen et al. (2011)
LIM homeo			
ISL1	Not known (critical role in heart development)	Rathke's pouch induction	Cai, Liang, Shi, et al. (2003), Takuma et al. (1998)
LHX2	Not known (regulates hematopoietic stem cells and head development)	Failure to form pituitary stalk and infundibulum. Small dysmorphic anterior lobe	Porter, Drago, Xu, et al. (1997)
LHX3	Hypopituitarism, variable effects on cervical spine	Pouch induction, functional overlap with LHX4	Bhangoo, Hunter, Savage, et al. (2006), Sheng, Zhadanov, Mosinger, et al. (1996)
LHX4	Hypopituitarism, variable cerebellar and skull defects	Pouch induction, functional overlap with LHX3	Machinis, Pantel, Netchine, et al. (2001), Sheng et al. (1997)

Continued

Table 1.2 Pituitary transcription factors—cont'd

Family, gene	Human disease	Pituitary function	References
T-box			
TBX2	Dose-dependent heart defects	Dispensable, but marks posterior lobe cells	Radio, Bernardini, Loddo, et al. (2010), Trowe et al. (2013)
TBX3	Ulnar-mammary syndrome	Required to establish *Tbx2* expression, repress *Shh*, pituitary stalk formation, growth Rathke's pouch	Bamshad, Lin, Law, et al. (1997), Trowe et al. (2013)
TPIT (Tbx19)	Adrenocorticotropic hormone deficiency	Signature factor for corticotrophs and melanotrophs	Lamolet et al. (2001), Pulichino, Vallette-Kasic, Couture, et al. (2003), Pulichino, Vallette-Kasic, Tsai, et al. (2003)
Helix-loop-helix			
NEUROD1	Allelic variants cause maturity onset diabetes of the young (MODY)	Signature factor for corticotrophs Delayed corticotroph development	Cherrington, Bailey, Diaz, and Mellon (2008), Lamolet et al. (2004), Lavoie, Budry, Balsalobre, and Drouin (2008), Malecki, Jhala, Antonellis, et al. (1999)
NEUROD4 (Math3)	Unknown	Stimulate somatotrophs expression of POU1F1, GH, and GHRHR	Zhu et al. (2006)
HES1	Chronic myelomonocytic leukemia	Notch target, regulates melanotroph cell fate	Klinakis, Lobry, Abdel-Wahab, et al. (2011), Raetzman et al. (2007)

Zn finger			
GATA2	Various hematopoietic defects tbx2	Suppresses gonadotroph and promotes thyrotroph fate, Pituitary KO has modest effects, Gata3 compensation	Charles, Saunders, Wood, et al. (2006), Dasen, O'Connell, Flynn, et al. (1999), Zhang, Ma, Huang, et al. (2008)
GATA3	Hypoparathyroidism, sensorineural deafness, and renal dysplasia	Not known, overlaps with *Gata2*	Charles et al. (2006), Van Esch, Groenen, Nesbit, et al. (2000)
Orphan nuclear receptor			
NR5A1	Hypogonadotropic hypogonadism, sex reversal, premature ovarian insufficiency, adrenal failure	Signature factor for gonadotrophs. Activates *Gnrhr*, *Lhb*, and *Fshb*	Achermann, Gu, et al. (1999), Achermann, Ito, Ito, Hindmarsh, and Jameson (1999), Luo, Ikeda, and Parker (1994)
High mobility group			
SOX2	Microphthalmia, anterior pituitary hypoplasia, hypogonadotropic hypogonadism	Anterior pituitary growth, stimulation of *Pou1f1*, *Gh*, and *Tshb* expression, progenitor proliferation	Jayakody, Andoniadou, Gaston-Massuet, et al. (2012), Kelberman, Rizzoti, Avilion, et al. (2006)
SOX3	Hypopituitarism, mental retardation	Neural ectoderm expression (pituitary organizer), Rathke's pouch growth and shape	Hamel et al. (1996), Rizzoti et al. (2004)
Kruppel			
GLI2	Holoprosencephaly, central incisor, hypopituitarism	Regulates expression of BMP and FGF necessary for Rathke's pouch induction	Franca, Jorge, Carvalho, et al. (2010), Wang et al. (2010)

Continued

Table 1.2 Pituitary transcription factors—cont'd

Family, gene	Human disease	Pituitary function	References
EGR1 (*Krox24*, *Ngf1a*)	Likely tumor suppressor in acute myeloid leukemia, myelodysplastic syndrome	*Gh* and *Lhb* expression	Joslin, Fernald, Tennant, et al. (2007), Topilko, Schneider-Maunoury, Levi, et al. (1998)
EGR2 (*Krox20*)	Charcot–Marie–Tooth disease, neuropathy	GH production	Bouchoucha, Charnay, and Gilardi-Hebenstreit (2013), Warner, Mancias, Butler, et al. (1998)

[a]Due to space constraints only selected references are listed. Additional references can be found in OMIM (http://omim.org) and MGI (http://www.informatics.jax.org).

(Pulichino, Vallette-Kasic, Couture, et al., 2003; Pulichino, Vallette-Kasic, Tsai, et al., 2003). HES1 deficiency also can cause ectopic differentiation in the intermediate lobe: instead of melanotrophs, the hypoplastic intermediate lobe contains POU1F1-dependent somatotrophs. Premature cell cycle exit appears to be the underlying permissive factor (Monahan, Rybak, & Raetzman, 2009).

The idea that single signature transcription factors direct cell specification is an overly simplistic one. Many transcription factors are required to produce the characteristic features of specialized hormone-producing cell types. For example, additional factors, both positive and negative, are implicated in driving POU1F1-expressing cells toward the specialization in production of GH, PRL, or TSH (Scully & Rosenfeld, 2002). The glucocorticoid and estrogen receptors and ETS factors are examples of factors that promote specialization (Budry, Couture, Balsalobre, & Drouin, 2011; Gutierrez-Hartmann, Duval, & Bradford, 2007; Vakili & Cattini, 2012). In addition, components of the combinatorial code of factors can change during development in order to achieve the fully differentiated hormone-producing cell or to maintain it (reviewed in Davis, Castinetti, Carvalho, et al., 2010).

The current state of the art requires an understanding of the epigenetic regulation that makes chromatin accessible for transcription factor binding and the mechanisms that initiate this state. Genome-wide analysis of DNase-sensitive open chromatin and transcription factor-binding sites are powerful tools for dissecting the differentiation steps for hormone-producing cells. A recent example comes from the study of PAX7, which is a pioneer transcription factor that binds enhancers associated with many different genes, and opens the chromatin to permit binding of TPIT or to suppress its binding (Budry, Balsalobre, Gauthier, et al., 2012). Transcription factors that set up chromatin accessibility for other transcription factors are known as selectors. In the absence of the PAX7 selector factor, intermediate lobe cells differentiate into corticotrophs instead of melanotrophs. An important future challenge is to understand these initiating, selector steps for other hormone-producing cell types. These types of experiments are particularly challenging unless there are cell culture systems that can recapitulate the differentiation process because the quantity of embryonic pituitary tissue is limited.

5. EARLY-ACTING TRANSCRIPTION FACTORS

PROP1 is the earliest pituitary-specific transcription factor expressed in development. This paired homeodomain protein is required for both activation and silencing of several genes that individually have important roles in organogenesis. It is critical for initial activation of POU1F1 and NOTCH2 and for temporally appropriate silencing HESX1 and OTX2 (Dasen, Martinez Barbera, Herman, et al., 2001; Gage et al., 1996; Mortensen, MacDonald, Ghosh, & Camper, 2011; Raetzman et al., 2004; Sornson, Wu, Dasen, et al., 1996). PROP1's switch from repressor to activator may be controlled by β-catenin, but other evidence suggests that β-catenin must be strongly suppressed for normal development (Brinkmeier et al., 2007; Gaston-Massuet et al., 2011; Olson et al., 2006). In humans, PROP1 deficiency can affect all hormone-producing cell types of the anterior lobe (Pernasetti, Toledo, Vasilyev, et al., 2000), and in mice, it causes a congenital deficiency of GH, TSH, PRL, and reduced levels of gonadotropins (Nasonkin, Ward, Raetzman, et al., 2004; Tang, Bartke, Gardiner, Wagner, & Yun, 1993). In the absence of PROP1, the proliferating cells located along Rathke's cleft fail to delaminate, mimicking failed epithelial to mesenchymal transition (Himes & Raetzman, 2009; Ward et al., 2005). This results in a highly dysmorphic and hypoplastic organ in mice and a variety of organ sizes in humans (Davis et al., 2010; Ward et al., 2005).

Many of the transcription factors that act early in pituitary development have effects on multiple pituitary cell types. In contrast to PROP1, most of these genes are not pituitary specific and affect multiple developing structures when mutated, resulting in syndromic hypopituitarism (Table 1.2; reviewed in Davis et al., 2010). Loss-of-function mutations in some genes are likely lethal in humans because of their pleiotropic effects. Highly variable craniofacial and pituitary phenotypes are observed, possibly because there is functional overlap between members of the same gene family such as PITX1, PITX2; LHX2, LHX3, LHX4; and OTX1, OTX2. Mutations in many of these genes are associated with reduced proliferation and increased apoptosis (Charles et al., 2005; Ellsworth, Butts, & Camper, 2008; Raetzman, Ward, & Camper, 2002; Sheng et al., 1997).

OTX1, OTX2, EMX1, and EMX2 have overlapping patterns of expression in the head of the developing mouse embryo (Acampora, Gulisano, & Simeone, 2000). Gene-targeting experiments revealed essential roles for each of these transcription factors and demonstrated compensation by members of the gene family during embryogenesis. *Otx2*$^{-/-}$ mice lack head structures anterior to rhombomere 3, while *Otx2*$^{+/-}$ heterozygotes have variable craniofacial phenotypes that range from pituitary aplasia to striking pituitary dysmorphology and hypoplasia (Matsuo, Kuratani, Kimura, Takeda, & Aizawa, 1995). *Otx1*$^{-/-}$ mutants have an even milder phenotype, resulting in a transient delay in growth and puberty (Acampora, Mazan, Tuorto, et al., 1998), supporting the idea that both *Otx1* and *Otx2* have roles in pituitary development, with *Otx2* being the most critical for normal organ morphology and function. *Otx2* is prominently expressed in the neural ectoderm, which produces FGF and stimulates the growth of Rathke's pouch (Mortensen et al., 2011). *Otx2* expression in Rathke's pouch is very low and transient, and little or no expression is apparent when *Pou1f1* transcription is initiated. This suggests that the hypopituitarism characteristic of *OTX2* mutations in humans and mice occurs because OTX2 is required for development of the posterior lobe and pituitary stalk. The anterior lobe hypoplasia is likely to be secondary to the neural ectoderm defect, resulting from reduced inductive signals that normally emanate from the organizing center.

Compensation and functional overlap are not limited to members of the same gene family. For example, in addition to the EMX and OTX genes, there are multiple genes that enhance or suppress the *Otx2* mutant phenotype, and they vary among different inbred strains (Matsuo et al., 1995). The C57BL/6 background enhances the susceptibility of *Otx2* heterozygotes to severe craniofacial defects, while CBA is protective (Hide, Hatakeyama, Kimura-Yoshida, et al., 2002; Yagi, Tokunaga, Furuta, et al., 1993). Late in gestation, on a mixed background of B6 and CBA, *Otx2* heterozygotes range from normal appearance to anencephaly and acephaly. Classic genetic mapping of the modifier genes revealed contributing loci on several chromosomes, although the specific genes are not yet known. These types of mouse studies may uncover genes that influence the severity of craniofacial defects in human carriers for *OTX2* mutations. In cases where the transmission of *OTX2* variants has been studied in pedigrees, the heterozygous mutations are not completely penetrant (Beby & Lamonerie, 2013). Sequencing the genomic DNA of these human patients could identify genes with deleterious variants that contribute to the penetrance of the *OTX2* mutant phenotype. This has been employed successfully in identifying genes that contribute to hypogonadotropic hypogonadism, and uncovered multiple examples of digenic or oligogenic disease (Miraoui, Dwyer, Sykiotis, et al., 2013; Sykiotis, Pitteloud, Seminara, Kaiser, & Crowley, 2010).

6. EMERGING ROLES FOR ADDITIONAL TRANSCRIPTION FACTOR FAMILIES: THE FORKHEADS

Transcriptome studies reveal that there are many different transcription factor genes expressed in the pituitary gland with unknown functions

(Brinkmeier et al., 2009). The SIX gene family has been implicated in pituitary development (Gaston-Massuet, Andoniadou, Signore, et al., 2008; Li, Perissi, Liu, Rose, & Rosenfeld, 2002), and the common effects of this family on eye and pituitary development support the idea that organs developing from placodes utilize similar regulatory pathways (McCabe, Manzo, Gammill, & Bronner-Fraser, 2004). There are many examples of homeobox, HMG box, helix-loop-helix, and orphan nuclear receptors that remain to be analyzed (Brinkmeier et al., 2009; Davis et al., 2010; Davis, Potok, Brinkmeier, et al., 2009). Developmental expression studies are the first step in understanding the role of these novel genes. The role of forkhead genes in pituitary growth is beginning to emerge, and it serves as an example of the complexity that may characterize other gene families that remain to be explored.

Forkheads are a family of transcription factors that contain a conserved, winged helix DNA-binding domain and were named from the phenotype of the *Drosophila* mutant that founded the group. Forkheads are implicated in many physiological processes including development, metabolism, cell cycle progression, and chromatin remodeling (Benayoun, Dipietromaria, Bazin, & Veitia, 2009; Lalmansingh, Karmakar, Jin, & Nagaich, 2012). To date, 50 forkhead factors have been identified in humans and 44 in mice. A unified nomenclature has been adopted for forkhead factors: FOX (for *f*orkhead b*o*x) a letter to designate the subfamily to which the factor belongs and a number to identify each member of the subfamily (Jackson, Carpenter, Nebert, & Vasiliou, 2010; Kaestner, Knochel, & Martinez, 2000). Mutations in forkhead genes often result in autosomal dominant conditions in humans, with haploinsufficiency likely. Several forkhead genes are expressed in the pituitary gland, and the best known is FOXL2 (Table 1.3).

FOXL2, also known as *Pfrk*, was the first forkhead to be described in the pituitary (Treier et al., 1998). FOXL2 is important for ovarian development and function (Kuo, Fan, Bentsi-Barnes, Barlow, & Pisarska, 2012; Richards & Pangas, 2010), and it promotes female sex determination (Benayoun et al., 2009; Biason-Lauber, 2012; Veitia, 2010). Mutations in the human *FOXL2* gene result in an autosomal dominant, loss-of-function disease called blepharophimosis, ptosis, and epicanthus inversus syndrome (BPES), which causes eyelid abnormalities and premature ovarian failure (Verdin & De Baere, 2012). Humans with BPES do not exhibit pituitary abnormalities, but homozygous mutant mice reveal the role of FOXL2 in pituitary function, suggesting that loss of both *Foxl2* alleles is required to alter pituitary function (Schmidt, Ovitt, Anlag, et al., 2004; Tran, Zhou, Lafleur,

Table 1.3 Forkhead genes and pituitary function

	Mutation phenotypes		Pituitary expression	
Gene	**Human**	**Mouse**	**Adult cell specificity**	**Other comments**
FOXL2	BPES, POI, dominant, haploinsufficient normal pituitary[a]	Systemic and pituitary-specific knockouts Homozygous loss of function causes hypogonadotropic hypogonadism	Thyrotroph and gonadotroph	First detected in CGA-positive cells around e10.5–e11.5
FOXO1	Unknown 1/90 women with POI had potentially deleterious variants in FOXO1 (Watkins, Umbers, Woad, et al., 2006)	Knockouts die by e10.5 due to placental defects (Ferdous et al., 2011)	Somatotroph and gonadotroph	Expressed in quiescent pituitary cells
FOXE1 (TTF2, TITF2, FKHL15)	Bamforth–Lazarus syndrome: thyroid agenesis, cleft palate, choanal atresia, spiky hair (De Felice et al., 1998)	No pituitary phenotype, thyroid agenesis, cleft palate	Unknown	Expressed in oral ectoderm, e9.5–e10.5
FOXP3	IPEX, X-linked severe autoimmunity that can be fatal	Regulatory T-cells, autoimmune disorder, infertility. Reduced pituitary *Lhb*, *Fshb*, *Cga* expression	None detected	None detected
FOXD1	Unknown	Perinatal lethal, renal failure. Abnormal sella turcica, reduced *Lhb* expression	Adult pituitary (B.S. Ellsworth, unpublished)	Expressed in mesenchyme near Rathke's pouch at e10.5

[a]BPES, blepharophimosis, ptosis, and epicanthus inversus syndrome.
POI, premature ovarian insufficiency.

et al., 2013; Uda, Ottolenghi, Crisponi, et al., 2004; Uhlenhaut, Jakob, Anlag, et al., 2009).

FOXL2 expression is reported in mouse gonadotrophs and thyrotrophs and human gonadotrophs (Egashira, Takekoshi, Takei, Teramoto, & Osamura, 2011; Ellsworth, Egashira, Haller, et al., 2006). FOXL2 is expressed in most null cell and gonadotropin-subunit-producing adenomas, suggesting that FOXL2 contributes to gonadotroph differentiation and possibly influences proliferation, as FOXL2 cooperates with clusterin to regulate gonadotroph adenoma growth (Chesnokova, Zonis, Wawrowsky, et al., 2012). FOXL2 regulates activin responsiveness of follistatin (*Fst*) in cooperation with SMAD3 (Blount et al., 2009), and it stimulates the activin-responsive element of the *Gnrhr* gene promoter in αT3-1 cells (Ellsworth et al., 2003). FOXL2 is not necessary for *Gnrhr* expression, however, suggesting the possibility of genetic overlap and/or compensation (Tran et al., 2013). Ectopic FOXL2 expression in transgenic mice is sufficient to drive ectopic expression of the gene encoding the glycoprotein hormone α-subunit (αGSU), *Cga* (Ellsworth et al., 2006). The necessity of FOXL2 for *Cga* expression is unclear because *Cga* expression is reduced in *Foxl2* knockout mice, but transcripts are normal in a pituitary-specific deletion of *Foxl2* (Justice et al., 2011; Tran et al., 2013). This apparent discrepancy could be due to hypothalamic contributions of FOXL2 to regulation of *Cga* expression, or to the timing or efficiency of *Foxl2* deletion in conditional knockout animals.

The follicle-stimulating hormone (*Fshb*) gene is the most well studied FOXL2 target gene (reviewed in Bernard & Tran, 2013; Bilezikjian, Justice, Blackler, Wiater, & Vale, 2012; Coss, Mellon, & Thackray, 2010). Gonadotroph cell specification occurs in both systemic and pituitary-specific deletions of *Foxl2*, but basal and activin-stimulated FSH levels are severely impaired in both males and females. Pituitary-specific *Foxl2* knockout male mice have reduced testis size and spermatogenesis, and females have reduced ovarian weight and oogenesis (Tran et al., 2013). Consistent with these studies, activin does not stimulate FSH secretion from primary pituitary cells from *Foxl2* mutant mice (Justice et al., 2011). Several studies provide mechanistic insight into the regulation of *Fshb* expression by FOXL2. FOXL2 synergizes with SMADs to mediate activin stimulation of the murine and porcine *Fshb* genes (Lamba, Fortin, Tran, Wang, & Bernard, 2009; Lamba, Wang, Tran, et al., 2010; Tran, Lamba, Wang, & Bernard, 2011). FOXL2 is also involved in the synergy between activin and progestins on the *Fshb* promoter (Ghochani, Saini, Mellon, & Thackray, 2012).

Less is known about the roles of other forkhead transcription factors during pituitary development. FOXO1 is expressed in many tissues including pancreas, liver, brain, adipose, and ovary (Jackson et al., 2010; Richards, Sharma, Falender, & Lo, 2002). FOXO1 is present in quiescent cells of the developing pituitary, consistent with a role in suppressing cell cycle progression (Majumdar, Farris, Kabat, Jung, & Ellsworth, 2012). The cell specificity of FOXO1 expression in the pituitary is not clear. FOXO1 is reported in approximately half of somatotroph cells and one-tenth of gonadotrophs in one study (Majumdar et al., 2012), but another reports expression primarily in gonadotrophs and functional inhibition of *Lhb* expression (Arriola, Mayo, Skarra, Benson, & Thackray, 2012). Further studies are needed to establish the requirement for FOXO1 in pituitary development.

FOXE1 is important for thyroid organogenesis and exhibits transient expression at e9.5 and e10.5 in the oral ectoderm that will form Rathke's pouch (De Felice, Ovitt, Biffali, et al., 1998; Zannini, Avantaggiato, Biffali, et al., 1997). No pituitary defects have been detected in *Foxe1* null mice, however, suggesting that this gene may not be required for normal pituitary development (De Felice et al., 1998).

Autoimmune hypophysitis is a rare disease of pituitary inflammation that leads to reduced hormone production. Some forkhead genes affect the immune system and influence pituitary hormone production, but the mechanisms are not yet understood. For example, FOXP3 and FOXD1 are not expressed in the developing pituitary gland, but both affect pituitary hormone production (Gumbel, Patterson, Owusu, et al., 2012; Jung, Jasurda, Egashira, & Ellsworth, 2012). FOXP3 is necessary for normal development and function of regulatory T-cells, and FOXP3 deficiency causes severe autoimmune disease (Brunkow, Jeffery, Hjerrild, et al., 2001; Ziegler, 2006). Mice with an inactivating *Foxp3* mutation (scurfy mice) have reduced expression of the gonadotropins *Lhb*, *Fshb*, and *Cga*, suggesting that FOXP3 is indirectly important for gonadotroph function (Jung et al., 2012). Besides its renal expression, *Foxd1* is expressed in the kidney and the mesenchyme surrounding the developing pituitary at e10.5. Mice deficient in *Foxd1* die within 24 h after birth due to renal failure (Hatini, Huh, Herzlinger, Soares, & Lai, 1996; Levinson et al., 2005), and there is a significant reduction in *Lhb* expression, specifically. These mice also exhibit failure of the sella turcica to form properly (Gumbel et al., 2012). Thus, both FOXB3 and FOXD1 affect pituitary function.

These studies indicate that forkhead factors play an important role in pituitary development and function. There is much more to be done before we can truly appreciate the contribution of this family of factors to pituitary organogenesis and hormone production. Similar to many other transcription factor families, there may be functional overlap among the members of the forkhead family.

7. PITUITARY PROGENITORS: STEM CELLS AND THE NICHE

The study of anterior pituitary stem cells or progenitors has recently accelerated (reviewed in Castinetti, Davis, Brue, & Camper, 2011). Alvarez's group discovered that the growth factor receptor GFRa2, glial cell line-derived neurotrophic factor receptor alpha 2, is a pituitary stem cell marker (Garcia-Lavandeira, Quereda, Flores, et al., 2009). GFRa2 is expressed in 0.9% of adult pituitary cells lining the cleft and a few cells scattered in the anterior pituitary parenchyma. These cells express several stem markers, such as SOX2 and OCT4, and interestingly, they are positive for PROP1, the early-acting, pituitary-specific transcription factor that is essential for maintenance of all pituitary cell types in humans (Bottner, Keller, Kratzsch, et al., 2004). GFPRa2-positive, PROP1-positive cells are slowly proliferating cells that can form spheres *in vitro*, generate secondary pituispheres, and differentiate into the five pituitary lineages. PROP1 may be an important player in establishing and/or maintaining a pool of pituitary progenitors. PROP1 expression overlaps with several stem cell markers including SOX2, OCT4, and GFRa2 (Garcia-Lavandeira et al., 2009; Susa et al., 2012; Yoshida, Kato, Yako, et al., 2011). In addition, *Prop1* is expressed in a transitional zone in between proliferating and differentiating cells during fetal pituitary organogenesis. This transitional zone is also marked by expression of CYCLIN E and NOTCH2 (Bilodeau et al., 2009; Raetzman et al., 2004). Definitive studies are needed to assess the potential of *Prop1*-expressing cells to form pituispheres and to trace the lineages of cells that derive from *Prop1*-expressing progenitors.

Evidence has emerged supporting the idea that SOX2, SOX9, and OCT4 are markers of pituitary progenitors with many characteristics of stem cells (reviewed in Alatzoglou, Kelberman, & Dattani, 2009).

A small (0.03%) population of progenitors in the adult pituitary gland, that are SOX2-positve, SOX9-negative, hormone-negative, can form pituispheres *in vitro*, which can self-renew, giving to rise to secondary spheres, and they can differentiate into all of the five endocrine cells of the AP, as well as folliculostellate cells (Fauquier, Rizzoti, Dattani,

Lovell-Badge, & Robinson, 2008). While these characteristics comply with most of the criteria for labeling them as stem cells, the classical definition requires at least five passages to demonstrate self-renew clearly. It is possible that they have the capacity for multiple passages of self-renewal if cultured in a milieu that better mimics the niche and/or the inductive factors produced by the organizing center in the ventral diencephalon.

Folliculostellate cells are nongranular cells with long cytoplasmic projections that confer a starlike morphology. They are located in the parenchymal tissue of the adult anterior pituitary gland. A subset of the folliculostellate cells may be involved in creating a niche for progenitors in the adult organ. Thomas's group have confirmed the progenitor potential of a subpopulation of folliculostellate cells, based on their ability to form colonies *in vitro* and to differentiate *in vivo* (Lepore et al., 2005). These colony-forming cells represent 0.2% of the anterior pituitary cells and express the folliculostellate markers S100 and GFAP. In addition, they have the capacity to take up the fluorescent dipeptide β-Ala-Lys-Ne-AMCA (Lepore et al., 2005).

Advances are being made in identifying the niche, which appears to be associated with Rathke's cleft in humans and mice (Garcia-Lavandeira et al., 2009). During embryonic development, proliferating cells are enriched around the remnants of Rathke's cleft. This multilayer zone is described as the marginal zone or the niche for potential pituitary stem cells. The ability of individual progenitors to produce all anterior pituitary cell types has been demonstrated in pituisphere cultures (Fauquier et al., 2008). Different groups, using diverse approaches, have demonstrated the presence of cells in the pituitary with progenitor or stem cell capacities such as self-renewal and differentiation into multiple cell types. More work needs to be done to characterize the pituitary stem cells, progenitors, and transit-amplifying cells and to understand the regulation of progression through these steps.

8. CELL CYCLE REGULATION

Normal organ development requires the regulation of the transition from proliferation to differentiation and the maintenance of progenitors in a quiescent state while preserving the ability to recruit them to differentiation, while avoiding excess growth and adenoma formation. The events of the cell cycle have fundamental similarities in eukaryotic cells from the yeast to vertebrates, and a brief overview is valuable for the interpretation of normal pituitary development and disease states (Lodish, Berk, Kaiser, et al., 2008; Sacco, Hasan, Alberghina, & Vanoni, 2012). The genetic

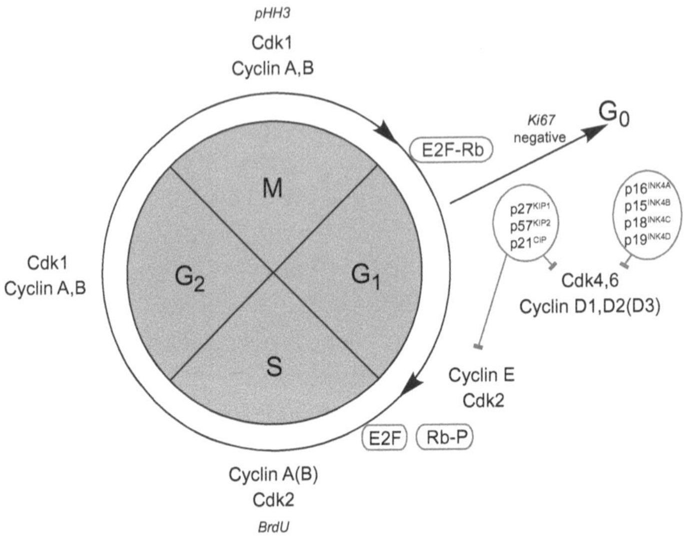

Figure 1.2 Regulation of the cell cycle. Immunostaining for phospho-histone H3 (PHH3) and incorporation of bromodeoxyuridine (BrdU) marks the mitosis (M) and synthesis (S) phases of the cell cycle. Ki67 staining is absent in quiescent cells. Association of E2F with Rb is disrupted when Rb is phosphorylated at the juncture between G1 and S phase. A variety of cyclins and cyclin-dependent kinases are expressed and have critical roles at specific points in the cell cycle.

material is copied in the synthetic (S) phase and divided between two daughter cells in the mitosis (M) phase. These two phases are separated by gaps (G_1 and G_2) as the cell prepares for the next phase (Fig. 1.2). Cell differentiation typically occurs concomitant with cell cycle exit, from G_1 to the G_0 phase. In some cases, stimulation can recruit quiescent cells to reenter the cell cycle. Such recruitment can occur during the normal tissue homeostasis, response to physiological challenges, and regeneration or wound healing (Pajalunga, Mazzola, Franchitto, Puggioni, & Crescenzi, 2008).

Cell cycle progression is regulated by critical checkpoint surveillance mechanisms. These are related to mitotic spindle assembly and position, and DNA integrity, including complete replication and proofreading for DNA damage (Curtin, 2012; Foley & Kapoor, 2013). Checkpoint blockage causes cell cycle arrest, but surveillance failure can permit uncontrolled proliferation. The length of a cell cycle varies considerably. Embryonic stem cells have a shorter cycle time than adult tissue stem cells or differentiated cell types (Lange & Calegari, 2010). Generally, the G1-phase lengthens substantially over time while cells progress toward differentiation, as demonstrated by sorting cells labeled with the DNA-binding dye propidium iodide (Behbehani, Bendall, Clutter, Fantl, & Nolan, 2012; Darzynkiewicz & Huang, 2004; Darzynkiewicz, Juan, & Bedner, 2001).

Mammalian cells have evolved a high degree of molecular regulation of the cell cycle in which multiple controller protein heterodimers provide checks and balances (Lodish et al., 2008). The presence of cyclins (*Ccn*) and cyclin-dependent kinases (*Cdk*) can dominate distinct cell cycle phases. Multiple signaling pathways regulate these proteins for endogenous checkpoint surveillance and response to external stimuli (Duronio & Xiong, 2013). Phosphorylation and dephosphorylation events and controlled protein degradation are significant parts of this process. One example is the phosphorylation of the tumor-suppressor retinoblastoma protein in late G_1 phase, which allows dissociation from the E2F1 transcription factor, and induction of gene expression necessary for G_1 to S phase progression (Talluri & Dick, 2012). Retinoblastoma remains phosphorylated and E2F1 dissociated, until M-phase is completed.

In humans and mice, there are at least 30 cyclin and 25 cyclin-dependent kinase and kinase-like genes known (http://www.ncbi.nlm.nih.gov/gene/), which illustrates the complexity and the potential for redundancy in cell cycle regulation. The cyclin-dependent kinase inhibitors (*Cdkns*) form inhibitory protein complexes with their phase representative counterparts. Many of these *Cdkns* affect at least G1-specific *Ccns* and *Cdks*, including *Ccnd1-3*, *Ccne1-2*, and *Cdk2/4/6* (Depamphilis, de Renty, Ullah, & Lee, 2012; Enders, 2012; Lodish et al., 2008). The CIP/KIP group includes *Cdkn1a*, *Cdkn1b*, and *Cdkn1c*, which are also known as *p21*, *p27*, and *p57*, respectively. A different group comprises INK4 (inhibitor of CDK4) and *Cdkn2a*, *Cdkn2b*, *Cdkn2c*, and *Cdkn2d* or *p16*, *p15*, *p18*, and *p19*, respectively (Bilodeau et al., 2009).

During pituitary development, $p57^{Kip2}$ (*Cdkn1c*) and cyclin E (*Ccne*) mark the exit of proliferating progenitors from the cell cycle, yielding non-cycling, undifferentiated precursors (Bilodeau et al., 2009). Differentiation is accompanied by extinction of $p57^{Kip2}$ and cyclin E expression and activation of $p27^{Kip1}$ (*Cdkn1b*) expression. $p57^{Kip2}$ deficiency causes pituitary overgrowth, possibly because it normally limits progenitor expansion, thereby controlling the size of the progenitor niche and the organ. $p57^{Kip2}$ and $p27^{Kip1}$ probably have redundant activities as inhibitors of the cyclin E complex. The redundancy of cell cycle regulators poses a challenge in understanding how the gateway to differentiation is regulated, and how transcription factor deficiencies result in mis-regulation of this process.

About one-third of the global knockouts of individual cell cycle regulatory genes lead to a pituitary phenotype (Table 1.4; reviewed in Quereda & Malumbres, 2009). There are two genes that cause generalized hypoplasia and three genes that cause intermediate lobe overgrowth. The intermediate lobe contains melanotrophs in mice and is rudimentary in humans (Budry et al., 2012). Anterior pituitary hyperplasia is detected in $p57^{-/-}$ mice during embryonic life, in contrast to other knockouts in which the hyperplasia appears

Table 1.4 Function of cell cycle regulators in pituitary gland growth

Pituitary phenotype	Targeted gene	References
AL hyperplasia	*Cdkn1c (p57)*	Bilodeau et al. (2009)
IL tumor	*Cdkn1b (p27)*	Fero, Rivkin, Tasch, et al. (1996), Nakayama, Ishida, Shirane, et al. (1996)
	Cdkn2c (p18)	Franklin et al. (1998)
	Rb	Hu et al. (1994), Jacks et al. (1992)
Hypoplasia	*Pttg1*	Chesnokova et al. (2005, 2008)
	Cdk4	Jirawatnotai et al. (2004), Moons et al. (2002)
None	*Cdk6,2,1*	Malumbres, Sotillo, Santamaria, et al. (2004), Martin, Odajima, Hunt, et al. (2005), Santamaria et al. (2007)
	Cyclin A, B, E, D	Brandeis, Rosewell, Carrington, et al. (1998), Carthon, Neumann, Das, et al. (2005), Fantl, Stamp, Andrews, Rosewell, and Dickson (1995), Kozar, Ciemerych, Rebel, et al. (2004), Minella, Loeb, Knecht, et al. (2008), van der Meer, Chan, Palazon, et al. (2004)
	Cdkn1a (p21)	Deng, Zhang, Harper, Elledge, and Leder (1995), Jackson et al. (2003)
	Cdkn2a (p16)	Serrano et al. (1996)
	Cdkn2b (p15)	Latres, Malumbres, Sotillo, et al. (2000)
	Cdkn2d (p19)	Zindy, van Deursen, Grosveld, Sherr, and Roussel (2000)
	E2f1	Yamasaki et al. (1996)
	Trp53	Salvador, Hollander, Nguyen, et al. (2002)

much later. The study of pituitary adenomas in mice can be confounded by the normally high incidence of adenomas at advanced ages, a characteristic that is genetic background dependent (Brayton, Treuting, & Ward, 2012). Pituitary hypoplasia is characteristic of *Pttg1* and *Cdk4* knockouts (Chesnokova, Kovacs, Castro, Zonis, & Melmed, 2005; Jirawatnotai, Aziyu, Osmundson, et al., 2004; Moons et al., 2002). Because cell cycle regulators have overlapping functions, double and triple loss-of-function mutations usually exhibit more severe phenotypes (Almeida, Muchow, Boikos, et al., 2010; Bilodeau et al., 2009; Chesnokova et al., 2005; Chesnokova, Zonis, Kovacs, et al., 2008; Franklin, Godfrey, Lee, et al., 1998; Monahan, Himes, Parfieniuk, & Raetzman, 2012). For example, triple knockouts of *Cdk2*, *Cdk4*, and *Cdk6* die at E14.5 (Santamaria, Barriere, Cerqueira, et al., 2007). There is a great deal of functional overlap and compensation among cell cycle regulators because these pathways are so important.

The role of cell cycle regulators, oncogenes, and tumor suppressors in pituitary adenomas is beginning to emerge (Salehi, Agur, et al., 2010; Salehi, Kovacs, et al., 2010). Most pituitary adenomas are benign and sporadic, although some familial types exist (Lee & Pellegata, 2013; Melmed, 2011). These include multiple endocrine neoplasia, due to mutations in menin (MEN1) or cyclin-dependent kinase inhibitor 1B (p27, MEN4), Carney Complex caused by mutations in protein kinase A regulatory subunit-1-alpha, *PRKAR1A*, and aryl hydrocarbon receptor-interacting protein. An active area of investigation involves studying the therapeutic potential of drugs that affect cell cycle regulators, like the histone deacylase inhibitors that affect the *p53*, *p21* DNA-damage pathway (Curtin, 2012; Ocker & Schneider-Stock, 2007).

9. VASCULARIZATION AND THE HYPOPHYSEAL PORTAL SYSTEM

Normal pituitary function is dependent upon development of the hypophyseal portal system. The steps in anatomical development have been cataloged using India ink, fluorescent gelatin, and immunostaining for markers such as endothelial cell adhesion molecule, yet the molecular mechanisms that regulate development of the vascular system are mostly unknown (Nakakura, Yoshida, Dohra, Suzuki, & Tanaka, 2006; Szabo & Csanyi, 1982; Ward, Stone, Raetzman, & Camper, 2006). Moreover, it is not clear whether the invasion of the vasculature has a direct role in stimulating pituitary differentiation. A variety of angiogenic and antiangiogenic factors are expressed in normal pituitary gland, and VEGF and FGF are among the best-studied (Lloyd, Vidal, Horvath, Kovacs, & Scheithauer, 2003). VEGFA is expressed at the appropriate time to have a role in stimulating the vascularization of the pars distalis. Expression coincides with penetration of the portal vessels into the pars distalis and connection with the secondary capillary plexus, at e15.5 in the rat (Nakakura et al., 2006). VEGFA expression is detectable in folliculostellate cells and some hormone-positive cells of the pars distalis. Normally, the pars distalis is much more vascularized than the pars intermedia. Ectopic expression of VEGFA in the pars intermedia causes reduced expression of the differentiation markers MSH and prohormone convertase 2 and increased growth of the lobe (Tanaka, Nakakura, Jansen, et al., 2013). Does vascularization of the pars distalis affect its differentiation? Treatment with an anti-VEGFA antibody reduces pituitary growth and serum PRL levels in mice predisposed to multiple endocrine neoplasia (Korsisaari, Ross, Wu, et al., 2008). Radiologic studies suggest that development of the hypophyseal arteries and portal system may be abnormal in some children with hypopituitarism, but it is not clear whether this is the cause or the effect (Maghnie, Genovese, Arico, et al., 1994). VEGFA expression is not sufficient for normal angiogenesis because *Prop1* mutant pituitaries express VEGFA, but they have poor vascularization, failed differentiation, and increased apoptosis (Ward et al., 2006). More research is necessary to identify the mechanisms that regulate normal vascularization and to decipher the influence of vascularization on pituitary differentiation.

10. CONCLUSION

Exploiting new technologies and diverse model systems will undoubtedly advance our understanding or pituitary development. Next-generation sequencing and bioinformatics make it feasible to monitor developmental and cell-specific changes in gene expression and chromatin accessibility on a genome-wide scale. Zebrafish provides the opportunity to enhance and suppress gene expression at various developmental times (Lohr & Hammerschmidt, 2011). The chick and frog offer the possibility of tissue transplantation during development (De Groef, Grommen, & Darras, 2008; Jenkins & Porter, 2004; Kanamori & Brown, 1996). The mouse excels in genetic engineering and has recently delivered breakthroughs in manipulating stem cells to differentiate into hormone-producing cells (Suga,

Kadoshima, Minaguchi, et al., 2011). Finally, human patients always identify the genes of relevance.

REFERENCES

Acampora, D., Gulisano, M., & Simeone, A. (2000). Genetic and molecular roles of Otx homeodomain proteins in head development. *Gene, 246*(1–2), 23–35.

Acampora, D., Mazan, S., Tuorto, F., et al. (1998). Transient dwarfism and hypogonadism in mice lacking Otx1 reveal prepubescent stage-specific control of pituitary levels of GH, FSH and LH. *Development, 125*(7), 1229–1239.

Achermann, J. C., Gu, W. X., Kotlar, T. J., et al. (1999). Mutational analysis of DAX1 in patients with hypogonadotropic hypogonadism or pubertal delay. *The Journal of Clinical Endocrinology and Metabolism, 84*(12), 4497–4500.

Achermann, J. C., Ito, M., Ito, M., Hindmarsh, P. C., & Jameson, J. L. (1999). A mutation in the gene encoding steroidogenic factor-1 causes XY sex reversal and adrenal failure in humans. *Nature Genetics, 22*(2), 125–126.

Alatzoglou, K. S., Kelberman, D., & Dattani, M. T. (2009). The role of SOX proteins in normal pituitary development. *The Journal of Endocrinology, 200*(3), 245–258.

Almeida, M. Q., Muchow, M., Boikos, S., et al. (2010). Mouse Prkar1a haploinsufficiency leads to an increase in tumors in the $Trp53^{+/-}$ or $Rb1^{+/-}$ backgrounds and chemically induced skin papillomas by dysregulation of the cell cycle and Wnt signaling. *Human Molecular Genetics, 19*(8), 1387–1398.

Arriola, D. J., Mayo, S. L., Skarra, D. V., Benson, C. A., & Thackray, V. G. (2012). FOXO1 transcription factor inhibits luteinizing hormone beta gene expression in pituitary gonadotrope cells. *The Journal of Biological Chemistry, 287*(40), 33424–33435.

Bamshad, M., Lin, R. C., Law, D. J., et al. (1997). Mutations in human TBX3 alter limb, apocrine and genital development in ulnar-mammary syndrome. *Nature Genetics, 16*(3), 311–315.

Bancalari, R. E., Gregory, L. C., McCabe, M. J., & Dattani, M. T. (2012). Pituitary gland development: An update. *Endocrine Development, 23*, 1–15.

Beby, F., & Lamonerie, T. (2013). The homeobox gene Otx2 in development and disease. *Experimental Eye Research, 111*, 9–16.

Behbehani, G. K., Bendall, S. C., Clutter, M. R., Fantl, W. J., & Nolan, G. P. (2012). Single-cell mass cytometry adapted to measurements of the cell cycle. *Cytometry Part A, 81*(7), 552–566.

Benayoun, B. A., Dipietromaria, A., Bazin, C., & Veitia, R. A. (2009). FOXL2: At the crossroads of female sex determination and ovarian function. *Advances in Experimental Medicine and Biology, 665*, 207–226.

Bentley, C. A., Zidehsarai, M. P., Grindley, J. C., Parlow, A. F., Barth-Hall, S., & Roberts, V. J. (1999). Pax6 is implicated in murine pituitary endocrine function. *Endocrine, 10*(2), 171–177.

Bernard, D. J., & Tran, S. (2013). Mechanisms of activin-stimulated FSH synthesis: The story of a pig and a FOX. *Biology of Reproduction, 88*(3), 78.

Bhangoo, A. P., Hunter, C. S., Savage, J. J., et al. (2006). Clinical case seminar: A novel LHX3 mutation presenting as combined pituitary hormonal deficiency. *The Journal of Clinical Endocrinology and Metabolism, 91*(3), 747–753.

Biason-Lauber, A. (2012). WNT4, RSPO1, and FOXL2 in sex development. *Seminars in Reproductive Medicine, 30*(5), 387–395.

Bilezikjian, L. M., Justice, N. J., Blackler, A. N., Wiater, E., & Vale, W. W. (2012). Cell-type specific modulation of pituitary cells by activin, inhibin and follistatin. *Molecular and Cellular Endocrinology, 359*(1–2), 43–52.

Bilodeau, S., Roussel-Gervais, A., & Drouin, J. (2009). Distinct developmental roles of cell cycle inhibitors p57Kip2 and p27Kip1 distinguish pituitary progenitor cell cycle exit from cell cycle reentry of differentiated cells. *Molecular and Cellular Biology, 29*(7), 1895–1908.

Blount, A. L., Schmidt, K., Justice, N. J., Vale, W. W., Fischer, W. H., & Bilezikjian, L. M. (2009). FoxL2 and Smad3 coordinately regulate follistatin gene transcription. *The Journal of Biological Chemistry, 284*(12), 7631–7645.

Bodner, M., Castrillo, J. L., Theill, L. E., Deerinck, T., Ellisman, M., & Karin, M. (1988). The pituitary-specific transcription factor GHF-1 is a homeobox-containing protein. *Cell, 55*(3), 505–518.

Bottner, A., Keller, E., Kratzsch, J., et al. (2004). PROP1 mutations cause progressive deterioration of anterior pituitary function including adrenal insufficiency: A longitudinal analysis. *The Journal of Clinical Endocrinology and Metabolism, 89*(10), 5256–5265.

Bouchoucha, Y. X., Charnay, P., & Gilardi-Hebenstreit, P. (2013). Ablation of Egr2-positive cells in male mouse anterior pituitary leads to atypical isolated GH deficiency. *Endocrinology, 154*(1), 270–282.

Brandeis, M., Rosewell, I., Carrington, M., et al. (1998). Cyclin B2-null mice develop normally and are fertile whereas cyclin B1-null mice die in utero. *Proceedings of the National Academy of Sciences of the United States of America, 95*(8), 4344–4349.

Brayton, C. F., Treuting, P. M., & Ward, J. M. (2012). Pathobiology of aging mice and GEM: Background strains and experimental design. *Veterinary Pathology, 49*(1), 85–105.

Brennan, D., Chen, X., Cheng, L., Mahoney, M., & Riobo, N. A. (2012). Noncanonical Hedgehog signaling. *Vitamins and Hormones, 88*, 55–72.

Brinkmeier, M. L., Davis, S. W., Carninci, P., et al. (2009). Discovery of transcriptional regulators and signaling pathways in the developing pituitary gland by bioinformatic and genomic approaches. *Genomics, 93*(5), 449–460.

Brinkmeier, M. L., Potok, M. A., Cha, K. B., et al. (2003). TCF and Groucho-related genes influence pituitary growth and development. *Molecular Endocrinology, 17*(11), 2152–2161.

Brinkmeier, M. L., Potok, M. A., Davis, S. W., & Camper, S. A. (2007). TCF4 deficiency expands ventral diencephalon signaling and increases induction of pituitary progenitors. *Developmental Biology, 311*(2), 396–407.

Brunkow, M. E., Jeffery, E. W., Hjerrild, K. A., et al. (2001). Disruption of a new forkhead/winged-helix protein, scurfin, results in the fatal lymphoproliferative disorder of the scurfy mouse. *Nature Genetics, 27*(1), 68–73.

Budry, L., Balsalobre, A., Gauthier, Y., et al. (2012). The selector gene Pax7 dictates alternate pituitary cell fates through its pioneer action on chromatin remodeling. *Genes & Development, 26*(20), 2299–2310.

Budry, L., Couture, C., Balsalobre, A., & Drouin, J. (2011). The Ets factor Etv1 interacts with Tpit protein for pituitary pro-opiomelanocortin (POMC) gene transcription. *The Journal of Biological Chemistry, 286*(28), 25387–25396.

Cai, C. L., Liang, X., Shi, Y., et al. (2003). Isl1 identifies a cardiac progenitor population that proliferates prior to differentiation and contributes a majority of cells to the heart. *Developmental Cell, 5*(6), 877–889.

Carthon, B. C., Neumann, C. A., Das, M., et al. (2005). Genetic replacement of cyclin D1 function in mouse development by cyclin D2. *Molecular and Cellular Biology, 25*(3), 1081–1088.

Castinetti, F., Davis, S. W., Brue, T., & Camper, S. A. (2011). Pituitary stem cell update and potential implications for treating hypopituitarism. *Endocrine Reviews, 32*(4), 453–471.

Castinetti, F., Regis, J., Dufour, H., & Brue, T. (2010). Role of stereotactic radiosurgery in the management of pituitary adenomas. *Nature Reviews Endocrinology, 6*(4), 214–223.

Cha, K. B., Douglas, K. R., Potok, M. A., Liang, H., Jones, S. N., & Camper, S. A. (2004). WNT5A signaling affects pituitary gland shape. *Mechanisms of Development, 121*(2), 183–194.

Charles, M. A., Saunders, T. L., Wood, W. M., et al. (2006). Pituitary-specific Gata2 knockout: Effects on gonadotrope and thyrotrope function. *Molecular Endocrinology, 20*(6), 1366–1377.

Charles, M. A., Suh, H., Hjalt, T. A., Drouin, J., Camper, S. A., & Gage, P. J. (2005). PITX genes are required for cell survival and Lhx3 activation. *Molecular Endocrinology, 19*(7), 1893–1903.

Cherrington, B. D., Bailey, J. S., Diaz, A. L., & Mellon, P. L. (2008). NeuroD1 and Mash1 temporally regulate GnRH receptor gene expression in immortalized mouse gonadotrope cells. *Molecular and Cellular Endocrinology, 295*(1–2), 106–114.

Chesnokova, V., Kovacs, K., Castro, A. V., Zonis, S., & Melmed, S. (2005). Pituitary hypoplasia in Pttg$^{-/-}$ mice is protective for Rb$^{+/-}$ pituitary tumorigenesis. *Molecular Endocrinology, 19*(9), 2371–2379.

Chesnokova, V., Zonis, S., Kovacs, K., et al. (2008). p21(Cip1) restrains pituitary tumor growth. *Proceedings of the National Academy of Sciences of the United States of America, 105*(45), 17498–17503.

Chesnokova, V., Zonis, S., Wawrowsky, K., et al. (2012). Clusterin and FOXL2 act concordantly to regulate pituitary gonadotroph adenoma growth. *Molecular Endocrinology, 26*(12), 2092–2103.

Cheung, L. Y., Rizzoti, K., Lovell-Badge, R., & Le Tissier, P. R. (2013). Pituitary phenotypes of mice lacking the notch signalling ligand delta-like 1 homologue. *Journal of Neuroendocrinology, 25*(4), 391–401.

Coss, D., Mellon, P. L., & Thackray, V. G. (2010). A FoxL in the Smad house: Activin regulation of FSH. *Trends in Endocrinology and Metabolism, 21*(9), 562–568.

Couly, G. F., & Le Douarin, N. M. (1985). Mapping of the early neural primordium in quail-chick chimeras. I. Developmental relationships between placodes, facial ectoderm, and prosencephalon. *Developmental Biology, 110*(2), 422–439.

Curtin, N. J. (2012). DNA repair dysregulation from cancer driver to therapeutic target. *Nature Reviews Cancer, 12*(12), 801–817.

Darzynkiewicz, Z., & Huang, X. (2004). Analysis of cellular DNA content by flow cytometry. *Current Protocols in Immunology*, Chapter 5: Unit 5.7.

Darzynkiewicz, Z., Juan, G., & Bedner, E. (2001). Determining cell cycle stages by flow cytometry. *Current Protocols in Cell Biology*, Chapter 8: Unit 8.4.

Dasen, J. S., Martinez Barbera, J. P., Herman, T. S., et al. (2001). Temporal regulation of a paired-like homeodomain repressor/TLE corepressor complex and a related activator is required for pituitary organogenesis. *Genes & Development, 15*, 3193–3207.

Dasen, J. S., O'Connell, S. M., Flynn, S. E., et al. (1999). Reciprocal interactions of Pit1 and GATA2 mediate signaling gradient-induced determination of pituitary cell types. *Cell, 97*(5), 587–598.

Dateki, S., Fukami, M., Sato, N., Muroya, K., Adachi, M., & Ogata, T. (2008). OTX2 mutation in a patient with anophthalmia, short stature, and partial growth hormone deficiency: Functional studies using the IRBP, HESX1, and POU1F1 promoters. *The Journal of Clinical Endocrinology and Metabolism, 93*(10), 3697–3702.

Dattani, M. T., Martinez-Barbera, J. P., Thomas, P. Q., et al. (1998). Mutations in the homeobox gene HESX1/Hesx1 associated with septo-optic dysplasia in human and mouse. *Nature Genetics, 19*(2), 125–133.

Davis, S. W., & Camper, S. A. (2007). Noggin regulates Bmp4 activity during pituitary induction. *Developmental Biology, 305*(1), 145–160.

Davis, S. W., Castinetti, F., Carvalho, L. R., et al. (2010). Molecular mechanisms of pituitary organogenesis: In search of novel regulatory genes. *Molecular and Cellular Endocrinology, 323*(1), 4–19.

Davis, R. J., D'Cruz, C. M., Lovell, M. A., Biegel, J. A., & Barr, F. G. (1994). Fusion of PAX7 to FKHR by the variant t(1;13)(p36;q14) translocation in alveolar rhabdomyosarcoma. *Cancer Research, 54*(11), 2869–2872.

Davis, S. W., Mortensen, A. H., & Camper, S. A. (2011). Birthdating studies reshape models for pituitary gland cell specification. *Developmental Biology, 352*(2), 215–227.
Davis, S. W., Potok, M. A., Brinkmeier, M. L., et al. (2009). Genetics, gene expression and bioinformatics of the pituitary gland. *Hormone Research, 71*(Suppl. 2), 101–115.
De Felice, M., Ovitt, C., Biffali, E., et al. (1998). A mouse model for hereditary thyroid dysgenesis and cleft palate. *Nature Genetics, 19*(4), 395–398.
De Groef, B., Grommen, S. V., & Darras, V. M. (2008). The chicken embryo as a model for developmental endocrinology: Development of the thyrotropic, corticotropic, and somatotropic axes. *Molecular and Cellular Endocrinology, 293*, 17–24.
De Moerlooze, L., Spencer-Dene, B., Revest, J. M., Hajihosseini, M., Rosewell, I., & Dickson, C. (2000). An important role for the IIIb isoform of fibroblast growth factor receptor 2 (FGFR2) in mesenchymal-epithelial signalling during mouse organogenesis. *Development, 127*(3), 483–492.
Deng, C., Zhang, P., Harper, J. W., Elledge, S. J., & Leder, P. (1995). Mice lacking p21CIP1/WAF1 undergo normal development, but are defective in G1 checkpoint control. *Cell, 82*(4), 675–684.
Depamphilis, M. L., de Renty, C. M., Ullah, Z., & Lee, C. Y. (2012). "The Octet": Eight protein kinases that control mammalian DNA replication. *Frontiers in Physiology, 3*, 368.
Diaczok, D., Romero, C., Zunich, J., Marshall, I., & Radovick, S. (2008). A novel dominant negative mutation of OTX2 associated with combined pituitary hormone deficiency. *The Journal of Clinical Endocrinology & Metabolism, 93*(11), 4351–4359.
Drouin, J., Bilodeau, S., & Roussel-Gervais, A. (2010). Stem cells, differentiation and cell cycle control in pituitary. *Frontiers of Hormone Research, 38*, 15–24.
Duronio, R. J., & Xiong, Y. (2013). Signaling pathways that control cell proliferation. *Cold Spring Harbor Perspectives in Biology, 5*(3), a008904.
Egashira, N., Takekoshi, S., Takei, M., Teramoto, A., & Osamura, R. Y. (2011). Expression of FOXL2 in human normal pituitaries and pituitary adenomas. *Modern Pathology, 24*(6), 765–773.
Ellsworth, B. S., Burns, A. T., Escudero, K. W., Duval, D. L., Nelson, S. E., & Clay, C. M. (2003). The gonadotropin releasing hormone (GnRH) receptor activating sequence (GRAS) is a composite regulatory element that interacts with multiple classes of transcription factors including Smads, AP-1 and a forkhead DNA binding protein. *Molecular and Cellular Endocrinology, 206*(1–2), 93–111.
Ellsworth, B. S., Butts, D. L., & Camper, S. A. (2008). Mechanisms underlying pituitary hypoplasia and failed cell specification in Lhx3-deficient mice. *Developmental Biology, 313*(1), 118–129.
Ellsworth, B. S., Egashira, N., Haller, J. L., et al. (2006). FOXL2 in the pituitary: Molecular, genetic, and developmental analysis. *Molecular Endocrinology, 20*(11), 2796–2805.
Enders, G. H. (2012). Mammalian interphase cdks: Dispensable master regulators of the cell cycle. *Genes & Cancer, 3*(11–12), 614–618.
Ericson, J., Norlin, S., Jessell, T. M., & Edlund, T. (1998). Integrated FGF and BMP signaling controls the progression of progenitor cell differentiation and the emergence of pattern in the embryonic anterior pituitary. *Development, 125*(6), 1005–1015.
Fantl, V., Stamp, G., Andrews, A., Rosewell, I., & Dickson, C. (1995). Mice lacking cyclin D1 are small and show defects in eye and mammary gland development. *Genes & Development, 9*(19), 2364–2372.
Fauquier, T., Rizzoti, K., Dattani, M., Lovell-Badge, R., & Robinson, I. C. (2008). SOX2-expressing progenitor cells generate all of the major cell types in the adult mouse pituitary gland. *Proceedings of the National Academy of Sciences of the United States of America, 105*(8), 2907–2912.
Ferdous, A., Morris, J., Abedin, M. J., Collins, S., Richardson, J. A., & Hill, J. A. (2011). Forkhead factor FoxO1 is essential for placental morphogenesis in the developing

embryo. *Proceedings of the National Academy of Sciences of the United States of America, 108*(39), 16307–16312.
Fero, M. L., Rivkin, M., Tasch, M., et al. (1996). A syndrome of multiorgan hyperplasia with features of gigantism, tumorigenesis, and female sterility in p27(Kip1)-deficient mice. *Cell, 85*(5), 733–744.
Foley, E. A., & Kapoor, T. M. (2013). Microtubule attachment and spindle assembly checkpoint signalling at the kinetochore. *Nature Reviews Molecular Cell Biology, 14*(1), 25–37.
Franca, M. M., Jorge, A. A., Carvalho, L. R., et al. (2010). Novel heterozygous nonsense GLI2 mutations in patients with hypopituitarism and ectopic posterior pituitary lobe without holoprosencephaly. *The Journal of Clinical Endocrinology and Metabolism, 95*(11), E384–E391.
Franklin, D. S., Godfrey, V. L., Lee, H., et al. (1998). CDK inhibitors p18(INK4c) and p27 (Kip1) mediate two separate pathways to collaboratively suppress pituitary tumorigenesis. *Genes & Development, 12*(18), 2899–2911.
Gage, P. J., Brinkmeier, M. L., Scarlett, L. M., Knapp, L. T., Camper, S. A., & Mahon, K. A. (1996). The Ames dwarf gene, df, is required early in pituitary ontogeny for the extinction of Rpx transcription and initiation of lineage-specific cell proliferation. *Molecular Endocrinology, 10*(12), 1570–1581.
Gage, P. J., Suh, H., & Camper, S. A. (1999). Dosage requirement of Pitx2 for development of multiple organs. *Development, 126*(20), 4643–4651.
Garcia-Lavandeira, M., Quereda, V., Flores, I., et al. (2009). A GRFa2/Prop1/stem (GPS) cell niche in the pituitary. *PLoS One, 4*(3), e4815.
Gaston-Massuet, C., Andoniadou, C. L., Signore, M., et al. (2008). Genetic interaction between the homeobox transcription factors HESX1 and SIX3 is required for normal pituitary development. *Developmental Biology, 324*(2), 322–333.
Gaston-Massuet, C., Andoniadou, C. L., Signore, M., et al. (2011). Increased Wingless (Wnt) signaling in pituitary progenitor/stem cells gives rise to pituitary tumors in mice and humans. *Proceedings of the National Academy of Sciences of the United States of America, 108*(28), 11482–11487.
Ghochani, Y., Saini, J. K., Mellon, P. L., & Thackray, V. G. (2012). FOXL2 is involved in the synergy between activin and progestins on the follicle-stimulating hormone beta-subunit promoter. *Endocrinology, 153*(4), 2023–2033.
Gleiberman, A. S., Fedtsova, N. G., & Rosenfeld, M. G. (1999). Tissue interactions in the induction of anterior pituitary: Role of the ventral diencephalon, mesenchyme, and notochord. *Developmental Biology, 213*(2), 340–353.
Goldberg, L. B., Aujla, P. K., & Raetzman, L. T. (2011). Persistent expression of activated Notch inhibits corticotrope and melanotrope differentiation and results in dysfunction of the HPA axis. *Developmental Biology, 358*(1), 23–32.
Gumbel, J. H., Patterson, E. M., Owusu, S. A., et al. (2012). The forkhead transcription factor, Foxd1, is necessary for pituitary luteinizing hormone expression in mice. *PLoS One*, 7(12), e52156.
Gurnett, C. A., Alaee, F., Kruse, L. M., et al. (2008). Asymmetric lower-limb malformations in individuals with homeobox PITX1 gene mutation. *American Journal of Human Genetics, 83*(5), 616–622.
Gutierrez-Hartmann, A., Duval, D. L., & Bradford, A. P. (2007). ETS transcription factors in endocrine systems. *Trends in Endocrinology and Metabolism, 18*(4), 150–158.
Hamel, B. C., Smits, A. P., Otten, B. J., van den Helm, B., Ropers, H. H., & Mariman, E. C. (1996). Familial X-linked mental retardation and isolated growth hormone deficiency: Clinical and molecular findings. *American Journal of Medical Genetics, 64*(1), 35–41.
Hatini, V., Huh, S. O., Herzlinger, D., Soares, V. C., & Lai, E. (1996). Essential role of stromal mesenchyme in kidney morphogenesis revealed by targeted disruption of Winged Helix transcription factor BF-2. *Genes & Development, 10*(12), 1467–1478.

Hermesz, E., Mackem, S., & Mahon, K. A. (1996). Rpx: A novel anterior-restricted homeobox gene progressively activated in the prechordal plate, anterior neural plate and Rathke's pouch of the mouse embryo. *Development, 122*(1), 41–52.

Hide, T., Hatakeyama, J., Kimura-Yoshida, C., et al. (2002). Genetic modifiers of otocephalic phenotypes in Otx2 heterozygous mutant mice. *Development, 129*(18), 4347–4357.

Himes, A. D., & Raetzman, L. T. (2009). Premature differentiation and aberrant movement of pituitary cells lacking both Hes1 and Prop1. *Developmental Biology, 325*(1), 151–161.

Ho, H. Y., Susman, M. W., Bikoff, J. B., et al. (2012). Wnt5a-Ror-Dishevelled signaling constitutes a core developmental pathway that controls tissue morphogenesis. *Proceedings of the National Academy of Sciences of the United States of America, 109*(11), 4044–4051.

Hu, N., Gutsmann, A., Herbert, D. C., Bradley, A., Lee, W. H., & Lee, E. Y. (1994). Heterozygous Rb-1 delta 20/+ mice are predisposed to tumors of the pituitary gland with a nearly complete penetrance. *Oncogene, 9*(4), 1021–1027.

Hui, C. C., & Angers, S. (2011). Gli proteins in development and disease. *Annual Review of Cell and Developmental Biology, 27*, 513–537.

Ingraham, H. A., Chen, R. P., Mangalam, H. J., et al. (1988). A tissue-specific transcription factor containing a homeodomain specifies a pituitary phenotype. *Cell, 55*(3), 519–529.

Ingraham, H. A., Lala, D. S., Ikeda, Y., et al. (1994). The nuclear receptor steroidogenic factor 1 acts at multiple levels of the reproductive axis. *Genes & Development, 8*(19), 2302–2312.

Jacks, T., Fazeli, A., Schmitt, E. M., Bronson, R. T., Goodell, M. A., & Weinberg, R. A. (1992). Effects of an Rb mutation in the mouse. *Nature, 359*(6393), 295–300.

Jackson, B. C., Carpenter, C., Nebert, D. W., & Vasiliou, V. (2010). Update of human and mouse forkhead box (FOX) gene families. *Human Genomics, 4*(5), 345–352.

Jackson, R. J., Engelman, R. W., Coppola, D., Cantor, A. B., Wharton, W., & Pledger, W. J. (2003). p21Cip1 nullizygosity increases tumor metastasis in irradiated mice. *Cancer Research, 63*(12), 3021–3025.

Japon, M. A., Rubinstein, M., & Low, M. J. (1994). In situ hybridization analysis of anterior pituitary hormone gene expression during fetal mouse development. *The Journal of Histochemistry and Cytochemistry, 42*(8), 1117–1125.

Jayakody, S. A., Andoniadou, C. L., Gaston-Massuet, C., et al. (2012). SOX2 regulates the hypothalamic-pituitary axis at multiple levels. *The Journal of Clinical Investigation, 122*(10), 3635–3646.

Jenkins, S. A., & Porter, T. E. (2004). Ontogeny of the hypothalamo-pituitary-adrenocortical axis in the chicken embryo: A review. *Domestic Animal Endocrinology, 26*(4), 267–275.

Jirawatnotai, S., Aziyu, A., Osmundson, E. C., et al. (2004). Cdk4 is indispensable for postnatal proliferation of the anterior pituitary. *The Journal of Biological Chemistry, 279*(49), 51100–51106.

Joslin, J. M., Fernald, A. A., Tennant, T. R., et al. (2007). Haploinsufficiency of EGR1, a candidate gene in the del(5q), leads to the development of myeloid disorders. *Blood, 110*(2), 719–726.

Jung, D. O., Jasurda, J. S., Egashira, N., & Ellsworth, B. S. (2012). The forkhead transcription factor, FOXP3, is required for normal pituitary gonadotropin expression in mice. *Biology of Reproduction, 86*(5), 144, 1–9.

Justice, N. J., Blount, A. L., Pelosi, E., Schlessinger, D., Vale, W., & Bilezikjian, L. M. (2011). Impaired FSHbeta expression in the pituitaries of Foxl2 mutant animals. *Molecular Endocrinology, 25*(8), 1404–1415.

Kaestner, K. H., Knochel, W., & Martinez, D. E. (2000). Unified nomenclature for the winged helix/forkhead transcription factors. *Genes & Development, 14*(2), 142–146.

Kanamori, A., & Brown, D. D. (1996). The analysis of complex developmental programmes: Amphibian metamorphosis. *Genes to Cells: Devoted to Molecular & Cellular Mechanisms, 1*(5), 429–435.

Karin, M., Theill, L., Castrillo, J. L., McCormick, A., & Brady, H. (1990). Tissue-specific expression of the growth hormone gene and its control by growth hormone factor-1. *Recent Progress in Hormone Research, 46*, 43–57, discussion 8.

Kelberman, D., Rizzoti, K., Avilion, A., et al. (2006). Mutations within Sox2/SOX2 are associated with abnormalities in the hypothalamo-pituitary-gonadal axis in mice and humans. *The Journal of Clinical Investigation, 116*(9), 2442–2455.

Kelberman, D., Rizzoti, K., Lovell-Badge, R., Robinson, I. C., & Dattani, M. T. (2009). Genetic regulation of pituitary gland development in human and mouse. *Endocrine Reviews, 30*(7), 790–829.

Kioussi, C., O'Connell, S., St-Onge, L., et al. (1999). Pax6 is essential for establishing ventral-dorsal cell boundaries in pituitary gland development. *Proceedings of the National Academy of Sciences of the United States of America, 96*(25), 14378–14382.

Klinakis, A., Lobry, C., Abdel-Wahab, O., et al. (2011). A novel tumour-suppressor function for the Notch pathway in myeloid leukaemia. *Nature, 473*(7346), 230–233.

Korsisaari, N., Ross, J., Wu, X., et al. (2008). Blocking vascular endothelial growth factor-A inhibits the growth of pituitary adenomas and lowers serum prolactin level in a mouse model of multiple endocrine neoplasia type 1. *Clinical Cancer Research: An Official Journal of the American Association for Cancer Research, 14*(1), 249–258.

Kozar, K., Ciemerych, M. A., Rebel, V. I., et al. (2004). Mouse development and cell proliferation in the absence of D-cyclins. *Cell, 118*(4), 477–491.

Kuo, F. T., Fan, K., Bentsi-Barnes, I., Barlow, G. M., & Pisarska, M. D. (2012). Mouse forkhead L2 maintains repression of FSH-dependent genes in the granulosa cell. *Reproduction, 144*(4), 485–494.

Lala, D. S., Rice, D. A., & Parker, K. L. (1992). Steroidogenic factor I, a key regulator of steroidogenic enzyme expression, is the mouse homolog of fushi tarazu-factor I. *Molecular Endocrinology, 6*(8), 1249–1258.

Lalmansingh, A. S., Karmakar, S., Jin, Y., & Nagaich, A. K. (2012). Multiple modes of chromatin remodeling by Forkhead box proteins. *Biochimica et Biophysica Acta, 1819*(7), 707–715.

Lamba, P., Fortin, J., Tran, S., Wang, Y., & Bernard, D. J. (2009). A novel role for the forkhead transcription factor FOXL2 in activin A-regulated follicle-stimulating hormone beta subunit transcription. *Molecular Endocrinology, 23*(7), 1001–1013.

Lamba, P., Wang, Y., Tran, S., et al. (2010). Activin A regulates porcine follicle-stimulating hormone beta-subunit transcription via cooperative actions of SMADs and FOXL2. *Endocrinology, 151*(11), 5456–5467.

Lamolet, B., Poulin, G., Chu, K., Guillemot, F., Tsai, M. J., & Drouin, J. (2004). Tpit-independent function of NeuroD1(BETA2) in pituitary corticotroph differentiation. *Molecular Endocrinology, 18*(4), 995–1003.

Lamolet, B., Pulichino, A. M., Lamonerie, T., et al. (2001). A pituitary cell-restricted T box factor, Tpit, activates POMC transcription in cooperation with Pitx homeoproteins. *Cell, 104*(6), 849–859.

Lamonerie, T., Tremblay, J. J., Lanctot, C., Therrien, M., Gauthier, Y., & Drouin, J. (1996). Ptx1, a bicoid-related homeo box transcription factor involved in transcription of the pro-opiomelanocortin gene. *Genes & Development, 10*(10), 1284–1295.

Lange, C., & Calegari, F. (2010). Cdks and cyclins link G1 length and differentiation of embryonic, neural and hematopoietic stem cells. *Cell Cycle, 9*(10), 1893–1900.

Latres, E., Malumbres, M., Sotillo, R., et al. (2000). Limited overlapping roles of P15 (INK4b) and P18(INK4c) cell cycle inhibitors in proliferation and tumorigenesis. *The EMBO Journal, 19*(13), 3496–3506.

Lavoie, P. L., Budry, L., Balsalobre, A., & Drouin, J. (2008). Developmental dependence on NurRE and EboxNeuro for expression of pituitary proopiomelanocortin. *Molecular Endocrinology*, *22*(7), 1647–1657.
Le Tissier, P. R., Hodson, D. J., Lafont, C., Fontanaud, P., Schaeffer, M., & Mollard, P. (2012). Anterior pituitary cell networks. *Frontiers in Neuroendocrinology*, *33*(3), 252–266.
Lee, M., & Pellegata, N. S. (2013). Multiple endocrine neoplasia type 4. *Frontiers of Hormone Research*, *41*, 63–78.
Lepore, D. A., Roeszler, K., Wagner, J., Ross, S. A., Bauer, K., & Thomas, P. Q. (2005). Identification and enrichment of colony-forming cells from the adult murine pituitary. *Experimental Cell Research*, *308*(1), 166–176.
Levinson, R. S., Batourina, E., Choi, C., Vorontchikhina, M., Kitajewski, J., & Mendelsohn, C. L. (2005). Foxd1-dependent signals control cellularity in the renal capsule, a structure required for normal renal development. *Development*, *132*(3), 529–539.
Li, S., Crenshaw, E. B., 3rd., Rawson, E. J., Simmons, D. M., Swanson, L. W., & Rosenfeld, M. G. (1990). Dwarf locus mutants lacking three pituitary cell types result from mutations in the POU-domain gene pit-1. *Nature*, *347*(6293), 528–533.
Li, X., Perissi, V., Liu, F., Rose, D. W., & Rosenfeld, M. G. (2002). Tissue-specific regulation of retinal and pituitary precursor cell proliferation. *Science*, *297*(5584), 1180–1183.
Lloyd, R. V., Vidal, S., Horvath, E., Kovacs, K., & Scheithauer, B. (2003). Angiogenesis in normal and neoplastic pituitary tissues. *Microscopy Research and Technique*, *60*(2), 244–250.
Lodish, H., Berk, A., Kaiser, C. A., et al. (2008). Regulating the eukaryotic cell cycle. In *Molecular cell biology* (6th ed., pp. 847–904). New York, NY: W.H. Freeman and Company.
Lohr, H., & Hammerschmidt, M. (2011). Zebrafish in endocrine systems: Recent advances and implications for human disease. *Annual Review of Physiology*, *73*, 183–211.
Luo, X., Ikeda, Y., & Parker, K. L. (1994). A cell-specific nuclear receptor is essential for adrenal and gonadal development and sexual differentiation. *Cell*, 77(4), 481–490.
Machinis, K., Pantel, J., Netchine, I., et al. (2001). Syndromic short stature in patients with a germline mutation in the LIM homeobox LHX4. *American Journal of Human Genetics*, *69*(5), 961–968.
Maghnie, M., Genovese, E., Arico, M., et al. (1994). Evolving pituitary hormone deficiency is associated with pituitary vasculopathy: Dynamic MR study in children with hypopituitarism, diabetes insipidus, and Langerhans cell histiocytosis. *Radiology*, *193*(2), 493–499.
Majumdar, S., Farris, C. L., Kabat, B. E., Jung, D. O., & Ellsworth, B. S. (2012). Forkhead Box O1 is present in quiescent pituitary cells during development and is increased in the absence of p27 Kip1. *PLoS One*, 7(12), e52136.
Malecki, M. T., Jhala, U. S., Antonellis, A., et al. (1999). Mutations in NEUROD1 are associated with the development of type 2 diabetes mellitus. *Nature Genetics*, *23*(3), 323–328.
Malumbres, M., Sotillo, R., Santamaria, D., et al. (2004). Mammalian cells cycle without the D-type cyclin-dependent kinases Cdk4 and Cdk6. *Cell*, *118*(4), 493–504.
Mangalam, H. J., Albert, V. R., Ingraham, H. A., et al. (1989). A pituitary POU domain protein, Pit-1, activates both growth hormone and prolactin promoters transcriptionally. *Genes & Development*, *3*(7), 946–958.
Martin, A., Odajima, J., Hunt, S. L., et al. (2005). Cdk2 is dispensable for cell cycle inhibition and tumor suppression mediated by p27(Kip1) and p21(Cip1). *Cancer Cell*, 7(6), 591–598.
Maruoka, Y., Ohbayashi, N., Hoshikawa, M., Itoh, N., Hogan, B. L., & Furuta, Y. (1998). Comparison of the expression of three highly related genes, Fgf8, Fgf17 and Fgf18, in the mouse embryo. *Mechanisms of Development*, *74*(1–2), 175–177.
Matsuo, I., Kuratani, S., Kimura, C., Takeda, N., & Aizawa, S. (1995). Mouse Otx2 functions in the formation and patterning of rostral head. *Genes & Development*, *9*(21), 2646–2658.

McCabe, M. J., Gaston-Massuet, C., Tziaferi, V., et al. (2011). Novel FGF8 mutations associated with recessive holoprosencephaly, craniofacial defects, and hypothalamo-pituitary dysfunction. *The Journal of Clinical Endocrinology and Metabolism, 96*(10), E1709–E1718.

McCabe, K. L., Manzo, A., Gammill, L. S., & Bronner-Fraser, M. (2004). Discovery of genes implicated in placode formation. *Developmental Biology, 274*(2), 462–477.

Medina-Martinez, O., Amaya-Manzanares, F., Liu, C., et al. (2009). Cell-autonomous requirement for rx function in the mammalian retina and posterior pituitary. *PLoS One, 4*(2), e4513.

Melmed, S. (2011). Pathogenesis of pituitary tumors. *Nature Reviews Endocrinology,* 7(5), 257–266.

Meyers, E. N., Lewandoski, M., & Martin, G. R. (1998). An Fgf8 mutant allelic series generated by Cre- and Flp-mediated recombination. *Nature Genetics, 18*(2), 136–141.

Mikami, S. (1992). Hypophysis. In A. Matsumoto & S. Ishii (Eds.), *Atlas of endocrine organs, vertebrates and invertebrates* (pp. 39–62). Berlin: Springer-Verlag.

Minella, A. C., Loeb, K. R., Knecht, A., et al. (2008). Cyclin E phosphorylation regulates cell proliferation in hematopoietic and epithelial lineages in vivo. *Genes & Development, 22*(12), 1677–1689.

Miraoui, H., Dwyer, A. A., Sykiotis, G. P., et al. (2013). Mutations in FGF17, IL17RD, DUSP6, SPRY4, and FLRT3 are identified in individuals with congenital hypogonadotropic hypogonadism. *American Journal of Human Genetics, 92*(5), 725–743.

Mollard, P., Hodson, D. J., Lafont, C., Rizzoti, K., & Drouin, J. (2012). A tridimensional view of pituitary development and function. *Trends in Endocrinology and Metabolism, 23*(6), 261–269.

Monahan, P., Himes, A. D., Parfieniuk, A., & Raetzman, L. T. (2012). p21, an important mediator of quiescence during pituitary tumor formation, is dispensable for normal pituitary development during embryogenesis. *Mechanisms of Development, 128*(11–12), 640–652.

Monahan, P., Rybak, S., & Raetzman, L. T. (2009). The notch target gene HES1 regulates cell cycle inhibitor expression in the developing pituitary. *Endocrinology, 150*(9), 4386–4394.

Moons, D. S., Jirawatnotai, S., Parlow, A. F., Gibori, G., Kineman, R. D., & Kiyokawa, H. (2002). Pituitary hypoplasia and lactotroph dysfunction in mice deficient for cyclin-dependent kinase-4. *Endocrinology, 143*(8), 3001–3008.

Mortensen, A. H., MacDonald, J. W., Ghosh, D., & Camper, S. A. (2011). Candidate genes for panhypopituitarism identified by gene expression profiling. *Physiological Genomics, 43*(19), 1105–1116.

Nakakura, T., Yoshida, M., Dohra, H., Suzuki, M., & Tanaka, S. (2006). Gene expression of vascular endothelial growth factor-A in the pituitary during formation of the vascular system in the hypothalamic-pituitary axis of the rat. *Cell and Tissue Research, 324*(1), 87–95.

Nakayama, K., Ishida, N., Shirane, M., et al. (1996). Mice lacking p27(Kip1) display increased body size, multiple organ hyperplasia, retinal dysplasia, and pituitary tumors. *Cell, 85*(5), 707–720.

Nasonkin, I. O., Ward, R. D., Raetzman, L. T., et al. (2004). Pituitary hypoplasia and respiratory distress syndrome in Prop1 knockout mice. *Human Molecular Genetics, 13*(22), 2727–2735.

Ocker, M., & Schneider-Stock, R. (2007). Histone deacetylase inhibitors: Signalling towards p21cip1/waf1. *The International Journal of Biochemistry & Cell Biology, 39*(7–8), 1367–1374.

Ohbayashi, N., Hoshikawa, M., Kimura, S., Yamasaki, M., Fukui, S., & Itoh, N. (1998). Structure and expression of the mRNA encoding a novel fibroblast growth factor, FGF-18. *The Journal of Biological Chemistry, 273*(29), 18161–18164.

Ohuchi, H., Hori, Y., Yamasaki, M., et al. (2000). FGF10 acts as a major ligand for FGF receptor 2 IIIb in mouse multi-organ development. *Biochemical and Biophysical Research Communications, 277*(3), 643–649.

Olson, L. E., Tollkuhn, J., Scafoglio, C., et al. (2006). Homeodomain-mediated beta-catenin-dependent switching events dictate cell-lineage determination. *Cell, 125*(3), 593–605.

Pajalunga, D., Mazzola, A., Franchitto, A., Puggioni, E., & Crescenzi, M. (2008). The logic and regulation of cell cycle exit and reentry. *Cellular and Molecular Life Sciences, 65*(1), 8–15.

Pernasetti, F., Toledo, S. P., Vasilyev, V. V., et al. (2000). Impaired adrenocorticotropin-adrenal axis in combined pituitary hormone deficiency caused by a two-base pair deletion (301-302delAG) in the prophet of Pit-1 gene. *The Journal of Clinical Endocrinology and Metabolism, 85*(1), 390–397.

Porter, F. D., Drago, J., Xu, Y., et al. (1997). Lhx2, a LIM homeobox gene, is required for eye, forebrain, and definitive erythrocyte development. *Development, 124*(15), 2935–2944.

Potok, M. A., Cha, K. B., Hunt, A., et al. (2008). WNT signaling affects gene expression in the ventral diencephalon and pituitary gland growth. *Developmental Dynamics, 237*(4), 1006–1020.

Puertas-Avendano, R. A., Gonzalez-Gomez, M. J., Ruvira, M. D., et al. (2011). Role of the non-canonical notch ligand delta-like protein 1 in hormone-producing cells of the adult male mouse pituitary. *Journal of Neuroendocrinology, 23*(9), 849–859.

Pulichino, A. M., Lamolet, B., Vallette-Kasic, S., et al. (2004). $Tpit^{-/-}NeuroD1^{-/-}$ mice reveal novel aspects of corticotroph development. *Endocrine Research, 30*(4), 551–552.

Pulichino, A. M., Vallette-Kasic, S., Couture, C., et al. (2003). Human and mouse TPIT gene mutations cause early onset pituitary ACTH deficiency. *Genes & Development, 17*(6), 711–716.

Pulichino, A. M., Vallette-Kasic, S., Tsai, J. P., Couture, C., Gauthier, Y., & Drouin, J. (2003). Tpit determines alternate fates during pituitary cell differentiation. *Genes & Development, 17*(6), 738–747.

Quereda, V., & Malumbres, M. (2009). Cell cycle control of pituitary development and disease. *Journal of Molecular Endocrinology, 42*(2), 75–86.

Radio, F. C., Bernardini, L., Loddo, S., et al. (2010). TBX2 gene duplication associated with complex heart defect and skeletal malformations. *American Journal of Medical Genetics Part A, 152A*(8), 2061–2066.

Raetzman, L. T., Cai, J. X., & Camper, S. A. (2007). Hes1 is required for pituitary growth and melanotrope specification. *Developmental Biology, 304*(2), 455–466.

Raetzman, L. T., Ross, S. A., Cook, S., Dunwoodie, S. L., Camper, S. A., & Thomas, P. Q. (2004). Developmental regulation of Notch signaling genes in the embryonic pituitary: Prop1 deficiency affects Notch2 expression. *Developmental Biology, 265*(2), 329–340.

Raetzman, L. T., Ward, R., & Camper, S. A. (2002). Lhx4 and Prop1 are required for cell survival and expansion of the pituitary primordia. *Development, 129*(18), 4229–4239.

Rathke, M. H. (1839). *Entwicklungsgeschichte der Natter (Coluber natrix).* Königsberg: Bornträger.

Richards, J. S., & Pangas, S. A. (2010). The ovary: Basic biology and clinical implications. *The Journal of Clinical Investigation, 120*(4), 963–972.

Richards, J. S., Sharma, S. C., Falender, A. E., & Lo, Y. H. (2002). Expression of FKHR, FKHRL1, and AFX genes in the rodent ovary: Evidence for regulation by IGF-I, estrogen, and the gonadotropins. *Molecular Endocrinology, 16*(3), 580–599.

Rizzoti, K., Brunelli, S., Carmignac, D., Thomas, P. Q., Robinson, I. C., & Lovell-Badge, R. (2004). SOX3 is required during the formation of the hypothalamo-pituitary axis. *Nature Genetics, 36*(3), 247–255.

Sacco, E., Hasan, M. M., Alberghina, L., & Vanoni, M. (2012). Comparative analysis of the molecular mechanisms controlling the initiation of chromosomal DNA replication in yeast and in mammalian cells. *Biotechnology Advances, 30*(1), 73–98.

Salehi, F., Agur, A., Scheithauer, B. W., Kovacs, K., Lloyd, R. V., & Cusimano, M. (2010). Biomarkers of pituitary neoplasms: A review (Part II). *Neurosurgery, 67*(6), 1790–1798, discussion 8.

Salehi, F., Kovacs, K., Scheithauer, B. W., et al. (2010). Immunohistochemical expression of pituitary tumor transforming gene (PTTG) in pituitary adenomas: A correlative study of tumor subtypes. *International Journal of Surgical Pathology, 18*(1), 5–13.

Salvador, J. M., Hollander, M. C., Nguyen, A. T., et al. (2002). Mice lacking the p53-effector gene Gadd45a develop a lupus-like syndrome. *Immunity, 16*(4), 499–508.

Santamaria, D., Barriere, C., Cerqueira, A., et al. (2007). Cdk1 is sufficient to drive the mammalian cell cycle. *Nature, 448*(7155), 811–815.

Schlosser, G. (2006). Induction and specification of cranial placodes. *Developmental Biology, 294*(2), 303–351.

Schmidt, D., Ovitt, C. E., Anlag, K., et al. (2004). The murine winged-helix transcription factor Foxl2 is required for granulosa cell differentiation and ovary maintenance. *Development, 131*(4), 933–942.

Schwind, J. L. (1928). The development of the hypophysis cerebri of the albino rat. *American Journal of Anatomy, 41*, 295–315.

Scully, K. M., & Rosenfeld, M. G. (2002). Pituitary development: Regulatory codes in mammalian organogenesis. *Science, 295*(5563), 2231–2235.

Semina, E. V., Reiter, R., Leysens, N. J., et al. (1996). Cloning and characterization of a novel bicoid-related homeobox transcription factor gene, RIEG, involved in Rieger syndrome. *Nature Genetics, 14*(4), 392–399.

Serrano, M., Lee, H., Chin, L., Cordon-Cardo, C., Beach, D., & DePinho, R. A. (1996). Role of the INK4a locus in tumor suppression and cell mortality. *Cell, 85*(1), 27–37.

Seuntjens, E., & Denef, C. (1999). Progenitor cells in the embryonic anterior pituitary abruptly and concurrently depress mitotic rate before progressing to terminal differentiation. *Molecular and Cellular Endocrinology, 150*(1–2), 57–63.

Sheng, H. Z., Moriyama, K., Yamashita, T., et al. (1997). Multistep control of pituitary organogenesis. *Science, 278*(5344), 1809–1812.

Sheng, H. Z., Zhadanov, A. B., Mosinger, B., Jr., et al. (1996). Specification of pituitary cell lineages by the LIM homeobox gene Lhx3. *Science, 272*(5264), 1004–1007.

Simmons, D. M., Voss, J. W., Ingraham, H. A., et al. (1990). Pituitary cell phenotypes involve cell-specific Pit-1 mRNA translation and synergistic interactions with other classes of transcription factors. *Genes & Development, 4*(5), 695–711.

Solomon, B. D., Pineda-Alvarez, D. E., Balog, J. Z., et al. (2009). Compound heterozygosity for mutations in PAX6 in a patient with complex brain anomaly, neonatal diabetes mellitus, and microophthalmia. *American Journal of Medical Genetics Part A, 149A*(11), 2543–2546.

Sornson, M. W., Wu, W., Dasen, J. S., et al. (1996). Pituitary lineage determination by the Prophet of Pit-1 homeodomain factor defective in Ames dwarfism. *Nature, 384*(6607), 327–333.

Spielmann, M., Brancati, F., Krawitz, P. M., et al. (2012). Homeotic arm-to-leg transformation associated with genomic rearrangements at the PITX1 locus. *American Journal of Human Genetics, 91*(4), 629–635.

Suga, H., Kadoshima, T., Minaguchi, M., et al. (2011). Self-formation of functional adenohypophysis in three-dimensional culture. *Nature, 480*(7375), 57–62.

Susa, T., Kato, T., Yoshida, S., Yako, H., Higuchi, M., & Kato, Y. (2012). Paired-related homeodomain proteins Prx1 and Prx2 are expressed in embryonic pituitary stem/progenitor cells and may be involved in the early stage of pituitary differentiation. *Journal of Neuroendocrinology, 24*(9), 1201–1212.

Sykiotis, G. P., Pitteloud, N., Seminara, S. B., Kaiser, U. B., & Crowley, W. F., Jr. (2010). Deciphering genetic disease in the genomic era: The model of GnRH deficiency. *Science Translational Medicine*, *2*(32), 32rv2.

Szabo, K., & Csanyi, K. (1982). The vascular architecture of the developing pituitary-median eminence complex in the rat. *Cell and Tissue Research*, *224*, 563–577.

Szeto, D. P., Rodriguez-Esteban, C., Ryan, A. K., et al. (1999). Role of the Bicoid-related homeodomain factor Pitx1 in specifying hindlimb morphogenesis and pituitary development. *Genes & Development*, *13*(4), 484–494.

Takuma, N., Sheng, H. Z., Furuta, Y., et al. (1998). Formation of Rathke's pouch requires dual induction from the diencephalon. *Development*, *125*(23), 4835–4840.

Talluri, S., & Dick, F. A. (2012). Regulation of transcription and chromatin structure by pRB: Here, there and everywhere. *Cell Cycle*, *11*(17), 3189–3198.

Tanaka, S., Nakakura, T., Jansen, E. J., et al. (2013). Angiogenesis in the intermediate lobe of the pituitary gland alters its structure and function. *General and Comparative Endocrinology*, *185*, 10–18.

Tang, K., Bartke, A., Gardiner, C. S., Wagner, T. E., & Yun, J. S. (1993). Gonadotropin secretion, synthesis, and gene expression in two types of bovine growth hormone transgenic mice. *Biology of Reproduction*, *49*(2), 346–353.

Tatsumi, K., Miyai, K., Notomi, T., et al. (1992). Cretinism with combined hormone deficiency caused by a mutation in the PIT1 gene. *Nature Genetics*, *1*(1), 56–58.

Topilko, P., Schneider-Maunoury, S., Levi, G., et al. (1998). Multiple pituitary and ovarian defects in Krox-24 (NGFI-A, Egr-1)-targeted mice. *Molecular Endocrinology*, *12*(1), 107–122.

Tran, S., Lamba, P., Wang, Y., & Bernard, D. J. (2011). SMADs and FOXL2 synergistically regulate murine FSHbeta transcription via a conserved proximal promoter element. *Molecular Endocrinology*, *25*(7), 1170–1183.

Tran, S., Zhou, X., Lafleur, C., et al. (2013). Impaired fertility and FSH synthesis in gonadotrope-specific Foxl2 knockout mice. *Molecular Endocrinology*, *27*(3), 407–421.

Treier, M., Gleiberman, A. S., O'Connell, S. M., et al. (1998). Multistep signaling requirements for pituitary organogenesis in vivo. *Genes & Development*, *12*(11), 1691–1704.

Treier, M., O'Connell, S., Gleiberman, A., et al. (2001). Hedgehog signaling is required for pituitary gland development. *Development*, *128*(3), 377–386.

Trowe, M. O., Zhao, L., Weiss, A. C., Christoffels, V., Epstein, D. J., & Kispert, A. (2013). Inhibition of Sox2-dependent activation of Shh in the ventral diencephalon by Tbx3 is required for formation of the neurohypophysis. *Development*, *140*(11), 2299–2309.

Uda, M., Ottolenghi, C., Crisponi, L., et al. (2004). Foxl2 disruption causes mouse ovarian failure by pervasive blockage of follicle development. *Human Molecular Genetics*, *13*(11), 1171–1181.

Uhlenhaut, N. H., Jakob, S., Anlag, K., et al. (2009). Somatic sex reprogramming of adult ovaries to testes by FOXL2 ablation. *Cell*, *139*(6), 1130–1142.

Vakili, H., & Cattini, P. A. (2012). The hidden but positive role for glucocorticoids in the regulation of growth hormone-producing cells. *Molecular and Cellular Endocrinology*, *363*(1–2), 1–9.

van der Meer, T., Chan, W. Y., Palazon, L. S., et al. (2004). Cyclin A1 protein shows haplo-insufficiency for normal fertility in male mice. *Reproduction*, *127*(4), 503–511.

Van Esch, H., Groenen, P., Nesbit, M. A., et al. (2000). GATA3 haplo-insufficiency causes human HDR syndrome. *Nature*, *406*(6794), 419–422.

Vandeva, S., Vasilev, V., Vroonen, L., et al. (2010). Familial pituitary adenomas. *Annales d'Endocrinologie*, *71*(6), 479–485.

Veitia, R. A. (2010). FOXL2 versus SOX9: A lifelong "battle of the sexes" *Bioessays*, *32*(5), 375–380.

Verdin, H., & De Baere, E. (2012). FOXL2 impairment in human disease. *Hormone Research in Paediatrics*, *77*(1), 2–11.

Wang, Y., Martin, J. F., & Bai, C. B. (2010). Direct and indirect requirements of Shh/Gli signaling in early pituitary development. *Developmental Biology*, *348*(2), 199–209.

Ward, R. D., Raetzman, L. T., Suh, H., Stone, B. M., Nasonkin, I. O., & Camper, S. A. (2005). Role of PROP1 in pituitary gland growth. *Molecular Endocrinology*, *19*(3), 698–710.

Ward, R. D., Stone, B. M., Raetzman, L. T., & Camper, S. A. (2006). Cell proliferation and vascularization in mouse models of pituitary hormone deficiency. *Molecular Endocrinology*, *20*(6), 1378–1390.

Warner, L. E., Mancias, P., Butler, I. J., et al. (1998). Mutations in the early growth response 2 (EGR2) gene are associated with hereditary myelinopathies. *Nature Genetics*, *18*(4), 382–384.

Watkins, W. J., Umbers, A. J., Woad, K. J., et al. (2006). Mutational screening of FOXO3A and FOXO1A in women with premature ovarian failure. *Fertility and Sterility*, *86*(5), 1518–1521.

Yagi, T., Tokunaga, T., Furuta, Y., et al. (1993). A novel ES cell line, TT2, with high germline-differentiating potency. *Analytical Biochemistry*, *214*(1), 70–76.

Yamaguchi, S., Yamazaki, Y., Ishikawa, Y., Kawaguchi, N., Mukai, H., & Nakamura, T. (2005). EWSR1 is fused to POU5F1 in a bone tumor with translocation t(6;22)(p21; q12). *Genes, Chromosomes & Cancer*, *43*(2), 217–222.

Yamasaki, L., Jacks, T., Bronson, R., Goillot, E., Harlow, E., & Dyson, N. J. (1996). Tumor induction and tissue atrophy in mice lacking E2F-1. *Cell*, *85*(4), 537–548.

Yoshida, S., Kato, T., Yako, H., et al. (2011). Significant quantitative and qualitative transition in pituitary stem/progenitor cells occurs during the postnatal development of the rat anterior pituitary. *Journal of Neuroendocrinology*, *23*(10), 933–943.

Zannini, M., Avantaggiato, V., Biffali, E., et al. (1997). TTF-2, a new forkhead protein, shows a temporal expression in the developing thyroid which is consistent with a role in controlling the onset of differentiation. *The EMBO Journal*, *16*(11), 3185–3197.

Zhang, S. J., Ma, L. Y., Huang, Q. H., et al. (2008). Gain-of-function mutation of GATA-2 in acute myeloid transformation of chronic myeloid leukemia. *Proceedings of the National Academy of Sciences of the United States of America*, *105*(6), 2076–2081.

Zhao, Y., Mailloux, C. M., Hermesz, E., Palkovits, M., & Westphal, H. (2010). A role of the LIM-homeobox gene Lhx2 in the regulation of pituitary development. *Developmental Biology*, *337*(2), 313–323.

Zhao, L., Zevallos, S. E., Rizzoti, K., Jeong, Y., Lovell-Badge, R., & Epstein, D. J. (2012). Disruption of SoxB1-dependent Sonic hedgehog expression in the hypothalamus causes septo-optic dysplasia. *Developmental Cell*, *22*(3), 585–596.

Zhu, X., Wang, J., Ju, B. G., & Rosenfeld, M. G. (2007). Signaling and epigenetic regulation of pituitary development. *Current Opinion in Cell Biology*, *19*(6), 605–611.

Zhu, X., Zhang, J., Tollkuhn, J., et al. (2006). Sustained Notch signaling in progenitors is required for sequential emergence of distinct cell lineages during organogenesis. *Genes & Development*, *20*(19), 2739–2753.

Ziegler, S. F. (2006). FOXP3: Of mice and men. *Annual Review of Immunology*, *24*, 209–226.

Zindy, F., van Deursen, J., Grosveld, G., Sherr, C. J., & Roussel, M. F. (2000). INK4d-deficient mice are fertile despite testicular atrophy. *Molecular and Cellular Biology*, *20*(1), 372–378.

CHAPTER TWO

Development of the Medial Hypothalamus: Forming a Functional Hypothalamic-Neurohypophyseal Interface

Caroline Alayne Pearson*, **Marysia Placzek**†,‡,1

*Department of Neurobiology and Broad Center for Regenerative Medicine and Stem Cell Research, David Geffen School of Medicine at UCLA, Los Angeles, California, USA

†MRC Centre for Developmental and Biomedical Genetics, University of Sheffield, Sheffield, United Kingdom

‡Department of Biomedical Science, University of Sheffield, Sheffield, United Kingdom

[1]Corresponding author: e-mail address: m.placzek@sheffield.ac.uk

Contents

Current Topics in Developmental Biology, Volume 106
ISSN 0070-2153
http://dx.doi.org/10.1016/B978-0-12-416021-7.00002-X

Abstract

The medial hypothalamus is composed of nuclei of the tuberal hypothalamus, the paraventricular nucleus of the anterior hypothalamus, and the neurohypophysis. Its arrangement, around the third ventricle of the brain, above the adenohypophysis, and in direct contact with the vasculature, means that it serves as an interface with circulating systems, providing a key conduit through which the brain can sample, and control, peripheral body systems. Through these interfaces, and interactions with other parts of the brain, the medial hypothalamus centrally governs diverse homeostatic processes, including energy and fluid balance, stress responses, growth, and reproductive behaviors. Here, we summarize recent studies that reveal how the diverse cell types within the medial hypothalamus are assembled in an integrated manner to enable its later function. In particular, we discuss how the temporally protracted operation of signaling pathways and transcription factors governs the appearance and regionalization of the hypothalamic primordium from the prosencephalic territory, the specification and differentiation of progenitors into neurons in organized nuclei, and the establishment of interfaces. Through analyses of mouse, chick, and zebrafish, a picture emerges of an evolutionarily conserved and highly coordinated developmental program. Early indications suggest that deregulation of this program may underlie complex human pathological conditions and dysfunctional behaviors, including stress and eating disorders.

ABBREVIATIONS

AgRP agouti-related protein
ARC arcuate nucleus
AVP arginine vasopressin
bHLH basic helix–loop–helix
BMP bone morphogenetic protein
CART cocaine- and amphetamine-regulated transcript
CRH corticotropin-releasing hormone
DA dopamine
Fgf fibroblast growth factor
GABA gamma-aminobutyric acid
GHRH growth hormone-releasing hormone
GnRH gonadotropin-releasing hormone
Gsh1 genomic screen homeobox 1
HD homeodomain
H-NH hypothalamic-neurohypophyseal
IHD intrahypothalamic diagonal
NPY neuropeptide Y
Otp Orthopedia
OXT oxytocin
POMC proopiomelanocortin
PVN paraventricular nucleus
Rax retina and anterior neural fold homeobox

SF-1 *steroidogenic factor 1*
Shh sonic hedgehog
Sim1 homolog of Drosophila *Single-minded*
SST somatostatin
SVZ subventricular zone
TRH thyrotropin-releasing hormone
VMN ventromedial nucleus
VZ ventricular zone

1. INTRODUCTION

The hypothalamus is an evolutionarily ancient part of the brain, occupying the ventral-most portion of the diencephalon, just above the anterior pituitary gland (adenohypophysis). The hypothalamus serves to centrally integrate sensory inputs, process them, and effect responses to regulate homeostatic processes that are essential to survival and species propagation. These include autonomic regulation of energy and fluid balance, stress responses, growth and reproductive behaviors, and ill-defined roles in emotional and social/behavioral homeostasis. Its complex adaptive functions are accomplished through interactions with numerous other regions of the brain. Additionally, interfaces with circulatory systems are pivotal to its ability to mediate homeostasis, enabling it to detect circulating metabolites and hormones and to operate as the central regulator of the pituitary endocrine (hypophyseal) system.

The diverse functions of the hypothalamus, together with its relatively small size, its complex three-dimensional architecture, the paucity of specific markers, and the complex migratory and morphological events that accompany its ontogeny, mean that our understanding of hypothalamic development lags behind that of other regions of the central nervous system. Many questions still remain as to the origin, regionalization, growth, and differentiation of defined hypothalamic territories and cells. Obtaining a description of these events is important, as it will allow us to understand the complex human pathological conditions and dysfunctional behaviors that are underlain by hypothalamic cells and circuits, including neurological conditions such as chronic stress and eating disorders (see reviews by Kelberman & Dattani, 2009; McCabe, Alatzoglou, & Dattani, 2011; Michaud, 2001; Sternson, 2013; Swaab, 2004). Here, we provide an overview of recent studies that address the development of cells of the medial hypothalamus.

We describe regionalization and neurogenesis within the medial hypothalamus and then review recent studies that show how developmental programs within medial hypothalamic glial cells construct the infundibulum and neurohypophysis. In discussing the neurohypophysis, we focus on our emerging understanding of the development of specialized cellular interfaces that support integrated functions between the brain and the peripheral body.

2. ARCHITECTURE OF THE ADULT MEDIAL HYPOTHALAMUS

Historically, the adult hypothalamus has been divided into four rostrocaudal divisions: preoptic, anterior, tuberal, and mammillary (Fig. 2.1A, inset; Swanson, 1987). Each area harbors defined nuclei, agglomerations of cell bodies that occupy stereotyped positions along the dorsoventral and mediolateral axes (Figs. 2.1 and 2.2A), whose regulatory functions have been principally assigned through lesion, or stimulation studies (e.g., Tokunaga, Fukushima, Kemnitz, & Bray, 1986), and, more recently, through genetic analyses (Lutz & Woods, 2012; Muller & Keck, 2002; Tschop & Heiman, 2001).

The medial hypothalamus, the focus of this chapter, encompasses nuclei of the tuberal hypothalamus including the arcuate (Arc), ventromedial (VMN), dorsomedial (DMN) and anterobasal (ABa) nuclei, and, additionally, the paraventricular nucleus (PVN) of the anterior hypothalamus (Fig. 2.1A and Fig. 2.2A). Each medial nucleus mediates a variety of physiologies, but they share a role and act together to regulate energy balance, fluid balance, stress responses, reproduction, and growth, integrating these with other hypothalamic functions such as circadian cycles (for details, see Table 2.1 and the table legend). Physiological function is orchestrated through multiple neuronal types that are present within each nucleus. For instance, in the arcuate nucleus, neuropeptide Y/agouti-related protein (NPY/AgRP) neurons and proopiomelanocortin/cocaine- and amphetamine-regulated transcript (POMC/CART) neurons modulate feeding, energy balance, and body composition. Arc and PVN are linked via the ventromedial nucleus (VMN), whose steroidogenic neurons are implicated in both energy homeostasis and stress. The DMN integrates inputs from these nuclei and from the suprachiasmatic nucleus to coordinating feeding with wakefulness. Gamma-aminobutyric acid (GABA) and glutamate interneurons exist widely within medial hypothalamic nuclei,

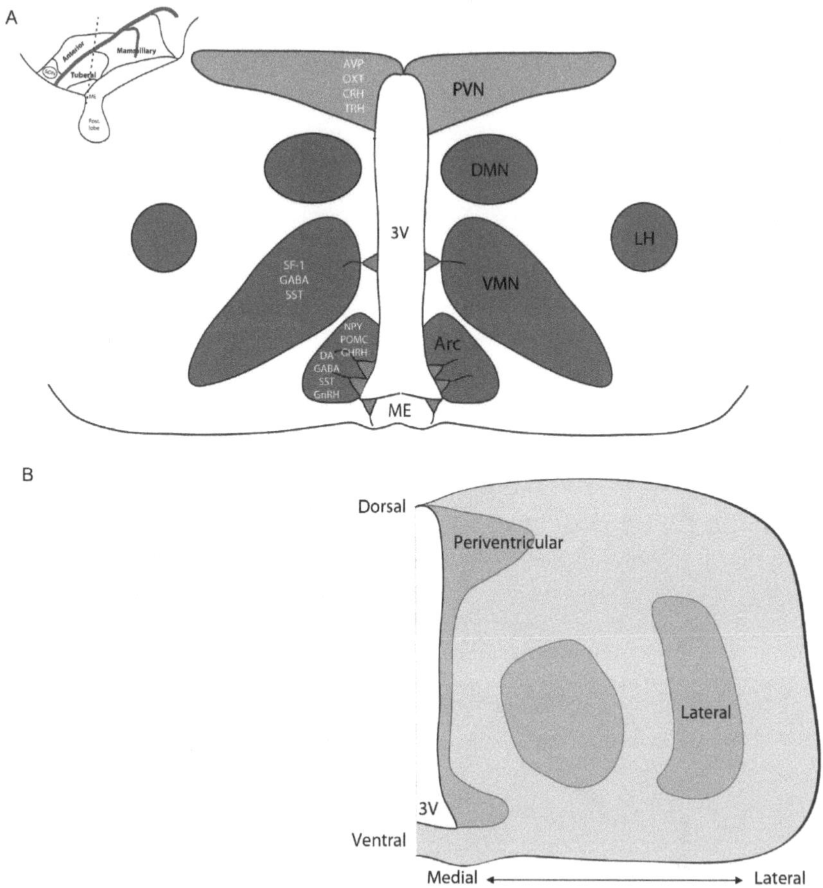

Figure 2.1 Organization of the medial hypothalamus. (A) Inset: divisions of adult hypothalamus; SON marks preoptic area. Main panel shows positions of medial hypothalamic nuclei relative to third ventricle and median eminence; plane of sections shown by dotted line in inset. Red shows tuberal nuclei, names on right hand side; orange shows anterior nuclei. Neuronal effectors (see text for details) shown on left-hand side. Tanycytes (green) are located in the ventral two-thirds of the ventricular zone (VZ). β-Tanycytes line the median eminence. (B) Mediolateral divisions of hypothalamus. SON, supraoptic nucleus; LH, lateral nucleus; PVN, paraventricular nucleus; DMN, dorsomedial nucleus; VMN, ventromedial nucleus; ARC, arcuate nucleus; ME, median eminence. (See color plate.)

supporting a connectivity that integrates afferent input and sculpts an integrated response to changing conditions (see Table 2.1 legend for details).

In addition to these neurons, medial hypothalamic nuclei contain neuroendocrine neurons that fall into two subsets. Magnocellular neuroendocrine neurons located in the PVN project axons into the posterior

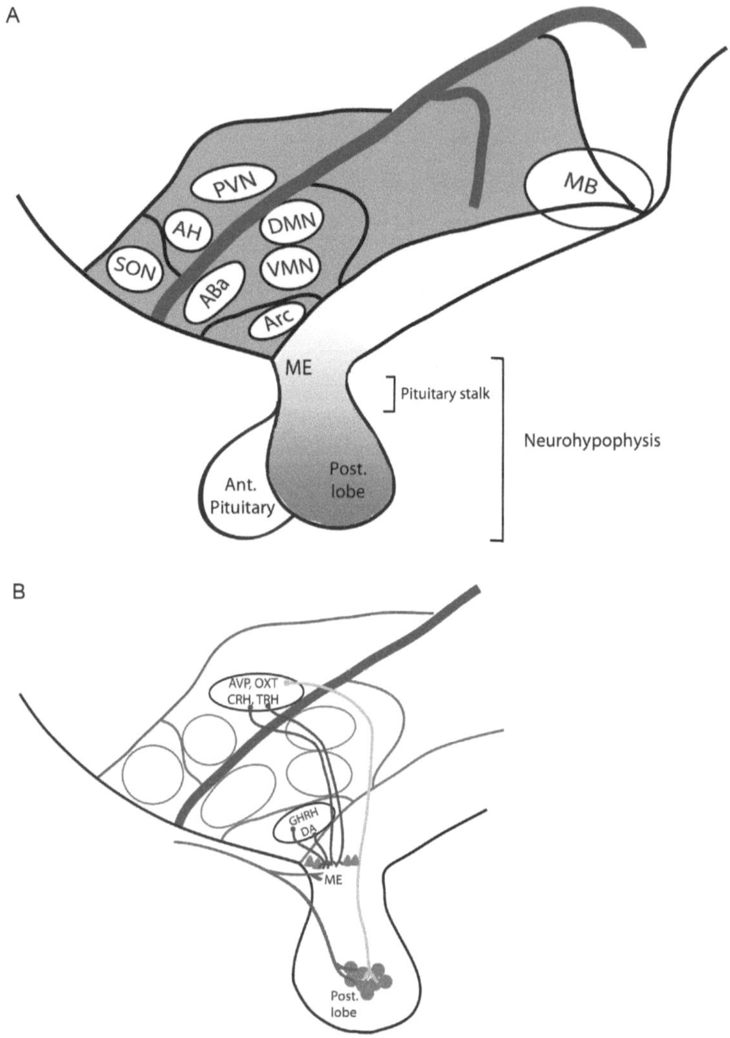

Figure 2.2 Structure of the adult neurohypophysis; side view. (A) Positions of hypothalamic nuclei and components of neurohypophysis. We include ME, pituitary stalk, and posterior lobe in the definition of the neurohypophysis. Originally named for being a "growth or attachment underneath," the neurohypophysis has been variously defined as the posterior lobe or the posterior lobe and pituitary stalk. Both exclude the ME from its definition. However, other studies include the ME as an integral part of the neurohypophysis (e.g., Wittkowski et al., 1999). The similar embryonic origins and features of cells in the ME and posterior lobe lead us to use this broader definition. (B) Cells and cellular components of hypothalamic-neurohypophyseal interfaces. Magnocellular and parvocellular axons project to the neurohypophysis, terminating in the posterior lobe (magnocellular neurons, turquoise) or median eminence (parvocellular neurons, blue). Axon termini are in intimate contact with blood capillaries (red) and glial cells

lobe of the pituitary, where they release arginine vasopressin (AVP) or oxytocin (OXT) into the systemic circulation to influence fluid balance and the reproductive axis. Parvocellular neuroendocrine neurons located within the Arc and PVN project to the median eminence where they release regulating peptides (including growth hormone-releasing hormone (GHRH), somatostatin (SST), dopamine (DA), thyroid-releasing hormone (TRH), and corticotropin-releasing hormone (CRH)) that pass through the pituitary portal system to govern the production of anterior pituitary hormones that control many physiological axes (Fig. 2.1, 2.2B; Table 2.1 legend; Bargmann, 1949; Fuxe & Hokfelt, 1972; Harris, 1955; Markakis, 2002).

The median eminence and posterior lobe, which we refer to, collectively, as the neurohypophysis (Fig. 2.2A and legend) provide crucial interfaces between the peripheral body and the hypothalamus. The best-understood components of these interfaces are the axon terminals of neuroendocrine neurons and the fenestrated portal blood capillaries of the median eminence and posterior lobe, whose intimacy enables hypothalamic neurohormones and neurotransmitters released from neuroendocrine axon terminals to reach the bloodstream (Fig. 2.2B). But additional cell types contribute to neurohypophyseal interfaces. In the posterior lobe, astrocytic-like cells termed pituicytes can modulate the release of neurohormones (reviewed in Wittkowski, 1998). In the median eminence, radial glial-like tanycytes are thought to support the bidirectional flow of biologically active components between the hypothalamus and circulating systems, allowing the hypothalamus to sense and respond to circulating metabolites/biological factors in both plasma and cerebrospinal fluid of the third ventricle (Figs. 2.1A; reviewed in Wittkowski, 1998; Wittkowski, Bockmann, Kreutz, & Bockers, 1999; Rodriguez et al., 2005). Tanycytes are diverse and have an enormous range of biological functions, from Ca^{2+} signaling to stem cell potential, enabling them to evoke changes in the hypothalamus across widely different temporal scales (Bolborea & Dale, 2013). Thus, three

(green) composed of tanycytes in the ME and pituicytes in the posterior lobe. SON, supraoptic nucleus; AH, anterior hypothalamic nucleus; PVN, paraventricular nucleus; Aba, anterobasal nucleus; VMN, ventromedial nucleus; Arc, arcuate nucleus; MB, mammillary bodies; ME, median eminence; post. lobe, posterior lobe of the pituitary; OXT, oxytocin; AVP, arginine vasopressin; CRH, corticotropin-releasing hormone; TRH, thyrotropin-releasing hormone; GHRH, growth hormone-releasing hormone; DA, dopamine. (See color plate.)

Table 2.1 Functions of neurons and nuclei in the medial hypothalamus

Nucleus	Region(s)	Zone(s)	Neurohormone	Function
Arcuate	Tuberal	Periventricular, medial	NPY	Energy balance
			POMC	Control of anterior pituitary
			DA	
			SST	
			GHRH	
Ventromedial	Tuberal	Medial		Emotion
Dorsomedial	Tuberal	Medial		Energy balance
				Emotion
				Integration with circadian cycle
Lateral	Tuberal	Lateral		Energy balance
Paraventricular	Anterior	Periventricular, medial	TRH	Control of anterior pituitary
			CRH	Fluid balance
			DA	
			OXT	
			AVP	

The Arc nucleus is a primary site for the sensing of energy balance, due to its orexigenic NPY/AgRP neurons and anorexigenic POMC/CART neurons that act opposingly to modulate feeding and body composition. Additionally, the Arc harbors a number of neurons, including GHRH, SST, kisspeptin, and dopaminergic neurons (the latter express tyrosine hydroxylase, the rate-limiting factor in the production of catecholamines, including dopamine) that control the anterior pituitary and hence physiological axes. Specifically, dopaminergic neurons regulate the reproductive axis, while GHRH and SST neurons regulate growth and development (see reviews by Hill, 2012; Jo & Chua, 2009; Markakis, 2002; Szarek et al., 2010). The VMN receives input from the Arc and, in parallel with the Arc, governs energy balance. VMN neurons express the SF-1/Nr5a1 transcription factor, encoded by the *Ftz-F1* gene. SF-1 coordinates the control of energy homeostasis: mice in which *SF-1* is ablated in the CNS are susceptible to high fat diet-induced obesity (Kim et al., 2011). In addition, conditional knockout SF-1 mice display increased anxiety-like behavior and show impaired female reproductive function, linking stress, feeding behavior, and reproduction (Kim et al., 2009, 2010; see Maniam & Morris, 2012). The DMN receives many projections from the suprachiasmatic nucleus, a major regulator of circadian rhythm, and appears to act as a node, the net effect of which is to coordinate wakefulness, feeding, and corticosteroid secretion during the active part of the circadian cycle (reviewed in Saper, 2006). The PVN receives inputs from the Arc and VMN and harbors many of the neuroendocrine neurons whose axons descend to the pituitary to effect or regulate endocrine control of the body. Magnocellular neuroendocrine neurons, located mainly ventrolaterally in the PVN, project axons into the posterior pituitary, where they release either AVP or OXT into the systemic circulation to influence fluid balance and the reproductive axis. Parvocellular neuroendocrine neurons located mainly medially synthesize CRH or TRH. Their axons project to the ME, where they release regulating peptides that pass through the pituitary portal system to the anterior pituitary, and govern the production of pituitary hormones to regulate the adrenal and thyroid axes (see Fig. 2.2B and reviews by Hill, 2012; Jo & Chua, 2009; Markakis, 2002; Szarek et al., 2010). Essential neurohormone and neurotransmitter components of the medial hypothalamus can be found in lower nonvertebrates (Tessmar-Raible et al., 2007) and are highly conserved in mammals, birds, and zebrafish (see Kuenzel & van Tienhoven, 1982; Lohr & Hammerschmidt, 2011; Machluf et al., 2011).

fundamentally different biological components constitute the adult neurohypophysis and contribute to its interface functions: neuroendocrine axon terminals, portal capillary vessels, and glial-like cells composed of tanycytes and pituicytes.

3. ESTABLISHING THE MEDIAL HYPOTHALAMUS

How, then, are the different subdivisions of the medial hypothalamus established? What are the cellular and molecular mechanisms that orchestrate its development and the differentiation of its constituent cells? What mechanisms operate during embryogenesis to dictate the complex topology of medial hypothalamic neurons? And what mechanisms operate to build hypothalamic interfaces in an integrated manner? Focused analyses in the last decade have begun to unravel details of these events. The process is protracted: new neurons are added throughout embryogenesis, postnatally and even into adulthood; similarly, aspects of gliogenesis/glial differentiation occur late in embryogenesis and continue into adulthood. Here, we restrict our discussion to a description of the events that occur within embryogenesis and that lay the foundations for formation of a functional neurohypophysis. We summarize recent discoveries, gained through analyses of mouse, chick, and zebrafish. Despite the large evolutionary distance between these species, a picture emerges of a conserved and highly coordinated developmental program that we break into three main events: (1) appearance and regionalization of the hypothalamic primordium from the prosencephalic territory, (2) specification and differentiation of progenitors into neurons in organized nuclei, and (3) establishment of interfaces. Each phase is enacted under the influence of signaling ligands and transcription factors, whose operative functions are being increasingly well defined.

3.1. Transcription factors define early regionalized territories

Systematic gene expression profiling and fate-mapping analyses are allowing us to precisely define the hypothalamic borders and establish a map of its early regional territories (e.g., Caqueret, Coumailleau, & Michaud, 2005; Morales-Delgado et al., 2011; Shimogori et al., 2010; Wolf & Ryu, 2013). These studies suggest, in fact, that the preoptic area (classically included in the definition of the adult hypothalamus) is at least partly derived from telencephalic tissue, and potentially, that preoptic nuclei should be considered as having subpallial telencephalic origins. Other than this, such studies have substantially advanced our appreciation of hypothalamic development, demonstrating that hypothalamic transcription factors and neurohormones have been highly conserved during evolution (reviewed in Aubert, Grumbach, & Kaplan, 1977; Lohr & Hammerschmidt, 2011; Machluf, Gutnick, & Levkowitz, 2011), revealing how physiological stressors can integrate directly with transcriptional modulators (Amir-Zilberstein et al., 2012) and leading to a good understanding of the embryonic origins of defined hypothalamic nuclei and neurons (see succeeding text and reviews by Jo & Chua, 2009; Puelles, 1995, 2001; Puelles & Rubenstein, 2003; Rubenstein, Shimamura, Martinez, & Puelles, 1998; Szarek, Cheah, Schwartz, & Thomas, 2010).

Very few genes are expressed exclusively in specific regions of the developing hypothalamus. Nonetheless, it is becoming possible to use combinatorial patterns of gene expression to uniquely identify hypothalamic regions along the early dorsoventral and anterior–posterior (future rostrocaudal) axes (Fig. 2.3). Early developmentally expressed genes including the HD proteins, *Pax6*, *Nkx2.1* (also known as *Thyroid transcription factor 1*, *Ttf1*), *Nkx2.2*, *Aristaless* (*Arx*), *Distal-less 5* (*Dlx*), *Six3*, *Vax1*, and *Orthopedia* (*Otp*); LIM domain *Lhx* family members; basic helix–loop–helix (bHLH) factors such as *Sim1*; the Forkhead domain protein *Foxb1*; and T-box family (*Tbx*) members suggest a dorsoventral regionalization of the hypothalamus into alar, basal, and floor plate longitudinal subdivisions that is conserved in all species examined (Fig. 2.3A) and established in response to early dorsoventral signaling events (see following text). For instance, at early stages *Sim1* and *Pax6* are restricted to the alar plate, whereas *Nkx2.1* characterizes basal and floor plate regions. *Nkx2.2* is expressed in a longitudinal band that overlaps the alar and basal plate boundaries. A parallel strip of cells termed the IHD expresses *Arx3* (and the glutamate decarboxylase, *Gad67*) and occupies ventral-most parts of the alar plate. *Tbx2* and *Tbx3* are nested within ventral-most *Nkx2.1*-expressing cells and mark the hypothalamic floor plate. Along the rostrocaudal axis, territories in the alar plate, IHD, basal plate, and floor plate can be further subdivided by additional markers, including *Nkx6.2*, *Lymphoid enhancer-binding factor 1 (Lef1)*, and *Iroquois5* (Shimogori et al., 2010). Together, dorsoventral and rostrocaudal subdivisions suggest an early regional map of the developing hypothalamus (Fig. 2.3A and B).

Two general points emerge from such transcriptional mapping. First, the accepted rostrocaudal divisions of the adult hypothalamus (anterior, tuberal, and mammillary) derive from dorsal and ventral embryonic anlage. Thus, tuberal (and mammillary) areas are ventral (basal and floor plate)-derived, whereas anterior hypothalamic regions are dorsal (alar plate)-derived.

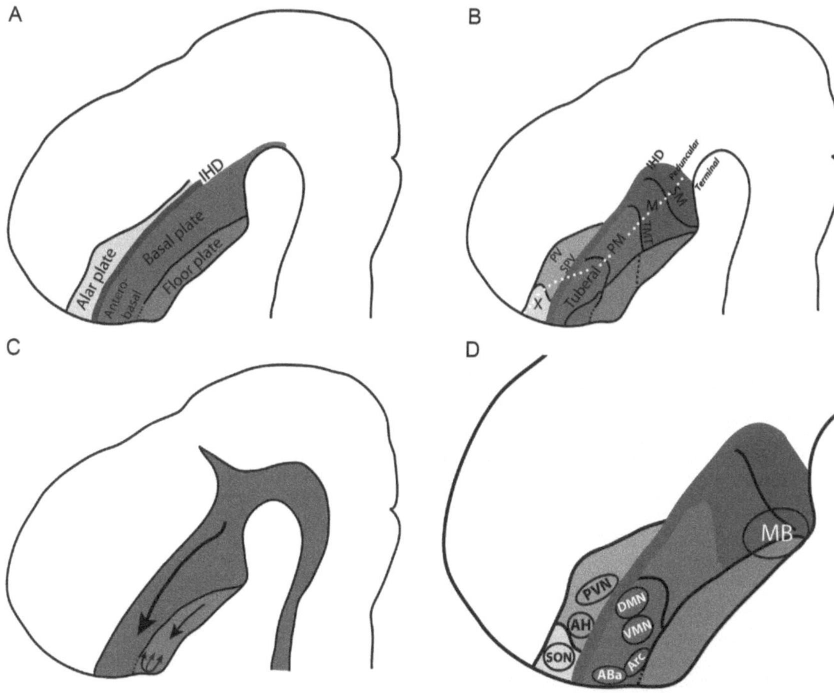

Figure 2.3 Combinatorial patterns of gene expression identify developing hypothalamic regions that prefigure adult nuclei. (A) The developing hypothalamus can be broadly divided along the dorsoventral axis into alar plate, basal plate, and floor plate. The alar and basal plates are separated by the intrahypothalamic diagonal (IHD). Each region can be identified by expression of a specific gene or combination of genes. Floor plate: *Six3* (high), *Shh* (transient), *Tbx1, 2,* and *3, Nkx2.1*. Basal plate: *Six3* (low), *Nkx2.1, Shh*. Alar/basal plate boundary: *Nkx2.2*. IHD: *Arx3, Gad67*. Alar plate: *Pax6, Sim1* (excluding rostral-most). (B) The developing hypothalamus can be further subdivided along the anterior–posterior (future rostrocaudal) axis by additional markers that define specific territories in the floor, basal, and alar plates. The alar plate is divided into two regions, an unnamed region (X) that expresses Nkx6.2 and the paraventricular and subparaventricular regions that express Pax6 and Sim1. Within the basal plate, there is a high degree of subregional heterogeneity in the early embryo, identified through particular markers. Tuberal: *Pomc/SF-1*. Premamillary: *Lef1*. Tuberomammillary terminal: *Arx*. Mammillary: *Foxb1*. Supramammillary: *Irx3*. The posterior IHD expresses *Lhx6*, whereas the anterior IHD expresses *Lhx8*. Note that anterior, tuberal, premammillary, and mammillary areas have constituent parts in the terminal hypothalamus and associated retro counterparts in the more caudal peduncular hypothalamus that are not included in this figure (see Morales-Delgado et al., 2011). Note also that cartoon should be treated as a "working model." Most of the transcription factors shown show highly dynamic profiles. (C) Fate-mapping studies show that populations of hypothalamic cells migrate along the rostrocaudal axis. Basal plate cells from a more caudal position undergo a rostral migration to intermingle with cells of the hypothalamic floor plate. A small subpopulation of rostral floor plate cells migrate caudolaterally to surround midline cells and form a collar (see Fig. 2.5). (D) Schematic shows general position of nascent hypothalamic nuclei and indicates their mapping onto embryonic regional territories. SON, supraoptic nucleus; AH, anterior hypothalamus; PVN, paraventricular nucleus; ABa, anterior basal nucleus; DMN, dorsomedial nucleus; VMN, ventromedial nucleus; Arc, arcuate nucleus; MB, mammillary bodies. (See color plate.)

Second, the borders of some territories remain ill-defined. In part, this remains due to the paucity of early unique transcriptional identifiers. In addition, however, extensive cell migrations occur during regionalization of the hypothalamus, including opposing rostrocaudal movements of adjacent cell populations (Fig. 2.3C; Manning et al., 2006; Morales-Delgado et al., 2011; Pearson et al., 2011; Zhao et al., 2008). A potential fountain-like movement of cells in the anterior-most floor and basal plate regions means that the anterior-most tuberal regions harbor cells transiently and of mixed origin, precluding a clear definition of the floor and basal plate boundary in the tuberal hypothalamus (Fig. 2.3C).

4. SIGNALING LIGANDS IN INDUCTION AND REGIONALIZATION

Many of the secreted signaling factors that confer regional identity elsewhere in the CNS, in particular Sonic hedgehog (Shh), bone morphogenetic proteins (BMPs), and Wnts (Briscoe & Novitch, 2008; Edlund & Jessell, 1999; Shirasaki & Pfaff, 2002), operate within the hypothalamus. Here, they impart early dorsoventral and rostrocaudal identity and, together with additional ligands, notably fibroblast growth factors (FGFs), specify later aspects of neuronal fate and orchestrate the continued refinement of hypothalamic territories. Three issues confound our understanding of the exact role of each signal. First, signaling ligands display dynamic expression profiles in the hypothalamus (Fig. 2.4) and many regulatory interactions exist between them (reviewed in Monuki, 2007). Second, over the period of regionalization and neuronal specification, the hypothalamus grows and undergoes complex morphological changes: at present, the contributions of proliferation and morphogenesis to regionalization/specification are poorly understood. Finally, in some regions of the hypothalamus, notably in the midline, lineage-distinct compartments appear to be established (Kapsimali, Caneparo, Houart, & Wilson, 2004; Manning et al., 2006; Pearson et al., 2011). These presumably dictate the emergence of distinct adult subregions and cells, but their significance remains ill-defined. Nonetheless, despite these issues, a picture is emerging for a temporally protracted role for each ligand in the establishment of the medial hypothalamus and its resident cells.

4.1. Wnt signaling in anterior-posterior regionalization

Along the general anterior–posterior (future rostrocaudal) hypothalamic axis, the opposing actions of Wnt signaling and Wnt antagonists regionalize

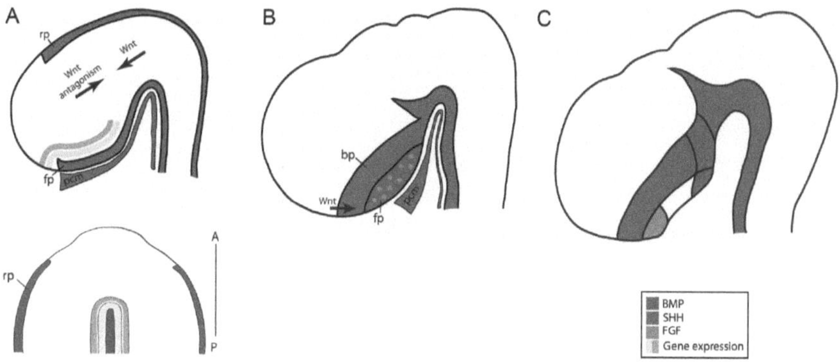

Figure 2.4 Dynamic expression of secreted signaling ligands at three successive times, all prior to formation of the infundibulum/neurohypophysis. (A) Side (top) or ventral (bottom) views illustrating early patterning events. Wnt antagonists promote anterior neural/hypothalamic fate by repressing Wnt activity (black arrows). Shh from the prechordal mesoderm/nascent hypothalamic floor plate is opposed by BMPs from the telencephalic roof plate. Shh from the prechordal mesoderm and hypothalamic floor plate establishes arcs of gene expression around the hypothalamic ventral midline. (B) The onset of BMP activity in the prechordal mesoderm leads to the downregulation of *Shh* and the upregulation of *Bmp* and *Fgf* in the hypothalamic floor plate. *Shh* is secondarily induced in the basal plate. Rostrally derived Wnt signals restrict the *Fgf/Bmp* expression domain. (C) *Fgf* and *Bmp* expression domains resolve, concomitant with the relative backward shift of the prechordal mesoderm. *Fgfs* become restricted to the tuberal region of the hypothalamic floor, where they will play a role in the development of the infundibulum/neurohypophysis. *Bmps* become restricted to the supramammillary hypothalamus. The *Shh*-expressing basal plate surrounds the hypothalamic floor plate. rp, roof plate; fp, floor plate; pcm, prechordal mesoderm; bp, basal plate. (See color plate.)

hypothalamic territories (Fig. 2.4). Many studies have demonstrated an early role for Wnt antagonists in the promotion of anterior neural plate/prosencephalic identities and a role for Wnt signaling in the promotion of posterior fates (Altmann & Brivanlou, 2001). The homeodomain (HD) factor Six3, an early determinant of anterior neural fate, is upregulated in a manner that depends upon low Wnt activity (Braun, Etheridge, Bernard, Robertson, & Roelink, 2003; Lagutin et al., 2003) and marks the hypothalamus, including future tuberal regions. Six3 exerts both cell-autonomous effects (Kobayashi et al., 2002; Kobayashi, Nishikawa, Suzuki, & Yamamoto, 2001; Lavado, Lagutin, & Oliver, 2008) and, through its ability to govern *Shh* expression (Geng et al., 2008; Jeong et al., 2008), nonautonomous effects. Wnt signal antagonism, similarly, is required for hypothalamic floor plate (rather than posterior floor plate) identities, contributing to the upregulation of Fgfs and Bmps in the hypothalamic floor plate (Fig. 2.4B) (Kapsimali et al., 2004;

Manning et al., 2006; and section 4.2). Thus, Wnt antagonism imparts hypothalamic fate early in development (Fig. 2.4A).

A short time later, Wnt signaling plays a forming role in defining the anterior expression limits of *Fgf10* and *Bmp4* in the floor plate of the tuberal hypothalamus. Canonical Wnt signaling exerts its effects by transforming Lef/Tcf transcription factors from repressors to activators of target genes. *Tcf4* is expressed within the anterior-most basal plate (termed the anterobasal plate), rostral to the *Fgf/Bmp*-expressing floor plate (see Fig. 2.4B). In the absence of *Tcf4*, and therefore Wnt-responsiveness, the expression domains of *Fgf* and *Bmp* expand rostrally into the anterobasal territory (Brinkmeier, Potok, Davis, & Camper, 2007). Similarly, *Wnt5a* mutant mice exhibit expanded *Fgf10* and *Bmp* domains (Potok et al., 2008). Thus, Wnt5a signaling via Tcf4 defines the rostral extent of *Fgf* and *Bmp* expression in the tuberal hypothalamic floor plate (Fig. 2.4B).

In summary, early Wnt/Wnt antagonism plays a role in the establishment of discrete zones of signaling ligand expression, in particular, influencing the profiles of *Shh*, *Bmps*, and *Fgfs*. As described in the succeeding text, each of these signaling ligands has significant roles in the development of the tuberal hypothalamus.

4.2. Shh in induction and ventral regionalization

Regulated Wnt signalling operates together with ventral signals, that derive initially from axial mesoderm cells of the prechordal plate/prechordal mesoderm that underlie the nascent hypothalamus (Fig. 2.4A). Physical ablation experiments reveal that the prechordal mesoderm induces both floor and basal plate territories in the gastrulating embryo (Adelmann, 1922; Garcia-Calero, Fernandez-Garre, Martinez, & Puelles, 2008; Patten, Kulesa, Shen, Fraser, & Placzek, 2003). Prevention of prechordal mesoderm formation or inhibition of its maintenance, through genetic manipulation, results in dorsalization of the hypothalamic anlage and cyclopic, holoprosencephalic phenotypes (Chiang et al., 1996; Warr et al., 2008; reviewed in Krauss, 2007; Lipinski, Godin, O'Leary-Moore, Parnell, & Sulik, 2010; Muenke & Beachy, 2000; Wallis & Muenke, 2000).

A number of secreted ligands are expressed within prechordal mesoderm, and their cross-regulatory interactions mediate its activities (reviewed in Monuki, 2007). During gastrulation, Nodal and Shh from prechordal mesoderm cooperate through an as-yet undefined pathway to induce *Shh*-expressing hypothalamic midline floor plate cells (reviewed in

Placzek & Briscoe, 2005; Fig. 2.4A). In many species, including humans, deregulation of Shh and Nodal pathways leads to holoprosencephaly, due to a failure of hypothalamic midline floor plate induction and optic field separation (reviewed in Roessler & Muenke, 2010).

In contrast to the posterior neuraxis, where floor plate cells are largely nonproliferative (Altman & Bayer, 1986), hypothalamic floor plate cells show highly regulated proliferation (Bosco et al., 2013; Manning et al., 2006; Pearson et al., 2011) that is likely to contribute to a significant widening of the hypothalamic floor plate. What factors might mediate this? Signaling factors deriving from the prechordal mesoderm exert their effects on a neural canvas already stamped with prosencephalic identity, including expression of the prosencephalic-specific Six3 (Geng et al., 2008). Loss- and gain-of-function approaches demonstrate that Six3 is directly required to activate *Shh* expression in ventral midline floor plate cells (Geng et al., 2008; Jeong et al., 2008). Elsewhere in the CNS, Six3 promotes the proliferation of forebrain progenitors by antagonizing Geminin, a DNA replication inhibitor (del Bene, Tessmar-Raible, & Wittbrodt, 2004). Conceivably, then, Six3 contributes to the widening of the hypothalamic floor plate by driving cell cycle within progenitors.

Although not studied systematically, evidence supports the idea that Shh deriving from the prechordal plate mesoderm and hypothalamic floor plate establishes arcs of progenitor gene expression in the floor and basal plates, centered around the hypothalamic ventral midline (Fig. 2.4A). Increased Shh activity leads to ectopic expression of hypothalamic markers (Barth & Wilson, 1995; Hauptmann & Gerster, 1996; Ohyama, Ellis, Kimura, & Placzek, 2005; Pabst, Herbrand, Takuma, & Arnold, 2000; Rohr, Barth, Varga, & Wilson, 2001; Shimamura & Rubenstein, 1997). Conversely, blockade of Shh from prechordal mesoderm or hypothalamic floor plate leads to a reduction in expression of floor plate and basal plate markers, including *Nkx2.1*, *Nkx2.2*, and *Dlx2*, and a ventral expansion of the dorsal marker *Pax6* (Marcucio, Cordero, Hu, & Helms, 2005; Ohyama, Das, & Placzek, 2008; Ohyama et al., 2005; Shimamura & Rubenstein, 1997). The outcome of Shh signaling is governed by additional factors: thus, the HD transcription factor Six3, which is itself maintained by Shh (Geng et al., 2008; Sanek, Taylor, Nyholm, & Grinblat, 2009), cooperates with Shh to induce expression of Nkx2.1 (Marin, Baker, Puelles, & Rubenstein, 2002; Ohyama et al., 2005; Szabo et al., 2009). An unanswered question is the extent to which Shh acts as a morphogen to pattern the hypothalamus, in the manner in which it operates in the spinal cord (reviewed in Briscoe & Novitch, 2008; Jessell, 2000; Ulloa & Briscoe, 2007). As in the posterior CNS, Shh-activated genes can attenuate the effects of Shh signaling (Sanek et al., 2009), presumably confining the regional and temporal expression and actions of Shh (Szabo et al., 2009).

Concomitant with the induction of HD transcription factors within the basal plate, secondary and sustained sites of *Shh* expression are established there, initially seamlessly interfacing with Shh-expressing floor plate cells (Fig. 2.4B). Shh deriving from the hypothalamic floor plate appears to be required for Shh expression in most of the basal plate, suggesting a homeogenetic induction (Echelard et al., 1993; Mathieu, Barth, Rosa, Wilson, & Peyrieras, 2002; Ohyama et al., 2008). What, though, might be the significance of sustained expression of Shh within the basal hypothalamus? Conditional ablation studies in mice are beginning to dissect this, revealing a role in late and local events. Thus, loss of Shh from the anterobasal and basal plate from E10.5 does not downregulate progenitor transcription factors in the tuberal region, but instead is required for late aspects of neuronal differentiation, namely, for expression of *Pomc* and steroidogenic factor 1 (*SF-1*; aka *Nr5a1*), markers of Arc and VMN-specific neurons, respectively (Shimogori et al., 2010 and see succeeding text).

Two aspects of these collective studies are worth highlighting. First, they imply that Shh may exert relatively local effects within distinct hypothalamic territories. Second, they imply that local Shh signaling plays a role in late stages of tuberal neuronal differentiation. It will be intriguing, in future, to establish how restricted responses to Shh might be achieved. Do cells show different competence to Shh, or is Shh diffusion limited in particular directions? And what is the precise character of cells that sustain Shh expression: do they act as later suborganizers, actively building local hypothalamic complexity?

4.3. Spatiotemporal antagonism of Shh and Shh signaling by BMPs

Ligands that confer dorsal identities in the posterior CNS (reviewed in Chichikov & Millen, 2005), in particular BMPs, are expressed by telencephalic roof plate cells (Fig. 2.4A) and appear to establish and maintain early hypothalamic and alar identities. Pax6 and Pax7, for instance, are expressed in similar relative domains to those observed more posteriorly (O'Leary, Chou, & Sahara, 2007), while mice deficient for factors involved in BMP signaling exhibit abnormal hypothalamic development (e.g., Cheng, Zhang, & Reed, 2007).

In contrast to the developing spinal cord, however, where BMPs are largely dorsally maintained and antagonize Shh/Shh signaling in a spatial fashion (Ulloa & Briscoe, 2007), in the hypothalamus, BMP activity is initiated within the prechordal mesoderm shortly after the onset of hypothalamic floor plate induction, at least in chick (Dale et al., 1997; Fig. 2.4B). BMP signaling from prechordal mesoderm is instrumental in the development of ventral hypothalamic pattern, acting as a temporal antagonist of Shh (Manning et al., 2006; Ohyama et al., 2008). BMPs deriving from the prechordal mesoderm upregulate the transcriptional repressor, Tbx2, in the hypothalamic floor plate, resulting in the downregulation of *Shh* in a cell-autonomous manner (Manning et al., 2006; Fig. 2.3B).

Elements of this pathway are conserved in mammals and zebrafish. *Tbx2/3* and *Tbx2b* are expressed in the hypothalamic floor plate of mice and zebrafish, respectively (Pontecorvi, Goding, Richardson, & Kessaris, 2008; Thisse et al., 2004). Recent experiments in mice show that Tbx2 and Tbx3 can repress Shh by sequestering the SRY box-containing transcription factor Sox2 away from a Shh forebrain enhancer (SBE2): mice in which Tbx3 is removed fail to repress transcription of Shh (Trowe et al., 2013). Thus, T-box family (Tbx) proteins can directly repress *Shh* expression. Why is this significant? The extinction of Shh/Shh signaling from hypothalamic floor plate cells appears to be instrumental in their further development, including the onset of expression of the HD transcription factor, *Emx2*, and a synchronized proliferation of hypothalamic floor plate cells (Manning et al., 2006). In addition, the onset of *Tbx2* expression coincides with upregulation of *Fgfs*, initially throughout most of the hypothalamic floor plate and then confined to the forming tuberal hypothalamic floor (Fig. 2.4B and C; Geng et al., 2008; Ohuchi et al., 2000; Ohyama et al., 2008; Pearson et al., 2011; Tsai, Brooks, Rochester, Kavanaugh, & Chung, 2011). This event is key to the further development of the tuberal hypothalamic floor plate, in particular formation of the infundibulum and neurohypophysis (see succeeding text).

5. NEURONAL DIFFERENTIATION IN THE MEDIAL HYPOTHALAMUS

Once hypothalamic regional territories are established, progenitor cells begin to differentiate. Waves of neurogenesis within cycling stem/progenitors within the ventricular zone (VZ)/subventricular zone (SVZ) give rise to nascent neurons that migrate laterally and terminally differentiate. Birth-dating studies suggest neurons in the lateral hypothalamus are born first, followed by those in medial, and then periventricular nuclei (Fig. 2.1B; Altman & Bayer, 1986; Caqueret, Boucher, & Michaud, 2006), although in the medial hypothalamus, such "outside-in" temporal differentiation may only partly explain the final patterns of neuronal differentiation (see succeeding text). Functional neurons emerge through a hierarchical sequence of events involving lineage commitment, migration, survival, neurotransmitter selection, and axonal growth, the molecular determinants of which are beginning to be uncovered. These studies begin to show how regulatory transcription factors within the medial hypothalamus control different stages of neuronal fate decisions.

5.1. Neurogenesis and lineage commitment: Proneural gene activity

As in other regions of the CNS, cell cycle exit and the upregulation of neurogenic programs in the medial hypothalamus require the activity of proneural genes (Guillemot, 2007). Achaete–scute-like 1 (Mash1), a bHLH proneural transcription factor, is widely expressed in the VZ and SVZ in basal and floor plate regions of the tuberal hypothalamus. Lineage-tracing studies show that Mash1-expressing progenitors contribute to multiple hypothalamic regions (Kim, Battiste, Nakagawa, & Johnson, 2008), while knockout studies in mouse indicate an essential role in formation of the Arc and VMN. Thus, loss of Mash1 leads to hypoplasia of the Arc and VMN due to a failure of neurogenesis (McNay, Pelling, Claxton, Guillemot, & Ang, 2006). All lineages of the Arc and VMN are affected, but not all equally. Thus, there is a complete absence of neurons expressing GHRH. By contrast, there is a dramatic reduction, but not a complete loss, of SF-1, POMC, NPY, and dopaminergic neurons in $Mash1^{-/-}$ mutant mice (see Fig. 2.1A for positions of neurons within specific nuclei). Why might these neurons persist, albeit in reduced number? The most likely explanation is that other proneural genes operate in parallel with Mash1. Studies in zebrafish, in fact, suggest a role for the *atonal*-like gene *Neurogenin1* (Ngn1) in dopaminergic neuronal specification (Yang, Dong, & Guo, 2012).

The ability of the hypothalamus to control homeostasis is dependent on neurons that exert opposite effects. For instance, in the Arc, NPY neurons and POMC neurons act opposingly to modulate feeding and body composition. Recent studies suggest that proneural genes may govern this balance. *Neurogenin3* (*Ngn3*) is expressed in mitotic progenitors that contribute to subsets of POMC, NPY, dopaminergic, and SF-1 neurons in the Arc and VMN and plays a role in their specification (Pelling et al., 2011, 2013). However, while Ngn3 promotes POMC and SF-1 neurons, it inhibits the development of NPY and dopaminergic neurons (Pelling et al., 2011). How might this occur? Recent studies point to one possibility, showing that *Pomc* is expressed in the vast majority of neurons in the presumptive Arc but that its transcription is extinguished in more than half, some of which subsequently differentiate into NPY neurons (Padilla, Carmody, & Zeltser, 2010). Potentially, Ngn3 acts to determine fate in a common progenitor of POMC and NPY neurons but must be eliminated or downregulated to support differentiation of the progenitor to a NPY fate. Ultimately, POMC and NPY neuronal differentiation must be exquisitely controlled to achieve a balance of orexigenic/anorexigenic neurons that can act opposingly to modulate feeding and energy states. Their integrated development from a common precursor provides one way of achieving this balance.

5.2. Neurogenesis and lineage commitment: HD gene activity

The downstream programs initiated through the action of proneural genes in hypothalamic progenitors are not always clear, but increasing evidence suggests that, as in the posterior CNS (see reviews by Jessell, 2000; Shirasaki & Pfaff, 2002), proneural genes act in concert with HD proteins to determine lineage commitment/subtype-specific neuronal identities. For instance, Mash1 function is required for expression of the HD transcription factor, genomic screen homeobox 1 (Gsh1), which, in turn, is necessary for expression of GHRH, suggesting it acts to mediate the Mash1-dependent specification of GHRH neurons in the Arc (McNay et al., 2006; Mutsuga et al., 2001). The HD protein Nkx2.1 is expressed widely in progenitor cells that occupy the hypothalamic basal plate (Fig. 2.3A) and that differentiate into GABA, NPY, POMC, and dopaminergic neurons of the arcuate nucleus (ARC) (Yee, Wang, Anderson, Ekker, & Rubenstein, 2009). Loss of function of Nkx2.1 leads to an apparent failure to form or maintain the ARC and VMN (Kimura et al., 1996; Takuma et al., 1998), most likely due to a cell-autonomous requirement for Nkx2.1 in progenitor cells and/or differentiating neurons. Other HD proteins, such as the closely related genes, Hmx2 and Hmx3, appear to govern more specific differentiation programs. Hmx2 and Hmx3 overlap in the Arc and DMN (Wang, Grimmer, Van De Water, & Lufkin, 2004), where they regulate the expression of Gsh1. Consequently, in the absence of Hmx2/3, mice show a severe reduction of GHRH neurons (Wang & Lufkin, 2005). Studies in zebrafish suggest that proneural and HD activities are regulated in a concerted manner by the evolutionarily conserved zinc finger protein, Fezf2. Fezf2 is required to activate but not maintain the expression of bHLH genes and, additionally, is required to activate the HD genes *otpb* and *dlx2*, which, in fish, are involved in TH and GABA neuronal development, respectively (Blechman et al., 2007; Yang et al., 2012).

Until recently, many studies examining tuberal neurogenesis focused on progenitor cell selection and lineage commitment within the Arc, due mainly to the ability to follow POMC and NPY neuronal fates. New studies, though, have shed light on the manner in which early VMN cells are selected, indicating a crucial role for the paired-type HD gene, *Retina and anterior neural fold homeobox* (*Rax*). In mice, Rax is widely expressed in the floor and ventral basal plate of the hypothalamus, where it is confined to progenitor cells that lie at, or close to the VZ (Lu et al., 2013). In tuberal regions, expression is detected in progenitor cells that lie medial to the Arc and VMN. Genetic lineage-tracing studies demonstrate that *Rax*+ lineages give rise to SF-1-expressing VMN neurons and potentially to Arc neurons. Broad elimination of *Rax* in mice leads to a severe loss of both VMN and Arc neurons (Lu et al., 2013), corroborating earlier studies in zebrafish that demonstrate a reduction of neurons in *chokh* (rx3) mutants (Tessmar-Raible et al., 2007). Targeted ablation of *Rax* in a subset of VMN progenitors leads to a fate switch from a VMN identity to an alternate hypothalamic identity, suggesting that *Rax* selects VMN identity in a cell-autonomous manner.

5.3. Migration and survival

SF-1 is an orphan nuclear hormone receptor that is specifically expressed within the VMN. SF-1 appears to play a role, primarily, in the aggregation of the VMN nucleus and its terminal differentiation: in SF-1-deficient mice, neurons fail to migrate or coalesce to form the ventromedial portion of the VMN (Ingraham et al., 1994; Shinoda et al., 1995; reviewed in Jo & Chua, 2009). This highlights an essential part of hypothalamic development, namely, the enormous contribution of cell migration to the final organization of hypothalamic nuclei. Indeed, increasing numbers of studies show that the simple laminar organization of the hypothalamus along the mediolateral axis, established through the radial migration of newborn cells, is transient, eroded through a subsequent reorganization of populations of cells through tangential migrations. In some cases, neurons born within one part of the tuberal hypothalamus migrate into another tuberal territory. Topographic mapping of SST gene expression reveals, for instance, that a subset of cells produced in the anterobasal nucleus disperse in to the Arc and VMN (Morales-Delgado et al., 2011). In other cases, neurons born outside the hypothalamus migrate into it. Gonadotropin-releasing hormone (GnRH) neurons originate in the olfactory placode, migrate into the forebrain, and position themselves within discrete nuclei, including the Arc (Hu et al., 2013; Wray, 2001, 2002).

The importance of cell migration and coalescence is particularly well illustrated through studies of *Sim1* (homolog of Drosophila single-minded) mutant mice. The transcription factor *Sim1* is regionally constrained within the alar plate and prefigures and is required for the emergence of adult anterior nuclei, including the PVN (Caqueret et al., 2005; Goshu et al., 2004; Michaud, De Rossi, May, Holdener, & Fan, 2000; Michaud, Rosenquist, May, & Fan, 1998; Wang & Lufkin, 2000). SIM1 and its obligate dimerization partner Aryl hydrocarbon receptor nuclear translocator 2 (ARNT2) are required for expression of TRH, CRH, OXT, and AVP (Hosoya et al., 2001; Keith, Adelman, & Simon, 2001). Replacement of SIM1 and ARNT2 coding sequences with a reporter gene suggests that these genes play a role in migration and neuronal coalescence, rather than a primary effect on proliferation, survival, or subtype selection (Michaud et al., 2000, 1998). Thus, transcription regulators are playing essential roles at different stages of hypothalamic neuronal differentiation programs.

5.4. Neurotransmitter selection

In cases where proneural and HD genes govern survival or subtype selection, it is difficult to know how neurotransmitter identity is regulated within the wider context of survival or the promotion of neuronal subtype identity. For instance, in both mouse and zebrafish, the HD protein OTP plays a critical role in the differentiation of many medial hypothalamic neurons (Acampora et al., 1999; Blechman et al., 2007; Ryu et al., 2007; Wang & Lufkin, 2000) and is thought to regulate terminal differentiation/neurotransmitter selection. OTP modulates expression of CRH: analysis of a zebrafish *otpb* mutant reveals that CRH-containing neurons develop normally and that OTP regulates CRH synthesis during stress adaptation (Amir-Zilberstein et al., 2012). Similarly, OTP is required for the production of both OXT and DA neurons (Acampora et al., 1999; Blechman et al., 2007; Ryu et al., 2007; Wang & Lufkin, 2000), where, again, it is believed to govern neuropeptide and/or neurotransmitter activation. The mechanism through which it operates in these neurons, though, is unclear: as yet, there is no evidence that OTP directly activates promoters involved in neuropeptide/neurotransmitter synthesis. The possibility that HD proteins govern neurotransmitter identity indirectly may explain the different requirement for particular HD proteins in different species: thus, in mouse, cells that express the HD protein, DLX, contribute to GABA and dopaminergic neurons, but only dopaminergic neurons appear affected in *Dlx1*$^{-/-}$ mutants (Yee et al., 2009). This contrasts with studies in fish, where *dlx2* is required for GABA neuronal differentiation (Yang et al., 2012).

By contrast, in cases where transcriptional regulators govern later aspects of neuronal differentiation, it is possible to begin to dissect the mechanisms that govern neurotransmitter identity. Expression of the POU-domain class 3 transcription factor Brain-2 (Brn2) is governed by the parallel activities of SIM1/ARNT2 and OTP and is likely to mediate some of their actions: knockout studies reveal that Brn2 is required for expression of a subset of neurotransmitters that are controlled by SIM1/ARNT2 and OTP, namely, for CRH, OXT, and AVP (Nakai et al., 1995; Schonemann et al., 1995). By contrast, SIM2, whose expression is governed by the activities of SIM1/ARNT2/OTP, appears to govern TRH expression (Acampora et al., 1999; Goshu et al., 2002; Wang & Lufkin, 2000; see also reviews by Jo & Chua, 2009; Szarek et al., 2010).

Intriguingly, replacement of OTP-coding sequence with a reporter gene reveals a loss of reporter-positive cells during embryonic development but, additionally, results in nonautonomous effects. Six3 is ectopically upregulated in *Otp*$^{-/-}$ mice, and proliferating cells within the VZ/SVZ that do not themselves express *Otp* are reduced (Wang & Lufkin, 2000). This indicates that *Otp*-expressing cells produce secreted factors involved in the proliferation or maintenance of progenitor cells and suggests that emerging neurons may feedback to progenitor cells, actively participating in building other neuronal components that enable their function or, indeed, that act antagonistically to them. The nature of these nonautonomous factors is not clear, but both morphogens, such as Shh, and trophic factors, such as brain-derived neurotrophic factor, are expressed in specific maturing nuclei and are likely to contribute to a protracted and integrated developmental program that builds local complexity over time.

5.5. Integrating neuronal and endocrine development

While no study has systematically dissected the steps that govern the differentiation program of any single neuron in the medial hypothalamus, it is clear that the same set of signaling factors are deployed to build hypothalamic neurons from neural ectoderm and endocrine cells from oral ectoderm. A wealth of studies have demonstrated the importance of Shh, BMPs, and FGFs as spatial and proliferative cues for progenitors within Rathke's pouch, the precursor of the anterior pituitary (Davis & Camper, 2007; Ericson, Norlin, Jessell, & Edlund, 1998; Herzog et al., 2004; Norlin, Nordstrom, & Edlund, 2000; Potok et al., 2008; Zhu, Gleiberman, & Rosenfeld, 2007), and for the emergence of defined endocrine lineages (Guner, Ozacar, Thomas, & Karlstrom, 2008; Liu et al., 2008; Sbrogna, Barresi, & Karlstrom, 2003). Similarly, common transcriptional factors appear to underpin the integrated specification of medial hypothalamic neurons and associated endocrine cells. For instance, Gsh1 and a zinc finger transcription factor, Ikaros, play a role in GHRH neuronal specification (Ezzat et al., 2006) but are also expressed in, and required for, somatotropes of the anterior pituitary (Li, Zeitler, Valerius, Small, & Potter, 1996). Similarly, SF-1 is required for both VMN formation and endocrine cell differentiation (Ikeda, Luo, Abbud, Nilson, & Parker, 1995; Kim, Zhao, & Parker, 2009; Shinoda et al., 1995). SF-1$^{-/-}$ mice exhibit a complex endocrine phenotype, but conditional genetics, specifically removal of SF-1 function from the anterior pituitary, has demonstrated that SF-1 is required for pituitary gonadotrope function (Zhao, Bakke, & Parker, 2001). We speculate that such integrated construction of neuronal and endocrine lineages is important for later integrated function, providing a mechanism to ensure effective architecture and communication of regulator neurons in the hypothalamus and effector cells in the hypophysis.

6. ESTABLISHMENT OF INTERFACES

6.1. Development of the infundibulum/neurohypophysis

As outlined earlier, three fundamentally different biological components constitute the adult neurohypophysis and form the active interfaces that support medial hypothalamic function: neuroendocrine axon terminals, portal capillary vessels, and glial-like cells composed of tanycytes and pituicytes (Fig. 2.2). How does each form in embryogenesis? Studies are beginning to reveal the developmental events that govern the development of each of these components, highlighting an orchestrated series of steps that supports the spatiotemporally integrated development, and hence future function, of the neurohypophysis. An embryonic structure termed the infundibulum holds the key to the emergence of such interfaces, providing the anlage on which to establish the neurohypophysis, the physical link between the developing nervous, endocrine, and blood systems. And molecularly, FGF ligands appear to act as master regulators of neurohypophyseal development, potentially playing a role in the establishment of each of its constitutive biological components.

Electron and light microscopy studies in many species show that the infundibulum arises from a rostral portion of the tuberal hypothalamic floor plate and appears as a ventrally projecting outgrowth, with multilayered walls, closely apposed to the nascent adenohypophysis (Fig. 2.5) that develops over the period in which neurogenesis occurs. Anamniotes do not have an obvious infundibulum, but show a rudimentary structure that shares molecular identity with the infundibulum of mice and chicks and forms around 36 hpf (Liu et al., 2013).

Tissue recombination studies and analyses of mouse mutants suggest that infundibular development may be triggered through early signaling events between the tuberal floor plate and Rathke's pouch, the precursor of the anterior pituitary (Dasen & Rosenfeld, 2001; Hermesz, Williams-Simons, & Mahon, 2003; Pelletier, 1991; Rizzoti et al., 2004). However, the tuberal floor plate is in register first with prechordal mesoderm and then (as prechordal mesoderm regresses) with Rathke's pouch (Fig. 2.4A and B), and it is more likely that Rathke's pouch maintains and refines, rather than initiates, the processes that drive development of some rostral-most hypothalamic floor plate cells into infundibular cells (see preceding text and Fig. 2.4).

6.2. The infundibulum is composed of multiple glial-like cells

Fate-mapping studies in chick show that the infundibulum is composed of multiple cell types, all derived from subregions of the tuberal floor plate (Fig. 2.5). A midline-situated subpopulation of floor plate begins to bifurcate and migrate caudally, forming a collar of cells around adjacent ventral midline cells (Fig. 2.5A and B). "Collar cells" (Pearson et al., 2011) exhibit

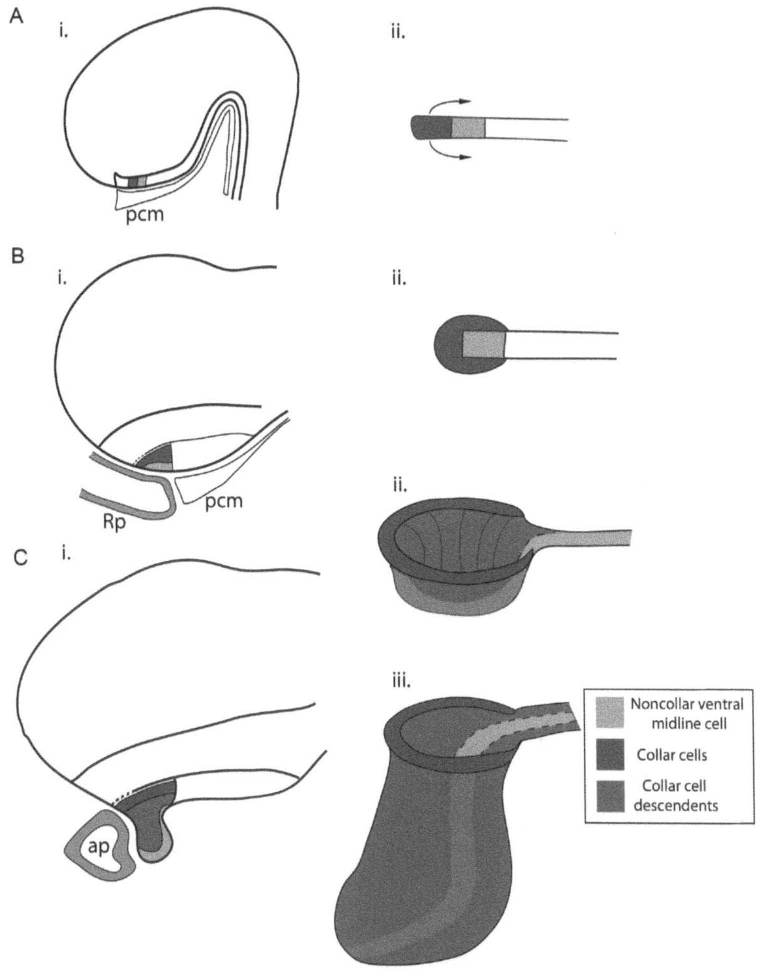

Figure 2.5 Multiple cell types form the infundibulum. The infundibulum is derived from at least three populations of cells termed collar cells, collar cell descendants, and noncollar rostral ventral midline cells. Collar cell descendants undergo a ventral migration, sculpting the infundibular walls. (A) Collar and noncollar rostral ventral midline cells are derived from two populations of floor plate cells directly adjacent to each other on the anterior–posterior axis. Lateral and dorsal views of the floor plate in the early hypothalamus (stage 10 in chick) indicate these two regions (collar cells in dark green and noncollar rostral ventral midline cells in light green) overlying the prechordal mesoderm. (i) is the lateral view and (ii) is the dorsal view. (B) Collar cells migrate caudally to surround the noncollar rostral ventral midline cells rostrally and laterally. During this time, the prechordal mesoderm retracts and part of the oral ectoderm, termed Rathke's pouch (in blue), comes to underlie collar cells and noncollar rostral ventral midline cells. Rathke's pouch is the presumptive anterior pituitary. (i) shows the lateral view and (ii) the dorsal view. (C) Collar cells show neural stem cell-like properties. Some appear

remarkable properties. They show neural stem-like properties and can self-renew or give rise to multipotent progenitors. Many progenitor cells migrate or grow ventrally, contributing to the multilayered walls of the infundibulum (Fig. 2.5C). Collar cell descendants populate much of the infundibulum, with the exception of a thin strip of caudoventral (posterior ventral) cells that derive from the noncollar rostral ventral midline. Thus, the infundibulum arises from at least three separate cell types: collar cells, collar cell descendants, and noncollar rostral midline floor plate cells. None express markers of early neuronal differentiation (Pearson et al., 2011). This, together with their floor plate origin, their stem-like potential, and their expression of glial markers (Placzek & Briscoe, 2005), suggests that the infundibulum is composed of multiple glial-like cell types.

6.3. Molecular pathways in infundibular formation

A number of transcription factors have been shown to be instrumental in infundibular formation. In mice mutants that lack *Hesx1, Lhx2, Nkx2.1, Rax,* and *Tbx3*, and in *Hes1 Hes5* double mutants, the infundibulum fails to form (Dattani et al., 1998; Kimura et al., 1996; Kita et al., 2007; Medina-Martinez et al., 2009; Takuma et al., 1998; Trowe et al., 2013; Zhao, Mailloux, Hermesz, Palkovits, & Westphal, 2010), while mice that are hypomorphic for Sox2 show aberrant infundibular development (Langer, Taranova, Sulik, & Pevny, 2012). At least some of these mutants affect early steps in the progression of hypothalamic floor plate cells to an infundibular fate: for instance, failure of infundibular formation in *Tbx3* mutant mice is likely due to failure to appropriately govern SOX2 and downregulate *Shh* expression (Langer et al., 2012; Trowe et al., 2013).

Additional studies suggest a pivotal role for FGFs in driving outgrowth of the infundibulum itself. In all vertebrates examined, *Fgfs* are upregulated throughout the tuberal floor plate (Manning et al., 2006; Ohuchi et al., 2000) and then become more restricted in and around the forming infundibulum (Fig. 2.4 and Fig. 2.5) (Herzog et al., 2004; Ohuchi et al., 2000;

Figure 2.5—Cont'd to self-renew; others proliferate and give rise to collar cell descendants. These migrate ventrally toward the anterior pituitary (blue) and sculpt the forming infundibulum (medium green). Additionally, noncollar rostral ventral midline cells migrate ventrally. Collar cells remain as a band around the developing infundibulum and continue to give rise to descendants into late embryogenesis. (i) shows the lateral view; (ii) and (iii) show a three-dimensional view of the sculpting of the infundibulum (between HH st 24 and 32 in chick). pcm, prechordal mesoderm; Rp, Rathke's pouch; ap, anterior pituitary. (See color plate.)

Pearson et al., 2011; Tsai et al., 2011). In *Fgf10*-null mice, the infundibulum fails to form properly and infundibular cells undergo apoptosis (Ohuchi et al., 2000). Similarly, in zebrafish, *fgf3* is required for maintenance of the infundibular-like structure (Liu et al., 2013). *In vitro* experiments in chick suggest that FGF signaling is required for the maintenance and proliferation of collar cells: inhibition of FGF signaling leads to the loss of both collar cells and collar-derived progenitors (Pearson et al., 2011).

Two transcription factors—the SoxB1 HMG-box transcription factor, SOX3, and the Lim HD transcription factor, LHX2—may operate downstream of FGF signaling. Both can be induced or maintained by FGF signaling (Pearson et al., 2011; Seth et al., 2006), and both are expressed in the tuberal floor plate. In *Lhx2* mutant mice, the infundibulum fails to form, possibly due to increased cell death (Zhao et al., 2010). Targeted conditional inactivation of *Sox3* in the mouse reduces proliferation rates resulting in thinning of the infundibulum and adjacent ventral hypothalamus (Rizzotti et al., 2004). Similarly, in humans, either reduced or elevated dosage of Sox3 leads to infundibular hypoplasia (di Iorgi et al., 2009; Woods et al., 2005). The broad expression of *Lhx2*, *Sox3*, and *Fgf10* within the tuberal hypothalamic floor plate makes it difficult to determine which cells are affected in mouse mutants and how their dysregulation results in aberrant infundibular formation. However, in chick, although *Sox3* is widely expressed in the rostral hypothalamic floor, SOX3 protein is detected only in collar cells and may play a direct role in the FGF-dependent maintenance and proliferation of collar cells (Pearson et al., 2011).

6.4. Extension of axons to the infundibulum/forming neurohypophysis

As detailed earlier, two major classes of neuroendocrine neurons are recognized, parvocellular and magnocellular. Both classes show stereotypic projection patterns that are conserved across species, projecting ventrally to terminate at the neurohypophysis (Fig. 2.2). What is known about the factors that govern these projection patterns?

Elsewhere in the CNS, axons project over long distances to intermediate and final targets under the influence of long-range chemoattractive and chemorepulsive cues. Along much of the rostrocaudal axis, two chemoattractants, Netrin and Shh, are expressed in the floor plate and guide axons to the midline (Charron, Stein, Jeong, McMahon, & Tessier-Lavigne, 2003; Sanchez-Camacho & Bovolenta, 2009; Chedotal, 2011). However, although briefly expressed in the hypothalamic floor plate (Fig. 2.4), both Netrin and Shh are rapidly downregulated and so are absent from the midline-situated infundibulum/neurohypophysis before H-NH axon pioneers begin to extend there. Thus, the alterations in midline axonal pathways in Netrin-1-deficient mice (Deiner & Stretavan, 1999) are more likely to reflect a sustained role in the basal plate, potentially an outgrowth-promoting effect, rather than a guiding effect (Low, Fiorini, Fisher, & Jasoni, 2012). What other cues might direct the early growth of H-NH axons toward their midline target(s)? Two studies suggest an essential role for FGFs in this event.

Early work in mouse first suggested a role for FGF signal reception in the migration of GnRH neurons from their birthplace in the olfactory placode to the hypothalamus (Tsai et al., 2011) and then in the subsequent growth of GnRH axons toward the median eminence (Gill & Tsai, 2006). A recent study extends this idea, demonstrating a requirement for FGFs in the pathfinding of magnocellular and parvocellular pioneers to the infundibulum. In chick, *in vitro* assays demonstrate that FGF3 and FGF10 deriving from the infundibulum can exert a direct guidance effect, stimulating and reorienting the growth of both magnocellular (AVP) and parvocellular (dopaminergic) subtypes. Similarly, in zebrafish, AVP and OXT magnocellular neurons, and dopaminergic parvocellular neurons differentiate in *fgf3* zebrafish mutants, but their axons fail to project to the neurohypophysis (Liu et al., 2013). Although, FGFs govern development of both the adenohypophysis and the infundibulum (the neurohypophyseal anlage), their role in H-NH axonal guidance can be separated from its roles in these events: temporal or tissue-specific manipulation of a dominant-negative Fgf-receptor (Fgf-R1) reveals that axons respond directly to FGF signaling (Liu et al., 2013).

The same studies suggest that FGF3/FGF10 exert long-range guidance effects to draw axons to the forming infundibulum but then prevent their further extension into more ventral parts of the forming neurohypophysis. Thus, whereas low concentrations of FGFs attract H-NH axons, high concentrations stall or repel their growth. Chick *Fgf10* and zebrafish *fgf3* are expressed in a graded fashion, with highest levels in the posterior ventral NH, and axons initially avoid regions that display highest and most persistent *Fgf3/10* expression levels, in-line with a dual (attractive, low; repellant, high) role for FGFs (Liu et al., 2013). Such dual activity may ensure that H-NH axons project to, but do not cross, the ventral midline: uniquely in this region of the CNS, axons are noncommissural. Conceivably, the stalling is part of a more intricate mechanism that ensures the subsequent correct projection of parvocellular axons to the median eminence and of magnocellular axons to the posterior lobe. Although we have little understanding of how this occurs, our knowledge of axonal sorting elsewhere in the embryo allows us to speculate on likely strategies that govern magnocellular and parvocellular sorting. Potentially, local cues, deriving from different cell populations in the median eminence and posterior lobe, orchestrate this. In mouse embryos that lack the Notch effector gene, *Hes1*, AVP axons project to the midline but then display abnormal trajectories in the region of the median eminence and posterior lobe (Aujila, Bora, Monahan, Sweedler, & Raetzman, 2011), suggesting a possible role for Notch signaling in the axonal targeting of this H-NH neuronal subtype.

6.5. Integrated establishment of axons and capillaries in the infundibulum/neurohypophysis

The rich vascularization of the neurohypophysis is critical to its function: the close proximity of H-NH axons and capillary vessels in the NH underlies the future functioning of the H-NH axis and homeostatic balance. What mechanisms, then, ensure that axons and capillaries develop in concert in a manner that enables them to interact with one another to form the functional neurohypophysis?

The same study that defined a role for FGF signaling in orienting the growth of neuroendocrine axons to the forming infundibulum showed that FGFs exert a weak, but direct, effect to promote local endothelial vasculogenesis, that is, the *de novo* formation of vessels via the assembly of endothelial cells. *In vivo*, endothelial cells/cell chains that prefigure neurohypophyseal capillaries extend from preexisting blood vessels toward the region of the forming infundibulum just prior to the arrival of early neuroendocrine axon pioneers. *In vitro* experiments show that endothelial cell outgrowth can be stimulated by the forming infundibulum *in vitro*, and this effect is mediated by FGFs. Likewise, in transgenic zebrafish, endothelial cells that express dominant-negative Fgf-R1 contribute to pituitary vessel formation with lower frequency than cells carrying a control transgene. Together, these studies point to a role of FGF signaling during the early modeling of endothelial cells and the initial formation of the H-NH capillary plexus (Liu et al., 2013). Mounting evidence shows that axonal growth cones and vascular sprouts share common receptors and are able to respond to the same guidance cues (reviewed by Adams & Eichmann, 2010). The demonstration that FGF-producing cells of the forming infundibulum govern both axons and nascent capillaries suggests that FGFs mediate long-range trophic/tropic activity to simultaneously attract both axons and endothelial cells and ensure coordinated growth of neurohypophyseal axons and capillaries to the same general target.

The effect of FGFs, though, is both weak and transient: the neurohypophysis can mimic the ability of the infundibulum to promote endothelial cell outgrowth, but, although neurohypophyseal cells continue to express *Fgfs*, they are not required for endothelial process formation. Instead, FGFs act in concert with other vascularization-promoting factors emanating from neurohypophyseal components.

One such factor is OXT, released from axon terminals of magnocellular neurons that have begun to innervate the posterior lobe. Elegant studies in zebrafish show that capillary vessels extend to the innervated neurohypophysis via angiogenesis (the sprouting and extension of new vessels from existing vessels). Vascular sprouting is controlled by sources of secreted angiogenic molecules, to which vascular tip cells are attracted. In zebrafish embryos, OXT-like acts as an angiogenic cue for nearby vascular sprouts, drawing them toward axon terminals, where they go on to form the hypophyseal arteries and veins (Gutnick et al., 2011). Thus, a local trophic action pulls capillary sprouts close to axonal terminals, ensuring the formation of intimate connections. It will be fascinating to see whether similar fine-tuning of other axonal-capillary networks occurs and whether, in general, magnocellular and parvocellular axons will actively participate in building the vascular networks required for their activity.

6.6. Development of glial-like cells of the neurohypophysis

What of the development of the third cellular component of the neurohypophysis, the differentiating glial cells? Despite the inroads made recently into an understanding of tanycyte cell function (reviewed in Bolborea & Dale, 2013), our understanding of pituicyte function remains poor, and the development of these fascinating and functionally diverse cells remains an enigma. However, recent studies provide some clues as to when and how they originate.

The lineage relationship, if any, of tanycyte and pituicyte cells is unclear, but they share many common markers, including expression of cellular retinoic acid-binding protein, GFAP, and Nestin. The early expression of a subset of these markers (Lee, Wang, Anderson, Ekker, & Rubenstein, 2009; Nakamura et al., 2001) in the embryonic infundibulum, and their common glial nature, suggests that tanycytes and pituicytes derive from infundibular cells, but the relationships between the different cell types of the embryonic infundibulum (collar cells, collar cell descendants, and non-collar ventral midline cells) and tanycytes or pituicytes remain to be determined. However, recent studies indicate an ongoing role for Wnt signaling in the transition from infundibular to neurohypophyseal fates, via a radial glial cell fate. In zebrafish and mice, Wnt signaling must be inhibited for radial glial cells, and then tanycytes, to form. Thus, ectopic activation of Wnt signaling inhibits radial glial and tanycyte formation (Wang et al., 2012). Given the antagonism between Wnt and FGF signaling within the ventrobasal hypothalamus in early development (see earlier) and the finding that a transient expression of *Fgf-10* governs the formation of cortical radial glia (Sahara & O'Leary, 2009), it is tempting to speculate that ongoing FGF signaling may again play some role in the transition from infundibular precursors to neurohypophyseal radial glial cells.

7. CONCLUDING REMARKS

Far from being a developmental backwater, the medial hypothalamus is emerging as one of the most beautiful examples of the manner in which the embryo assembles diverse cell types in an integrated manner to enable function. A common transcription factor can be expressed in both neural and ectoderm lineages in response to midline-derived signaling ligands and orchestrate the differentiation of regulatory neuroendocrine neurons and effector endocrine cells that govern a particular physiological axis. Neurons that act opposingly to modulate a particular behavior can develop from a common precursor, ensuring an exquisite control of balance. A common signaling pathway shapes the infundibulum and subsequently ensures that its constituent parts can be assembled there. Intriguingly, the signaling ligands that initiate these events in early embryogenesis are maintained, even into adulthood, where they are critical for postnatal establishment of novel cells and circuits, and for the modulation of glial-like cells (outside the scope of this review, but see Lee & Blackshaw, 2012). Similarly, the same early signal transcription factor pathways that establish medial hypothalamic cells and circuits can be redeployed for their maintenance and function (again, outside the scope of this review, but see Elghazi et al., 2012; Wang et al., 2012). Together, these studies hint that the mechanisms deployed to build the hypothalamus are used throughout life to maintain, refine, and modulate its circuitry, potentially both enabling and responding to changing needs. Clinically prevalent disorders, including obesity, stress, and reproductive failure, are likely to be caused by dysfunction of medial hypothalamic cells: an understanding of their normal differentiation and maintenance could therefore aid in the diagnosis and treatment of neuroendocrine and psychiatric diseases.

ACKNOWLEDGMENTS

We thank Andrew Furley for helpful comments on the chapter. This work was supported by the Medical Research Council of Great Britain.

REFERENCES

Acampora, D., Postiglione, M. P., Avantaggiato, V., Di Bonito, M., Vaccarino, F. M., Michaud, J., et al. (1999). Progressive impairment of developing neuroendocrine cell lineages in the hypothalamus of mice lacking the Orthopedia gene. *Genes & Development, 13*, 2787–2800.

Adams, R. H., & Eichmann, A. (2010). Axon guidance molecules in vascular patterning. *Cold Spring Harbor Perspectives in Biology, 2*, a001875.

Adelmann, H. B. (1922). The significance of the prechordal plate: An interpretative study. *The American Journal of Anatomy, 1*, 55–101.

Altman, J., & Bayer, S. A. (1986). The development of the rat hypothalamus. *Advances in Anatomy, Embryology, and Cell Biology, 100*, 1–178.

Altmann, C. R., & Brivanlou, A. H. (2001). Neural patterning in the vertebrate embryo. *International Review of Cytology, 203*, 447–482.

Amir-Zilberstein, L., Blechman, J., Sztainberg, Y., Norton, W. H., Reuveny, A., Borodovsky, N., et al. (2012). Homeodomain protein otp and activity dependent splicing modulate neuronal adaptation to stress. *Neuron, 73*, 279–291.

Aubert, M. L., Grumbach, M. M., & Kaplan, S. L. (1977). The ontogenesis of human fetal hormones. IV. Somatostatin, luteinizing hormone releasing factor, and thyrotropin releasing factor in hypothalamus and cerebral cortex of human fetuses 10–22 weeks of age. *Journal of Clinical Endocrinology and Metabolism, 44*, 1130–1141.

Aujila, P. K., Bora, A., Monahan, P., Sweedler, J. V., & Raetzman, L. T. (2011). The Notch effector gene Hes1 regulates migration of hypothalamic neurons, neuropeptide content and axon targeting to the pituitary. *Developmental Biology, 353*, 61–71.

Bargmann, W. (1949). Über die neurosekretorische Verknüpfung von Hypothalamus und Neurohypophyse. *Zeitschrift für Zellforschung, 34*, 610–634.

Barth, K. A., & Wilson, S. W. (1995). Expression of zebrafish nk2.2 is influenced by sonic hedgehog/vertebrate hedgehog-1 and demarcates a zone of neuronal differentiation in the embryonic forebrain. *Development, 121*, 1755–1768.

Blechman, J., Borodovsky, N., Eisenberg, M., Nabel-Rosen, H., Grimm, J., & Levkowitz, G. (2007). Specification of hypothalamic neurons by dual regulation of the homeodomain protein Orthopedia. *Development, 134*, 4417–4426.

Bolborea, M., & Dale, N. (2013). Hypothalamic tanycytes: Potential roles in the control of feeding and energy balance. *Trends in Neurosciences, 36*, 91–100.

Bosco, A., Bureau, C., Affaticati, P., Gaspar, P., Bally-Cuif, L., & Lillesaar, C. (2013). Development of hypothalamic serotoninergic neurons requires Fgf signalling via the ETS-domain transcription factor Etv5b. *Development, 140*, 372–384.

Braun, M. M., Etheridge, A., Bernard, A., Robertson, C. P., & Roelink, H. (2003). Wnt signaling is required at distinct stages of development for the induction of the posterior forebrain. *Development, 130*, 5579–5587.

Brinkmeier, M. L., Potok, M. A., Davis, S. W., & Camper, S. A. (2007). TCF4 deficiency expands ventral diencephalon signaling and increases induction of pituitary progenitors. *Developmental Biology, 311*, 396–407.

Briscoe, J., & Novitch, B. G. (2008). Regulatory pathways linking progenitor patterning, cell fates and neurogenesis in the ventral neural tube. *Philosophical Transactions of the Royal Society of London Series B, Biological Sciences, 363*, 57–70.

Caqueret, A., Boucher, F., & Michaud, J. L. (2006). Laminar organization of the early developing anterior hypothalamus. *Developmental Biology, 298*, 95–106.
Caqueret, A., Coumailleau, P., & Michaud, J. L. (2005). Regionalization of the anterior hypothalamus in the chick embryo. *Developmental Dynamics, 233*, 652–658.
Charron, F., Stein, E., Jeong, J., McMahon, A. P., & Tessier-Lavigne, M. (2003). The morphogen sonic hedgehog is an axonal chemoattractant that collaborates with netrin-1 in midline axon guidance. *Cell, 113*, 11–23.
Chedotal, A. (2011). Further tales of the midline. *Current Opinion in Neurobiology, 21*, 68–75.
Cheng, L. E., Zhang, J., & Reed, R. R. (2007). The transcription factor Zfp423/OAZ is required for cerebellar development and CNS midline patterning. *Developmental Biology, 307*, 43–52.
Chiang, C., Litingtung, Y., Lee, E., Young, K. E., Corden, J. L., Westphal, H., et al. (1996). Cyclopia and defective axial patterning in mice lacking Sonic hedgehog gene function. *Nature, 383*, 407–413.
Chichikov, V. V., & Millen, K. J. (2005). Roof plate-dependent patterning of the vertebrate dorsal central nervous system. *Developmental Biology, 277*, 287–295.
Dale, J. K., Vesque, C., Lints, T. J., Sampath, T. K., Furley, A., Dodd, J., et al. (1997). Cooperation of BMP7 and SHH in the induction of forebrain ventral midline cells by prechordal mesoderm. *Cell, 90*, 257–269.
Dasen, J. S., & Rosenfeld, M. G. (2001). Signaling and transcriptional mechanisms in pituitary development. *Annual Review of Neuroscience, 24*, 327–355.
Dattani, M. T., Martinez-Barbera, J. P., Thomas, P. Q., Brickman, J. M., Gupta, R., Mårtensson, I. L., et al. (1998). Mutations in the homeobox gene HESX1/hesx1 associated with septo-optic dysplasia in human and mouse. *Nature Genetics, 19*, 125–133.
Davis, S. W., & Camper, S. A. (2007). Noggin regulates Bmp4 activity during pituitary induction. *Developmental Biology, 305*, 145–160.
Deiner, M. S., & Stretavan, D. W. (1999). Altered midline axon pathways and ectopic neurons in the developing hypothalamus of netrin-1- and DCC-deficient mice. *Journal of Neuroscience, 19*, 9900–9912.
del Bene, F., Tessmar-Raible, K., & Wittbrodt, J. (2004). Direct interaction of geminin and Six3 in eye development. *Nature, 427*, 745–749.
di Iorgi, N., Secco, A., Napoli, F., Calandra, E., Rossi, A., & Maghnie, M. (2009). Developmental abnormalities of the posterior pituitary gland. *Endocrine Development, 14*, 83–94.
Echelard, Y., Epstein, D. J., St-Jacques, B., Shen, L., Mohler, J., McMahon, J. A., et al. (1993). Sonic hedgehog, a member of a family of putative signalling molecules, is implicated in the regulation of CNS polarity. *Cell, 75*, 1417–1430.
Edlund, T., & Jessell, T. M. (1999). Progression from extrinsic to intrinsic signalling in cell fate specification: A view from the nervous system. *Cell, 96*, 211–224.
Elghazi, L., Gould, A. P., Weiss, A. J., Barker, D. J., Callaghan, J., Opland, D., et al. (2012). Importance of beta-Catenin in glucose and energy homeostasis. *Scientific Reports*, Reports 2, article number 693.
Ericson, J., Norlin, S., Jessell, T., & Edlund, T. (1998). Integrated FGF and BMP signaling controls the progression of progenitor cell differentiation and the emergence of pattern in the embryonic anterior pituitary. *Development, 125*, 1005–1015.
Ezzat, S., Mader, R., Fischer, S., Yu, S., Ackerly, C., & Asa, S. L. (2006). An essential role for the hematopoietic transcription factor Ikaros in hypothalamic-pituitary-mediated somatic growth. In: *Proceedings of the National Academy of Sciences of the United States of America, 103*, 2214–2219.
Fuxe, K., & Hokfelt, T. (1972). On the existence of dopaminergic and noradrenergic mechanisms in the control of hormone releasing and inhibitory factor secretion from the median eminence. *Journal of Anatomy, 111*, 475–476.

Garcia-Calero, E., Fernandez-Garre, P., Martinez, S., & Puelles, E. (2008). Early pouch specification in the course of prechordal ventralization of the forebrain tegmentum. *Developmental Biology, 320*, 366–377.

Geng, X., Speirs, C., Lagutin, O., Inbal, A., Liu, W., Solnica-Krezel, L., et al. (2008). Haploinsufficiency of Six3 fails to activate Sonic hedgehog expression in the ventral forebrain and causes holoprosencephaly. *Developmental Cell, 15*, 236–247.

Gill, J. C., & Tsai, P. S. (2006). Expression of a dominant negative FGF receptor in developing GNRH1 neurons disrupts axon outgrowth and targeting to the median eminence. *Biology of Reproduction, 74*, 463–472.

Goshu, E., Jin, H., Fasnacht, R., Sepenski, M., Michaud, J. L., & Fan, C. M. (2002). Sim2 mutants have developmental defects not overlapping with those of Sim1 mutants. *Molecular Cell Biology, 22*, 4147–4157.

Goshu, E., Jin, H., Lovejoy, J., Marion, J. F., Michaud, J. L., & Fan, C. M. (2004). Sim2 contributes to neuroendocrine hormone gene expression in the anterior hypothalamus. *Molecular Endocrinology, 18*, 1251–1262.

Guillemot, F. (2007). Spatial and temporal specification of neural fates by transcription factor codes. *Development, 134*, 3771–3780.

Guner, B., Ozacar, A. T., Thomas, J. E., & Karlstrom, R. O. (2008). Graded hedgehog and fibroblast growth factor signaling independently regulate pituitary cell fates and help establish the pars distalis and pars intermedia of the zebrafish adenohypophysis. *Endocrinology, 149*, 4435–4451.

Gutnick, A., Blechman, J., Kaslin, J., Herwig, L., Belting, H. G., Affolter, M., et al. (2011). The hypothalamic neuropeptide oxytocin is required for formation of the neurovascular interface of the pituitary. *Developmental Cell, 21*, 642–654.

Harris, G. W. (1955). Pituitary-hypothalamic mechanisms. *AMA Archives of Neurology & Psychiatry, 73*, 124–126.

Hauptmann, G., & Gerster, T. (1996). Complex expression of the zp-50 pou gene in the embryonic zebrafish brain is altered by overexpression of sonic hedgehog. *Development, 122*, 1769–1780.

Hermesz, E., Williams-Simons, L., & Mahon, K. A. (2003). A novel inducible element, activated by contact with Rathke's pouch, is present in the regulatory region of the Rpx/Hesx1 homeobox gene. *Developmental Biology, 260*, 68–78.

Herzog, W., Sonntag, C., von der Hardt, S., Roehl, H. H., Varga, Z. M., & Hammerschmidt, M. (2004). Fgf3 signaling from the ventral diencephalon is required for early specification and subsequent survival of the zebrafish adenohypophysis. *Development, 131*, 3681–3692.

Hill, J. W. (2012). PVN pathways controlling energy homeostasis. *Indian Journal of Endocrinology and Metabolism, 16*, S627–S636.

Hosoya, T., Oda, Y., Takahashi, S., Morita, M., Kawauchi, S., Ema, M., et al. (2001). Defective development of secretory neurones in the hypothalamus of Arnt2-knockout mice. *Genes to Cells, 6*, 361–374.

Hu, Y., Poopalasundaram, S., Graham, A., & Bouloux, P. M. (2013). GnRH neuronal migration and olfactory bulb neurite outgrowth are dependent on FGF receptor 1 signaling, specifically via the PI3K p110α isoform in chick embryo. *Endocrinology, 154*, 388–399.

Ikeda, Y., Luo, X., Abbud, R., Nilson, J. H., & Parker, K. L. (1995). The nuclear receptor steroidogenic factor 1 is essential for the formation of the ventromedial hypothalamic nucleus. *Molecular Endocrinology, 9*, 478–486.

Ingraham, H. A., Lala, D. S., Ikeda, Y., Luo, X., Shen, W. H., Nachtigal, M. W., et al. (1994). The nuclear receptor steroidogenic factor 1 acts at multiple levels of the reproductive axis. *Genes & Development, 8*, 2302–2312.

Jeong, Y., Leskow, F. G., El-Jaick, K., Roessler, E., Muenke, M., Yocum, A., et al. (2008). Regulation of a remote Shh forebrain enhancer by the Six3 homeoprotein. *Nature Genetics, 40*, 1348–1353.

Jessell, T. (2000). Neuronal specification in the spinal cord: Inductive signals and transcriptional codes. *Nature Reviews Genetics*, *1*, 20–29.

Jo, Y. H., & Chua, S., Jr. (2009). Transcription factors in the development of medial hypothalamic structures. *American Journal of Physiology, Endocrinology and Metabolism*, *297*, E563–E567.

Kapsimali, M., Caneparo, L., Houart, C., & Wilson, S. W. (2004). Inhibition of Wnt/Axin/beta-catenin pathway activity promotes ventral CNS midline tissue to adopt hypothalamic rather than floorplate identity. *Development*, *131*, 5923–5933.

Keith, B., Adelman, D. M., & Simon, M. C. (2001). Targeted mutation of the murine arylhydrocarbon receptor nuclear translocator 2 (Arnt2) gene reveals partial redundancy with Arnt. In: *Proceedings of the National Academy of Sciences of the United States of America*, *98*, 6692–6697.

Kelberman, D., & Dattani, M. T. (2009). Role of transcription factors in midline central nervous system and pituitary defects. *Endocrine Development*, *14*, 67–82.

Kim, E. J., Battiste, J., Nakagawa, Y., & Johnson, J. E. (2008). Ascl1 (Mash1) lineage cells contribute to discrete cell populations in CNS architecture. *Molecular and Cellular Neurosciences*, *38*, 595–606.

Kim, K. W., Zhao, L., Donato, J., Jr., Kohno, D., Xu, Y., Elias, C. F., et al. (2011). Steroidogenic factor 1 directs programs regulating diet-induced thermogenesis and leptin action in the ventral medial hypothalamic nucleus. In: *Proceedings of the National Academy of Sciences of the United States of America*, *108*, 10673–10678.

Kim, K. W., Zhao, L., & Parker, K. L. (2009). Central nervous system-specific knockout of steroidogenic factor 1. *Molecular and Cellular Endocrinology*, *300*, 132–136.

Kimura, S., Hara, Y., Pineau, T., Fernandez-Salguero, P., Fox, C. H., Ward, J. M., et al. (1996). The T/ebp null mouse: Thyroid-specific enhancer-binding protein is essential for the organogenesis of the thyroid, lung, ventral forebrain, and pituitary. *Genes & Development*, *10*, 60–69.

Kita, A., Imayoshi, I., Hojo, M., Kitagawa, M., Kokubu, H., Ohsawa, R., et al. (2007). Hes1 and Hes5 control the progenitor pool, intermediate lobe specification and posterior lobe formation in the pituitary development. *Molecular Endocrinology*, *21*, 1458–1466.

Kobayashi, D., Kobayashi, M., Mastsumoto, K., Ogura, T., Nakafuku, M., & Shimamura, K. (2002). Early subdivisions in the neural plate define distinct competence for inductive signals. *Development*, *129*, 83–93.

Kobayashi, M., Nishikawa, K., Suzuki, T., & Yamamoto, M. (2001). The homeobox protein Six3 interacts with the Groucho corepressor and acts as a transcriptional repressor in eye and forebrain formation. *Developmental Biology*, *232*, 315–326.

Krauss, R. S. (2007). Holoprosencephaly: New models, new insights. *Expert Reviews in Molecular Medicine*, *9*, 1–17.

Kuenzel, W. J., & van Tienhoven, A. (1982). Nomenclature and location of avian hypothalamic nuclei and associated circumventricular organs. *Journal of Comparative Neurology*, *206*, 293–313.

Lagutin, O. V., Zhu, C. C., Kobayashi, D., Topczewski, J., Shimamura, K., Puelles, L., et al. (2003). Six3 expression of Wnt signaling in the anterior neuroectoderm is essential for vertebrate forebrain development. *Genes & Development*, *17*, 368–379.

Langer, L., Taranova, O., Sulik, K., & Pevny, L. (2012). SOX2 hypomorphism disrupts development of the prechordal floor and optic cup. *Mechanisms of Development*, *129*, 1–12.

Lavado, A., Lagutin, O. V., & Oliver, G. (2008). Six3 inactivation causes progressive caudalization and aberrant patterning of the mammalian diencephalon. *Development*, *135*, 441–450.

Lee, D. A., & Blackshaw, S. (2012). Functional implications of hypothalamic neurogenesis in the adult mammalian brain. *International Journal of Developmental Neuroscience*, *30*, 615–621.

Lee, C. L., Wang, Y., Anderson, S., Ekker, M., & Rubenstein, J. L. R. (2009). Arcuate nucleus expression of NKX2.1 and DLX, and lineages expressing these transcription factors in NPY+, POMC+ and TH+ neurons in neonatal and adult mice. *Journal of Comparative Neurology, 517*, 37–50.

Li, H., Zeitler, P. S., Valerius, M. T., Small, K., & Potter, S. S. (1996). Gsh-1, an orphan Hox gene, is required for normal pituitary development. *The EMBO Journal, 15*, 714–724.

Lipinski, R. J., Godin, E. A., O'Leary-Moore, S. K., Parnell, S. E., & Sulik, K. K. (2010). Genesis of teratogen-induced holoprosencephaly in mice. *American Journal of Medical Genetics Part C: Seminars in Medical Genetics, 154*, 29–42.

Liu, F., Pogoda, H. M., Pearson, C. A., Ohyama, K., Lohr, H., Hammerschmidt, M., et al. (2013). Direct and indirect roles of Fgf3 and Fgf10 in innervation and vascularisation of the vertebrate hypothalamic neurohypophysis. *Development, 140*, 1111–1122.

Liu, N. A., Ren, M., Song, J., Rios, Y., Wawrowsky, K., Ben-Shlomo, A., et al. (2008). In vivo time-lapse imaging delineates the zebrafish pituitary proopiomelanocortin lineage boundary regulated by FGF3 signal. *Developmental Biology, 319*, 192–200.

Lohr, H., & Hammerschmidt, M. (2011). Zebrafish in endocrine systems: Recent advances and implications for human disease. *Annual Review of Physiology, 73*, 183–211.

Low, V. F., Fiorini, Z., Fisher, L., & Jasoni, C. L. (2012). Netrin-1 stimulates developing GnRH neurons to extend neuritis to the median eminence in a calcium-dependent manner. *PLoS One*, 7, e46999.

Lu, F., Kar, D., Gruenig, N., Zhang, Z. W., Cousins, N., Rodgers, H. M., et al. (2013). Rax is a selector gene for mediobasal hypothalamic cell types. *Journal of Neuroscience, 33*, 259–272.

Lutz, T. A., & Woods, S. C. (2012). Overview of animal models of obesity. *Current Protocols in Pharmacology, 5.61*, 259–272.

Machluf, Y., Gutnick, A., & Levkowitz, G. (2011). Development of the zebrafish hypothalamus. *Annals of the New York Academy of Sciences, 1220*, 93–105.

Maniam, J., & Morris, M. J. (2012). The link between stress and feeding behaviour. *Neuropharmacology, 63*, 97–110.

Manning, L., Ohyama, K., Saeger, B., Hatano, O., Wilson, S. A., Logan, M., et al. (2006). Regional morphogenesis in the hypothalamus: A BMP-Tbx2 pathway coordinates fate and proliferation through Shh downregulation. *Developmental Cell, 11*, 873–885.

Marcucio, R. S., Cordero, D. R., Hu, D., & Helms, J. A. (2005). Molecular interactions coordinating the development of the forebrain and face. *Developmental Biology, 284*, 48–61.

Marin, O., Baker, J., Puelles, L., & Rubenstein, J. L. (2002). Patterning of the basal telencephalon and hypothalamus is essential for guidance of cortical projections. *Development, 129*, 761–773.

Markakis, E. A. (2002). Development of the neuroendocrine hypothalamus. *Frontiers in Neuroendocrinology, 23*, 257–291.

Mathieu, J., Barth, A., Rosa, F. M., Wilson, S. W., & Peyrieras, N. (2002). Distinct and cooperative roles for Nodal and Hedgehog signals during hypothalamic development. *Development, 129*, 3055–3065.

McCabe, M. J., Alatzoglou, K. S., & Dattani, M. T. (2011). Septo-optic dysplasia and other midline defects: The role of transcription factors: HESX1 and beyond. *Best Practice & Research Clinical Endocrinology & Metabolism, 25*, 115–124.

McNay, D. E., Pelling, M., Claxton, S., Guillemot, F., & Ang, S. L. (2006). Mash1 is required for generic and subtype differentiation of hypothalamic neuroendocrine cells. *Molecular Endocrinology, 20*, 1623–1632.

Medina-Martinez, O., Amaya-Manzanares, F., Liu, C., Mendoza, M., Shah, R., Zhang, L., et al. (2009). Cell-autonomous requirement for rx function in the mammalian retina and posterior pituitary. *PLoS One, 4*, e4513.

Michaud, J. L. (2001). The developmental program of the hypothalamus and its disorders. *Clinical Genetics, 60*, 255–263.

Michaud, J. L., De Rossi, C., May, N. R., Holdener, B. C., & Fan, C. M. (2000). ARNT2 acts as the dimerization partner of SIM1 for the development of the hypothalamus. *Mechanisms of Development, 90*, 253–261.
Michaud, J. L., Rosenquist, T., May, N. R., & Fan, C. M. (1998). Development of neuroendocrine lineages requires the bHLH-PAS transcription factor SIM1. *Genes & Development, 12*, 3264–3275.
Monuki, E. S. (2007). The morphogen signaling network in forebrain development and holoprosencephaly. *Journal of Neuropathology and Experimental Neurology, 66*, 566–575.
Morales-Delgado, N., Merchan, P., Bardet, S. M., Ferran, J. L., Puelles, L., & Diaz, C. (2011). Topography of somatostatin gene expression relative to molecular progenitor domains during ontogeny of the mouse hypothalamus. *Frontiers in Neuroanatomy, 5*, 10.
Muenke, M., & Beachy, P. A. (2000). Genetics of ventral forebrain development and holoprosencephaly. *Current Opinion in Genetics and Development, 10*, 262–269.
Muller, M. B., & Keck, M. E. (2002). Genetically engineered mice for studies of stress-related clinical conditions. *Journal of Psychiatric Research, 36*, 53–76.
Mutsuga, N., Iwasaki, Y., Morishita, M., Nomura, A., Yamamori, E., Yoshida, M., et al. (2001). Homeobox protein Gsh-1-dependent regulation of the rat GHRH gene promoter. *Molecular Endocrinology, 15*, 2149–2156.
Nakai, S., Kawano, H., Yudate, T., Nishi, M., Kuno, J., Nagata, A., et al. (1995). The POU domain transcription factor Brn-2 is required for the determination of specific neuronal lineages in the hypothalamus of the mouse. *Genes & Development, 9*, 3109–3121.
Nakamura, K., Kimura, S., Yamazaki, M., Kawaguchi, A., Inoue, K., & Sakai, T. (2001). Immunohistochemical analyses of thyroid-specific enhancer-binding protein in the fetal and adult rat hypothalamic and pituitary glands. *Developmental Brain Research, 130*, 159–166.
Norlin, S., Nordstrom, U., & Edlund, T. (2000). Fibroblast growth factor signaling is required for the proliferation and patterning of progenitor cells in the developing anterior pituitary. *Mechanisms of Development, 96*, 175–182.
O'Leary, D. D., Chou, S. J., & Sahara, S. (2007). Area patterning of the mammalian cortex. *Neuron, 56*, 252–269.
Ohuchi, H., Hori, Y., Yamasaki, M., Harada, H., Sekine, K., Kato, S., et al. (2000). FGF10 acts as a major ligand for FGF receptor 2 IIIb in mouse multi organ development. *Biochemical and Biophysical Research Communications, 277*, 643–649.
Ohyama, K., Das, R., & Placzek, M. (2008). Temporal progression of hypothalamic patterning by a dual action of BMP. *Development, 135*, 3325–3331.
Ohyama, K., Ellis, P., Kimura, S., & Placzek, M. (2005). Directed differentiation of neural cells to hypothalamic dopaminergic neurons. *Development, 132*, 5185–5197.
Pabst, O., Herbrand, H., Takuma, N., & Arnold, H. H. (2000). NKX2 gene expression in neuroectoderm but not in mesendodermally derived structures depends on sonic hedgehog in mouse embryos. *Development Genes and Evolution, 210*, 47–50.
Padilla, S. L., Carmody, J. S., & Zeltser, L. M. (2010). POMC-expressing progenitors give rise to antagonistic neuronal populations in hypothalamic feeding circuits. *Nature Medicine, 16*, 403–405.
Patten, I., Kulesa, P., Shen, M., Fraser, S., & Placzek, M. (2003). Distinct modes of floor plate induction in the chick embryo. *Development, 130*, 4809–4821.
Pearson, C. A., Ohyama, K., Manning, L., Aghamohammadzadeh, S., Sang, H., & Placzek, M. (2011). FGF-dependent midline-derived progenitor cells in hypothalamic infundibular development. *Development, 138*, 2613–2624.
Pelletier, G. (1991). Anatomy of the hypothalamic–pituitary axis. *Methods and Achievements in Experimental Pathology, 14*, 1–22.

Pelling, M., Anthwal, N., McNay, D., Gradwohl, G., Leiter, A. B., Guillemot, F., et al. (2011). Differential requirement for neurogenin3 in the development of POMC and NPY neurons in the hypothalamus. *Developmental Biology, 349*, 406–416.

Pelling, A. N., Claxton, S., Mellitzer, G., Collin, C., Kessaris, N., Richardson, W. D., et al. (2013). Conditional deletion of Neurogenin3 using Nkx2.1Cre results in a mouse model for the central control of feeding, activity and obesity. *Disease Models & Mechanisms, 6*, 1133–1145.

Placzek, M., & Briscoe, J. (2005). The floor plate: Multiple cells, multiple signals. *Nature Reviews Neuroscience, 6*, 230–240.

Pontecorvi, M., Goding, C. R., Richardson, W. D., & Kessaris, N. (2008). Expression of Tbx2 and Tbx3 in the developing hypothalamic–pituitary axis. *Gene Expression Patterns, 8*, 411–417.

Potok, M. A., Cha, K. B., Hunt, A., Brinkmeier, M. L., Leitges, M., Kispert, A., et al. (2008). WNT signaling affects gene expression in the ventral diencephalon and pituitary gland growth. *Developmental Dynamics, 237*, 1006–1020.

Puelles, L. (1995). A segmental morphological paradigm for understanding vertebrate forebrains. *Brain, Behaviour and Evolution, 46*, 319–337.

Puelles, L. (2001). Brain segmentation and forebrain development in amniotes. *Brain Research Bulletin, 55*, 695–710.

Puelles, L., & Rubenstein, J. L. (2003). Forebrain gene expression domains and the evolving prosomeric model. *Trends in Neurosciences, 26*, 469–476.

Rizzoti, K., Brunelli, S., Carmignac, D., Thomas, P. Q., Robinson, I. C., & Lovell-Badge, R. (2004). SOX3 is required during the formation of the hypothalamo–pituitary axis. *Nature Genetics, 36*, 247–255.

Rodriguez, E. M., Blazquez, J. L., Pastor, F. E., Pelaez, B., Pena, P., Peruzzo, B., et al. (2005). Hypothalamic tanycytes: A key component of brain-endocrine interaction. *International Review of Cytology, 247*, 89–164.

Roessler, E., & Muenke, M. (2010). The molecular genetics of holoprosencephaly. *American Journal of Medical Genetics, 154*, 52–61.

Rohr, K. B., Barth, K. A., Varga, Z. M., & Wilson, S. W. (2001). The nodal pathway acts upstream of hedgehog signaling to specify ventral telencephalic identity. *Neuron, 29*, 341–351.

Rubenstein, J. L., Shimamura, K., Martinez, S., & Puelles, L. (1998). Regionalization of the prosencephalic neural plate. *Annual Review of Neuroscience, 21*, 445–477.

Ryu, S., Mahler, J., Acampora, D., Holzschuh, J., Erhardt, S., Omodei, D., et al. (2007). Orthopedia homeodomain protein is essential for diencephalic dopaminergic neuron development. *Current Biology, 17*, 873–880.

Sahara, S., & O'Leary, D. D. (2009). Fgf10 regulates transition period of cortical stem cell differentiation to radial glia controlling generation of neurons and basal progenitors. *Neuron, 63*, 48–62.

Sanchez-Camacho, C., & Bovolenta, P. (2009). Emerging mechanisms in morphogen-mediated axon guidance. *Bioessays, 31*, 1013–1025.

Sanek, N. A., Taylor, A. A., Nyholm, M. K., & Grinblat, Y. (2009). Zebrafish zic2a patterns the forebrain through modulation of Hedgehog-activated gene expression. *Development, 136*, 3791–3800.

Saper, C. B. (2006). Staying awake for dinner: Hypothalamic integration of sleep, feeding and circadian rhythms. *Progress in Brain Research, 153*, 243–253.

Sbrogna, J. L., Barresi, M. J., & Karlstrom, R. O. (2003). Multiple roles for signaling in zebrafish pituitary development. *Developmental Biology, 254*, 19–35.

Schonemann, M. D., Ryan, A. K., McEvilly, R. J., O'Connell, S. M., Arias, C. A., Kalla, K. A., et al. (1995). Development and survival of the endocrine hypothalamus and posterior pituitary gland requires the neuronal POU domain factor Brn-2. *Genes & Development, 9*, 3122–3135.

Seth, A., Culverwell, J., Walkowicz, M., Toro, S., Rick, J. M., Neuhauss, S. C., et al. (2006). belladonna/(Ihx2) is required for neural patterning and midline axon guidance in the zebrafish forebrain. *Development, 133*, 725–735.

Shimamura, K., & Rubenstein, J. L. (1997). Inductive interactions direct early regionalization of the mouse forebrain. *Development, 124*, 2709–2718.

Shimogori, T., Lee, D. A., Miranda-Angulo, A., Yang, Y., Wang, H., Jiang, L., et al. (2010). A genomic atlas of mouse hypothalamic development. *Nature Neuroscience, 13*, 767–775.

Shinoda, K., Lei, H., Yoshii, H., Nomura, M., Nagano, M., Shiba, H., et al. (1995). Developmental defects of the ventromedial hypothalamic nucleus and pituitary gonadotroph in the Ftz-F1 disrupted mice. *Developmental Dynamics, 204*, 22–29.

Shirasaki, R., & Pfaff, S. L. (2002). Transcriptional codes and the control of neuronal identity. *Annual Review of Neuroscience, 25*, 251–281.

Sternson, S. M. (2013). Hypothalamic survival circuits: Blueprints for purposive behaviors. *Neuron, 77*, 810–824.

Swaab, D. F. (2004). Neuropeptides in hypothalamic neuronal disorders. *International Review of Cytology, 240*, 305–375.

Swanson, L. W. (1987). The hypothalamus. In A. Bjorland & T. Hoxfelt (Eds.), *Handbook of chemical neuroanatomy* (pp. 1–124). Amsterdam: Elsevier.

Szabo, N. E., Zhao, T., Cankaya, M., Theil, T., Zhou, X., & Alvarez-Bolado, G. (2009). Role of neuroepithelial Sonic hedgehog in hypothalamic patterning. *Journal of Neuroscience, 29*, 6989–7002.

Szarek, E., Cheah, P. S., Schwartz, J., & Thomas, P. (2010). Molecular genetics of the developing neuroendocrine hypothalamus. *Molecular and Cellular Endocrinology, 323*, 115–123.

Takuma, N., Sheng, H. Z., Furuta, Y., Ward, J. M., Sharma, K., Hogan, B. L., et al. (1998). Formation of Rathke's pouch requires dual induction from the diencephalon. *Development, 125*, 4835–4840.

Tessmar-Raible, K., Raible, F., Christodoulou, F., Guy, K., Rembold, M., Hausen, H., et al. (2007). Conserved sensory-neurosecretory cell types in annelid and fish forebrain: Insights into hypothalamus evolution. *Cell, 129*, 1389–1400.

Thisse, B., Heyer, V., Lux, A., Alunni, V., Degrave, A., Seiliez, I., et al. (2004). Spatial and temporal expression of the zebrafish genome by large-scale in situ hybridization screening. *Methods in Cell Biology, 77*, 505–519.

Tokunaga, K., Fukushima, M., Kemnitz, J. W., & Bray, G. A. (1986). Comparison of ventromedial and paraventricular lesions in rats that become obese. *American Journal of Physiology, 251*, 1221–1227.

Trowe, M. P., Zhao, L., Weiss, A. C., Christoffels, V., Epstein, D. J., & Kispert, A. (2013). Inhibition of Sox2-dependent activation of Shh in the ventral diencephalon by Tbx3 is required for formation of the neurohypophysis. *Development, 140*, 2299–2309.

Tsai, P. S., Brooks, L. R., Rochester, J. R., Kavanaugh, S. I., & Chung, W. C. (2011). Fibroblast growth factor signaling in the developing neuroendocrine hypothalamus. *Frontiers in Neuroendocrinology, 32*, 95–107.

Tschop, M., & Heiman, M. L. (2001). Rodent obesity models: An overview. *Experimental and Clinical Endocrinology & Diabetes, 109*, 307–319.

Ulloa, F., & Briscoe, J. (2007). Morphogens and the control of cell proliferation and patterning in the spinal cord. *Cell Cycle, 6*, 2640–2649.

Wallis, D., & Muenke, M. (2000). Mutations in holoprosencephaly. *Human Mutation, 16*, 99–108.

Wang, W., Grimmer, J. F., Van De Water, T. R., & Lufkin, T. (2004). Hmx2 and Hmx3 homeobox genes direct development of the murine inner ear and hypothalamus and can be functionally replaced by Drosophila Hmx. *Developmental Cell, 7*, 439–453.

Wang, X., Kopinke, D., Lin, J., McPherson, A. D., Duncan, R. N., Otsuna, H., et al. (2012). Wnt signaling regulates postembryonic hypothalamic progenitor differentiation. *Developmental Cell, 23*, 624–636.

Wang, W., & Lufkin, T. (2000). The murine Otp homeobox gene plays an essential role in the specification of neuronal cell lineages in the developing hypothalamus. *Developmental Biology*, *227*, 432–449.

Wang, W., & Lufkin, T. (2005). Hmx homeobox gene function in inner ear and nervous system cell-type specification and development. *Experimental Cell Research*, *306*, 373–379.

Warr, N., Powles-Glover, N., Chappell, A., Robson, J., Norris, D., & Arkell, R. M. (2008). Zic2-associated holoprosencephaly is caused by a transient defect in the organizer region during gastrulation. *Human Molecular Genetics*, *17*, 2986–2996.

Wittkowski, W. (1998). Tanycytes and pituicytes: Morphological and functional aspects of neuroglial interaction. *Microscopy Research and Technique*, *41*, 29–42.

Wittkowski, W., Bockmann, J., Kreutz, M. R., & Bockers, T. M. (1999). Cell and molecular biology of the pars tuberalis of the pituitary. *International Review of Cytology*, *185*, 157–194.

Wolf, A., & Ryu, S. (2013). Specification of posterior hypothalamic neurons requires coordinated activities of Fezf2, Otp, Sim1a and Foxb1.2. *Development*, *140*, 1762–1773.

Woods, K. S., Cundall, M., Turton, J., Rizotti, K., Mehta, A., Palmer, R., et al. (2005). Over- and underdosage of SOX3 is associated with infundibular hypoplasia and hypopituitarism. *American Journal of Human Genetics*, *76*, 833–849.

Wray, S. (2001). Development of luteinizing hormone releasing hormone neurons. *Journal of Neuroendocrinology*, *13*, 3–11.

Wray, S. (2002). Development of gonadotropin-releasing hormone-1 neurons. *Frontiers in Neuroendocrinology*, *23*, 292–316.

Yang, N., Dong, Z., & Guo, S. (2012). Fezf2 regulates multilineage neuronal differentiation through activating basic helix loop helix and homeodomain genes in the Zebrafish ventral forebrain. *The Journal of Neuroscience: The Official Journal of the Society for Neuroscience*, *32*, 10940–10948.

Yee, C. L., Wang, Y., Anderson, S., Ekker, M., & Rubenstein, J. L. (2009). Arcuate nucleus expression of NKX2.1 and DLX and lineages expressing these transcription factors in neuropeptide Y(+), proopiomelanocortin(+), and tyrosine hydroxylase(+) neurons in neonatal and adult mice. *Journal of Comparative Neurology*, *517*, 37–50.

Zhao, L., Bakke, M., & Parker, K. L. (2001). Pituitary-specific knockout of steroidogenic factor 1. *Molecular and Cellular Endocrinology*, *185*, 27–32.

Zhao, Y., Mailloux, C. M., Hermesz, E., Palkovits, M., & Westphal, H. (2010). A role of the LIM-homeobox gene Lhx2 in the regulation of pituitary development. *Developmental Biology*, *337*, 313–323.

Zhao, T., Szabó, N., Ma, J., Luo, L., Zhou, X., & Alverez-Bolado, G. (2008). Genetic fate mapping of Foxb1-cell lineage shows migration from caudal diencephalon to telencephalon and lateral hypothalamus. *European Journal of Neuroscience*, *28*, 1941–1955.

Zhu, X., Gleiberman, A. S., & Rosenfeld, M. G. (2007). Molecular physiology of pituitary development: Signaling and transcriptional networks. *Physiological Reviews*, *87*, 933–963.

CHAPTER THREE

Mammalian Sex Determination and Gonad Development

Dagmar Wilhelm[*,1], **Jennifer X. Yang**[*], **Paul Thomas**[†,1]
[*]Department of Anatomy and Developmental Biology, Monash University, Clayton, Victoria, Australia
[†]School of Molecular and Biomedical Science, University of Adelaide, Adelaide, South Australia, Australia
[1]Corresponding authors: e-mail address: dagmar.wilhelm@monash.edu; paul.thomas@adelaide.edu.au

Contents

Abstract

From a developmental biology perspective, gonadogenesis is of particular interest because it provides a unique example of how distinct organs, the testis and ovary, can arise from a common bipotential primordium. Gonadogenesis is also highly relevant from a clinical perspective, as congenital disorders of sex development (DSDs) are not uncommon, occurring in approximately 1 in 4500 live births. In recent years, there has been significant progress in our understanding of the genes and pathways that control important aspects of gonadogenesis including the initial decision to develop as a testis or ovary (sex determination), elaboration and cross-repression of the testis and ovary developmental pathways, and maintenance of the gonadal phenotype after birth. In this chapter, we provide an overview of gonadal morphogenesis and cell lineage specification, focusing primarily on these processes in mice and humans. We also provide an update on the genetic mechanisms that control murine gonadogenesis and maintenance and, where possible, discuss new insights into the pathological mechanisms of DSDs associated with mutation of orthologous genes in mice and humans.

Current Topics in Developmental Biology, Volume 106
ISSN 0070-2153
http://dx.doi.org/10.1016/B978-0-12-416021-7.00003-1

1. INTRODUCTION

Sexual reproduction is one of the two basic processes by which organisms reproduce, asexual reproduction being the other. Sexual reproduction takes two, a female and a male, each of whom contributes a gamete, the egg and the sperm, which unite to create a zygote containing the genetic information from both parents. If sexual reproduction in plants and animals is a result of evolutionary processes, an amazing series of chance events must have occurred. The complex and very different, but yet complementary, male and female reproductive systems need to independently evolve at every stage in parallel. It is therefore surprising that sexual reproduction has not evolved just once with the processes of sex determination and differentiation being similar for all animals, but many times independently and with a broad variety of different mechanisms. Sex determination is poorly conserved between different species, ranging from environmental factors, such as the temperature determining the sex of the offspring, to genetically determined sex. In mammals, the decision of becoming male or female is genetically determined at the time of fertilization, with the acquisition of a Y or an X chromosome from the father. The bifurcation of the developmental pathway into male and female becomes apparent when the bipotential genital ridge, the gonadal anlage, differentiates into either a testis, in an XY individual, or an ovary, in an XX individual. Most, if not all, secondary sexual dimorphisms are a consequence of the endocrine function of the testis and ovary.

This chapter reviews the development and differentiation of testes and ovaries and the molecular pathways driving these processes. We focus primarily on mice, a model organism extensively used to study these changes, as well as human where perturbation of these pathways causes disorders of sex development (DSDs).

2. DEVELOPMENT OF THE BIPOTENTIAL GENITAL RIDGE

The anlagen of the testes and ovaries in mammals are the paired indifferent and bipotential genital ridges (Fig. 3.1). They were first described by Waldeyer (1870) to develop as a thickening of the ventromedial surface of the mesonephros, the middle segment of the urogenital ridges, which are composed of the pronephros, mesonephros, and the kidney anlage, the

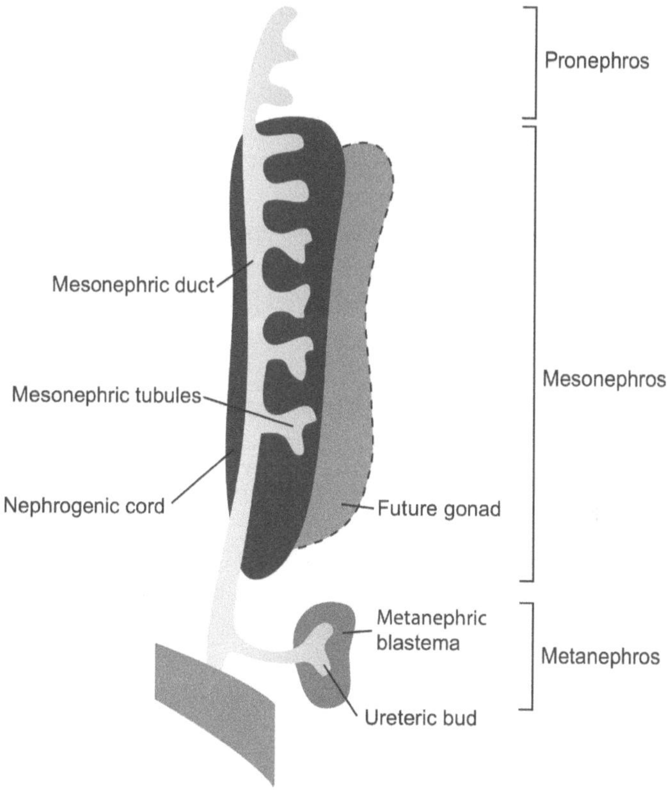

Figure 3.1 Schematic representation of a urogenital ridge showing position of the pronephros, mesonephros, and metanephros. The gonads develop from a thickening on the ventromedial surface of the mesonephros (shown in cyan). (See color plate.)

metanephros. The genital ridges are covered by the coelomic epithelium and are first visible at around 10 days *post coitum* (dpc) in mouse and during the fourth week postfertilization in human (Byskov, 1986; Satoh, 1991). Over the years, a variety of mechanisms have been proposed to explain the origin of the genital ridge. Initially, it was believed that the coelomic epithelium, termed "germinal epithelium," gives rise to the primordial germ cells (PGCs) and invades the underlying mesenchyme to form the genital ridge (Allen, 1904). It is clear now that PGCs are specified extragonadally and migrate to the developing genital ridges (see in the succeeding text). With respect to the somatic compartment of the gonads, it has been suggested that it is derived from mesenchymal cells (Fischel, 1930; Jirasek, 1971), from the mesonephros (Byskov & Lintern-Moore, 1973; Zamboni & Upadhyay, 1982), or from both the coelomic epithelium and the mesonephros, with the medulla being formed by the mesonephros and the cortex derived from the coelomic epithelium (Witschi, 1931). It is likely that some of these discrepancies are due to species-specific differences. In mouse, direct labeling of coelomic epithelial cells has shown that, at least in the developing testis, cells delaminate from the epithelium to give rise to Sertoli cells, the supporting cell lineage (Karl & Capel, 1998). In human, histological examination of serial sections of entire gonads suggests that the main constituent of somatic gonadal cells is derived from the mesonephros (Satoh, 1991).

In any case, the formation of the bipotential genital ridge is the obligatory first step for the generation of testes and ovaries, and a variety of nuclear/transcription factors and signaling proteins have been identified to have important roles in this process. In cases where the phenotype is not embryonic lethal, XY individuals that fail to fully generate genital ridges typically develop as females due to an inability to initiate or sustain the male differentiation pathway. In the succeeding text, we describe some of the well-known players in this process such as steroidogenic factor 1 (SF1) and Wilms' tumor suppressor 1 (WT1) as well as recently identified genes that give new insight into the signaling pathways that control this process.

SF1 (also known as Ftzf1, Ad4BP, and NR5A1) is a member of the orphan nuclear receptor family and is expressed in the genital ridge from 9.5 dpc and is maintained in the Sertoli and Leydig lineages. *Sf1*-null mutant mice exhibit complete gonadal agenesis resulting from an arrest in genital ridge development at 11.5 dpc and its subsequent regression (Luo, Ikeda, & Parker, 1994; Sadovsky et al., 1995). Although gonadal development is not compromised in *Sf1* heterozygous mice on most genetic backgrounds (Luo et al., 1994; Sadovsky et al., 1995), the gonads of humans with heterozygous *SF1* mutations fail to differentiate, causing XY female sex reversal (Achermann, Ito, Ito, Hindmarsh, & Jameson, 1999; Achermann et al., 2002; Correa et al., 2004; Mallet et al., 2004).

The Wilms' tumor suppressor gene 1 (*WT1*) encodes a zinc finger transcription factor protein and is expressed in the incipient gonad, mesonephros, and kidney. The expression of WT1 involves complex regulation and at least 24 protein isoforms are generated in mammals through differential splicing, alternative promoter usage, and RNA editing. Of particular importance to gonadal development are the +KTS and −KTS isoforms, which result from differential splicing of exon 9 to include (+KTS) or exclude (−KTS) the tripartite KTS amino acid motif. In humans, a reduction of the +KTS:−KTS ratio through point mutations in intron 9 that affects splicing leads to Fraser syndrome, a urogenital disorder that, in some cases, includes gonadal dysgenesis and XY female sex reversal. Mice lacking the WT1(+KTS) isoform also exhibit XY sex reversal and fail to upregulate *Sry* expression (Hammes et al., 2001), whereas ovaries develop normally in XX WT1(+KTS)-null embryos. In contrast, XY and XX embryos that lack the WT1(−KTS) isoform exhibit early gonadal degeneration indicating that the −KTS isoform is important for maintenance and differentiation of the gonadal primordium. This occurs, at least in part, through direct activation of the *Sf1* gene in concert with the homeoprotein LHX9 (Wilhelm & Englert, 2002) and the transcription cofactor CITED2 (Buaas, Val, & Swain, 2009). Accordingly, *Lhx9* mutants also display gonadal dysgenesis and exhibit a significant reduction in *Sf1* expression (Birk et al., 2000).

Insulin and the related growth factors insulin-like growth factor 1 (IGF1) and IGF2 are key regulators of cellular activity during embryonic and postnatal development. Three cell surface proteins mediate insulin/IGF signaling: insulin, IGF1, and insulin receptor-related receptors, all of which are expressed in the genital ridge (Nef et al., 2003). Gene dosage studies (Nef et al., 2003) including triple knockout (tKO) embryos lacking all three receptors, as well as a recently published analysis of *Insr* and *Igf1r* double mutants (dKO (Pitetti et al., 2013)), indicate that a threshold of IGF signaling is required for early gonad development with IR and IGF1R having the most critical roles. The primary gonadal defect in dKO and tKO mutant embryos is an early failure in genital ridge expansion resulting, at least in part, from decreased proliferation of somatic cells. XY dKO and tKO mutants exhibit male-to-female sex reversal due to a failure to activate the testis differentiation program. However, insulin/IGF signaling is not essential for the activation of the testis-determining gene *Sry per se* because *Sry* transcript and protein can be detected in XY dKO gonads at 11.5 and 12.5 dpc, albeit at significantly reduced levels. The failure to launch the male pathway eventually leads to upregulation of the ovarian pathway, although this is delayed and is associated with small gonads of ovarian morphology. XX dKO gonads also have genital ridge hypoplasia and delayed activation of female-specific genes. Together, these studies suggest that insulin/IGF1 signaling acts on somatic cells and is required to establish a "critical mass" of genital ridge tissue that is necessary for testicular and ovarian differentiation. However, it is worth noting that organ hypoplasia is a general feature of dKO embryos (indeed, dKO embryos are noticeably smaller from 11.5 dpc), indicating that the major role of insulin/IGF signaling in the developing gonad is to stimulate cell proliferation as is the case for other organs.

Other proteins with important roles in early gonad development include the homeoprotein EMX2 (Miyamoto, Yoshida, Kuratani, Matsuo, & Aizawa, 1997), the zinc finger transcription factor odd-skipped-related 1 (Wang, Lan, Cho, Maltby, & Jiang, 2005), and the homeobox factor PBX1 (Schnabel, Selleri, & Cleary, 2003). Mice lacking each of these proteins exhibit complete gonadal agenesis (Miyamoto et al., 1997; Schnabel et al., 2003; Wang et al., 2005). Deletion of the polycomb group gene *Cbx2* (chromobox homologue 2, also known as M33) causes hypoplasia of XX and XY gonads. While the latter develop as ovaries, this phenotype can be rescued by expression of *Sry* or *Sox9*. However, the resulting testes are hypoplastic, indicating that gonadal size and sex are controlled by different sets of genes (Katoh-Fukui et al., 2012). While these genetic experiments demonstrate the importance of these factors in early gonadogenesis, further studies are required to determine the target genes and pathways that are under their direct control.

3. PRIMORDIAL GERM CELLS

PGCs are the precursors of the gametes in both males and females. One of the first indications that these cells are specified outside of and migrate to the developing gonads came from experiments in frogs (Nussbaum, 1880). Around 30 years later, the same phenomenon was described in human embryos (Fuss, 1912). It is now generally accepted that in all mammals, PGCs arise during gastrulation through inductive signaling from neighboring cells. This specification takes place at the base of the allantois at around 7 dpc in mouse (Ginsburg, Snow, & McLaren, 1990; Lawson & Hage, 1994) and during the third week of gestation in the human embryo (Fuss, 1912). PGCs are then passively incorporated into the embryonic endoderm that gives rise to the hindgut (Tam & Snow, 1981). From 9.5 to 11.5 dpc in mouse (Clark & Eddy, 1975; Donovan, Stott, Cairns, Heasman, & Wylie, 1986), and during the 5th week in human (Fujimoto, Miyayama, & Fuyuta, 1977; McKay, Hertig, Adams, & Danziger, 1953; Witschi, 1948), PGCs migrate from the hindgut, through the dorsal mesentery to colonize the developing genital ridges. During migration and after colonization, in the 4- to 9-week period in the human embryo, PGCs proliferate and increase in number from about 1000 to

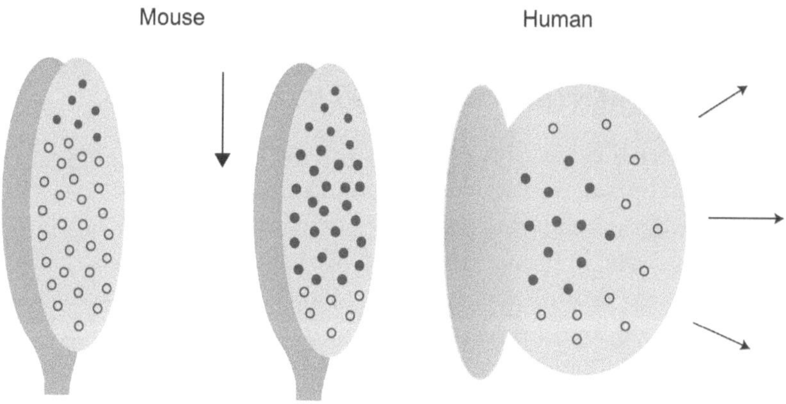

Figure 3.2 Comparison of germ cell meiosis in mouse and human ovaries. In mouse, germ cells (indicated by the red circles) undergo meiosis (closed circles) in an anterior-to-posterior wave in response to retinoic acid (RA) produced by the mesonephros (shown in yellow). In humans, the ovary produces RA and germ cell meiosis is initiated in the medulla and spreads radially into the cortex. (See color plate.)

450,000 in females and to around 150,000 in males (De Felici, 2009). In mouse, approximately 10–100 PGCs are present at 9.5 dpc, which increases to around 25,000 by 13.5 dpc. During this migration and proliferation period, PGCs maintain their bipotentiality. At 13.5 dpc in mouse and weeks 11–12 during human gestation, germ cell development starts to display a striking sexual dimorphism. In testes, PGCs or prospermatogonia enter mitotic arrest, whereas in ovaries, PGCs or oogonia start to enter the first meiotic prophase (Fig. 3.2; McCarrey, 1993). In the mouse ovary, it has been shown that retinoic acid produced by the mesonephros induces entry into meiosis in an anterior-to-posterior wave (Bowles et al., 2006; Koubova et al., 2006). However, this wave of entry into meiosis cannot be observed in the human ovary. Here, germ cell differentiation occurs radially, with PGCs in the medulla entering meiosis first and undifferentiated oogonia remain in the cortical region for longer (Anderson, Fulton, Cowan, Coutts, & Saunders, 2007; Stoop et al., 2005). Despite this difference, entry into meiosis in the human ovary is also thought to be initiated by retinoic acid, which, in this case, is produced by the ovary itself, and not the mesonephros (Childs, Cowan, Kinnell, Anderson, & Saunders, 2011). PGCs that do not complete their journey to the gonads are usually removed by apoptosis (Runyan et al., 2006).

4. LET US GET IT ON: ACTIVATION OF THE TESTIS DIFFERENTIATION PATHWAY

While sex differentiation of PGCs starts at 13.5 dpc and 11–12 weeks in mouse and human, respectively, differentiation of the somatic cells in an XY genital ridge is induced by the expression of the Y-linked *Sry*/*SRY* gene starting at 10.5 dpc in mouse (Bullejos & Koopman, 2001; Hacker, Capel, Goodfellow, & Lovell-Badge, 1995; Jeske, Bowles, Greenfield, & Koopman, 1995) and between 41 and 44 days postovulation in human (Hanley et al., 2000). Generation of the testis and its constituent male-specific lineages is critically dependent on *SRY*. *SRY* encodes a transcription factor that was identified in 1990 through the analysis of individuals with XY female sex reversal (who carried *SRY* mutations) and XX males in whom the *SRY* gene had translocated onto the X chromosome (reviewed in Sekido & Lovell-Badge, 2013). Subsequent experiments using *Sry*-transgenic mice and a spontaneous XY sex-reversing mouse mutant (Tdy) further showed that SRY is necessary and sufficient for male differentiation (Gubbay et al., 1990; Koopman, Gubbay, Vivian, Goodfellow, & Lovell-Badge, 1991). These landmark studies provided a molecular explanation for perplexing human sex reversal syndromes and heralded a new era in gonadogenesis research in which the hierarchy of genetic sex determination could be dissected from the top down. Interestingly, studies over the past two decades have revealed that the *SRY* sex determination switch is not as robust as might have been originally thought. Expression studies in mouse have shown that *Sry* is expressed at very low levels in the XY gonads beginning at 10.5 dpc, peaking at 11.5 dpc, and is extinguished at 12.5 dpc (Bullejos & Koopman, 2001; Hacker et al., 1995; Jeske et al., 1995). Elegant *Sry* induction experiments in transgenic mice indicate *Sry* expression must rise above a critical threshold within a 6-h window in order to irreversibly activate the testes pathway (Hiramatsu et al., 2009). Failure of *Sry* expression to reach this threshold can result in phenotypes such as ovotestis (ovarian and testicular tissue in the same gonad) or complete XY female sex reversal.

Given the importance of *Sry* expression levels for testis development, there has been considerable interest in identifying genetic factors that activate *SRY* expression. Recent analysis of the mitogen-activated protein kinase (MAPK) pathway indicates that it has an important role in this process (Fig. 3.3). Using a forward genetic screening approach, Bogani et al. (2009) identified a recessive mouse mutant with XY gonadal sex reversal named

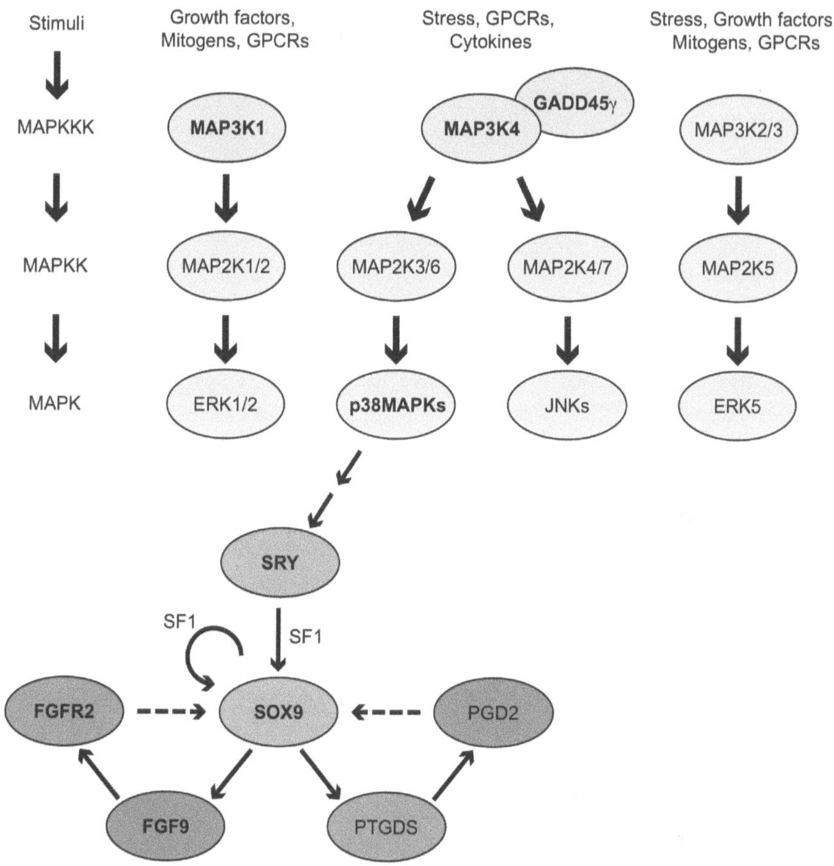

Figure 3.3 Overview of the testis differentiation pathway. Recent data in mice indicate that activation of the mitogen-activated protein kinase (MAPK) pathway is required for induction of the testis-determining gene *Sry*. SRY, in concert with SF1, directly activates *Sox9* expression, which is maintained via an autoregulatory loop requiring SF1, and two positive feedback circuits involving prostaglandin D_2 (PGD_2) and FGF9/FGFR2. See text for additional details. Color code: kinases (yellow), transcription factors (green), signaling factors (blue), enzymes (pink), and receptors (purple). Genes in bold exhibit gonadal phenotypes when mutated in mouse and/or human (See color plate.)

boygirl (*byg*) that contained a nonsense mutation in *Map3k4* gene. *Map3k4* encodes a MAPK that is widely expressed and functions in the MAPK and JNK (cJun N-terminal kinase) signaling pathways. On a C57BL/6J genetic background, *byg* mutant XY gonads fail to fully activate *Sry*, suggesting that MAPK signaling may regulate *Sry* expression. Recent analysis of *Gadd45γ* (growth arrest and DNA damage response) mutants supports this hypothesis (Gierl, Gruhn, von Seggern, Maltry, & Niehrs, 2012; Warr et al., 2012). *Gadd45γ* encodes a small, acidic protein that binds and activates MAP3K4 to promote MAPK signaling. Interestingly, unlike *Map3k4*, *Gadd45γ* expression in the developing XY gonads is remarkably similar to *Sry*. More importantly, *Gadd45γ*-null mutants display very poor *Sry* induction that is associated with a marked reduction phosphorylation (activation) of the MAPK signaling target protein p38 MAPK and GATA4, a transcription factor previously shown to transactivate *Sry* (Tevosian et al., 2002). These data, together with additional genetic and biochemical data (Gierl et al., 2012; Warr et al., 2012), provide definitive evidence that MAPK signaling is critical for *Sry* activation in mice. In humans, heterozygous mutations in *MAP3K1*, a closely related kinase, are associated with 46,XY DSD (Pearlman et al., 2010). Given the mouse data, surprisingly, these mutations appear to increase MAPK signaling activity in patient-derived lymphoblast cell lines, possibly reflecting species differences or context-dependent effects.

5. STAYING TURNED ON: ACTIVATION AND MAINTENANCE OF *SOX9*

SRY is the undisputed trigger for male development in the vast majority of mammalian species. However, although several putative direct targets have been identified (reviewed in Sekido & Lovell-Badge, 2013), the primary role of SRY in sex determination appears to be restricted to a single function: upregulation of *Sox9* (Fig. 3.3). *Sox9*, unlike *Sry*, is conserved throughout vertebrates and beyond and has been shown to have a central role in the establishment and maintenance of the male pathway in many species including humans, mouse, birds, fish, and flies (Foster et al., 1994; Kent, Wheatley, Andrews, Sinclair, & Koopman, 1996; Morais da Silva et al., 1996; Nanda et al., 2009; Spotila, Spotila, & Hall, 1998; Wagner et al., 1994; Western, Harry, Graves, & Sinclair, 1999). *Sox9* encodes a transcription factor belonging to the Sry-like HMG box family of which SRY is the founding member (Bowles, Schepers, & Koopman, 2000). In humans, heterozygous mutations in *SOX9* cause campomelic dysplasia, a bone disorder that also includes XY female sex reversal (or gonadal dysgenesis) in approximately 75% of affected individuals (Houston et al., 1983). Although *Sox9* heterozygous mice have normal gonad development, conditional homozygous deletion in the gonads causes XY ovary development, showing that SOX9 is required for testis differentiation (Barrionuevo et al., 2006; Chaboissier et al., 2004). To address whether *Sox9* is sufficient for male development, Vidal et al. (2001) generated transgenic mice with enforced expression of *Sox9* in the genital ridges. This resulted in complete XX male sex reversal that was phenotypically identical to the *Sry*-transgenic XX sex-reversed mice published by Koopman et al. (1991) 10 years earlier, which identified *Sry* as the testis-determining factor. *SOX9* gain of function (duplication) in humans also leads to XX male sex reversal (Huang, Wang, Ning, Lamb, & Bartley, 1999).

Careful comparison of *Sry* and *Sox9* expression during gonad development has shown that *Sox9* is upregulated in XY gonads a few hours after *Sry* expression is initiated at 10.5 dpc and follows the same central-to-polar pattern of activation (Bullejos & Koopman, 2001), suggesting that *Sox9* is directly regulated by SRY. This was confirmed in a landmark study in 2008, which identified a 3.2 kb testis enhancer sequence (TES) located approximately 14 kb upstream of *Sox9* (Sekido & Lovell-Badge, 2008). Using *in vivo* chromatin immunoprecipitation and transactivation analyses, it was shown that SRY and SF1 bind at several sites across the 1.4 kb core element of this enhancer (TESCO), resulting in synergistic upregulation of *Sox9*. The consequent increase in SOX9 levels appears to permit binding of SOX9 itself to enable autoregulation, thereby maintaining robust *Sox9* expression after *Sry* is downregulated at 12.5 dpc.

Despite the significant advance of TESCO identification, it has not yet been established whether or not TESCO is required for the activation of the testis differentiation pathway *in vivo*. To address this issue, mice with a targeted TESCO deletion are required. Further, it is unclear whether the role of TESCO is conserved in other species. TESCO sequences can be identified in humans and other eutherian mammals, but only partial conservation is evident in more distant species such as platypus, chicken, and frog (Bagheri-Fam, Sinclair, Koopman, & Harley, 2010). Reportedly, human TES transgenic mice do not exhibit testes-specific enhancer activity (Sekido & Lovell-Badge, 2013). Consistent with these data, molecular genetic analysis of humans with *SRY*-positive XY gonadal dysgenesis has thus far failed to detect any mutation or copy number change in the TES sequence. However, deletions and duplications upstream of TESCO have been identified in several individuals with 46,XY DSD and 46,XX DSD, respectively, which together provide substantial evidence for an alternative testis-specific enhancer located 517–595 kb upstream of *SOX9* (Fonseca et al., 2013). Further sequence analysis and functional studies are required to determine whether this putative enhancer has a human-specific or broadly conserved function and its regulatory relationship to TESCO.

In addition to autoregulation with SF1 via TESCO, studies in mice have shown that SOX9 initiates a least two regulatory loops that reinforce the testis determination pathway (Fig. 3.3). SOX9 has been shown to upregulate *Fgf9* (Kim et al., 2006), which, through generation of FGF9 protein, activates FGF signaling via FGF receptor 2 (FGFR2), which positively feeds back on *Sox9* expression. As a consequence, LOF mutations in *Fgf9* or *Fgfr2* lead to reduced *Sox9* expression and XY female sex reversal (Bagheri-Fam et al., 2008; Colvin, Green, Schmahl, Capel, & Ornitz, 2001; Kim et al., 2007). A second positive feedback loop in which SOX9 directly activates the prostaglandin D synthase (*Ptgds*) gene expression has also been identified (Moniot et al., 2009; Wilhelm et al., 2007, 2005). Increased PGD_2, acting via a paracrine and/or autocrine mechanism promotes nuclear translocation of SOX9 protein, thereby reinforcing Sertoli cell fate (Malki et al., 2005). Interestingly, both of these feed-forward loops can function non-cell autonomously to activate *Sox9* expression and promote Sertoli cell differentiation in non-SRY expressing supporting cells. Thus, expression of *Sry* is not a prerequisite for Sertoli cell differentiation (Sutton et al., 2011). These genetic circuits also provide an explanation for the recruitment of XX cells to the Sertoli lineage in chimeric gonads (Burgoyne, Buehr, Koopman, Rossant, & McLaren, 1988).

As a consequence of the expression of *Sry* and *Sox9*, a number of cellular and morphological changes occur, including the differentiation of Sertoli cells, the supporting cell lineage; a massive increase in testis size due to increased cell proliferation (Schmahl, Eicher, Washburn, & Capel, 2000) and migration of cells from the underlying mesonephros (Capel, Albrecht, Washburn, & Eicher, 1999; Martineau, Nordqvist, Tilmann, Lovell-Badge, & Capel, 1997); the formation of the testis-specific vasculature and testis cords, the precursors to the seminiferous tubules; comprising of clusters of germ cells, surrounded by Sertoli cells and a layer of long, flattened peritubular myoid (PM) cells (Jeanes et al., 2005), and the differentiation of steroidogenic Leydig cells in the interstitium, all of which are discussed in more detail in the following sections and are presented in Fig. 3.4.

5.1. Sertoli cells

Sry and *Sox9* are expressed in supporting cell precursors and their expression initiates the differentiation of the first testis-specific cell type, the Sertoli cell (Sekido, Bar, Narvaez, Penny, & Lovell-Badge, 2004; Wilhelm et al., 2005).

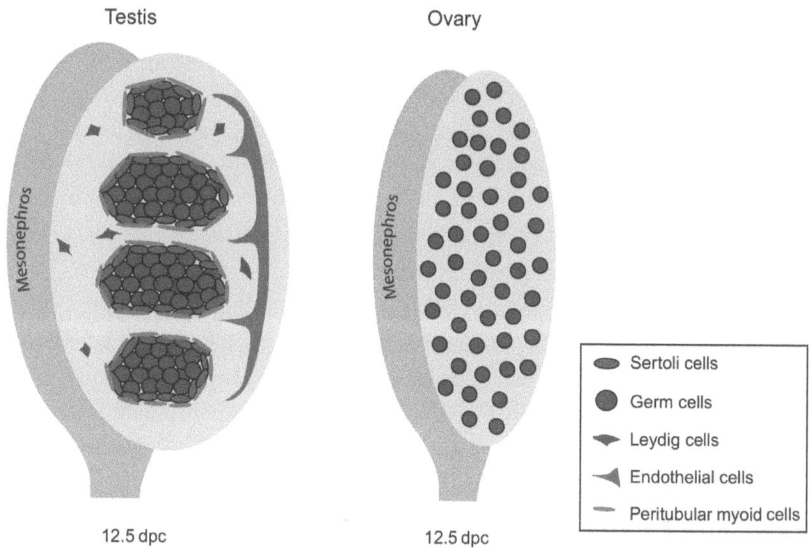

Figure 3.4 Comparison of the developing murine testis (left) and ovary (right). By 12.5 dpc, approximately 48 h after the onset of *Sry* expression, the testis has undergone significant morphogenesis. Cords composed of clusters of primordial germ cells enclosed by a layer of Sertoli and peritubular myoid cells. Testis-specific vasculature such as the coelomic vessel has also begun to form and a population of fetal Leydig cells is present. In contrast, the ovary at this stage appears morphologically undifferentiated, containing germ cells and somatic precursor cell lineages. (See color plate.)

Sertoli cells were named after Enrico Sertoli, an Italian physiologist who first described their role in supporting sperm development (Sertoli, 1865). In mouse, at least a subset of pre-Sertoli cells originate from the coelomic epithelium, which shows a higher proliferation rate in an XY genital ridge compared to the XX genital ridge (Karl & Capel, 1998; Schmahl et al., 2000). It has been suggested that signals emanating from differentiating pre-Sertoli cells are responsible for this increased proliferation, which in turn produces more Sertoli cells (Bradford et al., 2009; Schmahl et al., 2000) so that the Sertoli cell number reaches the threshold necessary for proper testis differentiation (Nagamine, Morohashi, Carlisle, & Chang, 1999; Palmer & Burgoyne, 1991). Sertoli cells only proliferate during fetal and neonatal development, although the time of proliferation varies depending on the species (O'Shaughnessy et al., 2007). The final number of Sertoli cells in the adult will determine the number of germ cells that they can support and therefore regulates the level of sperm production and hence fertility (Orth, 1982; Sharpe, McKinnell, Kivlin, & Fisher, 2003).

The differentiation of Sertoli cells is marked by changes in their shape and structure. Starting with a more mesenchymal shape, differentiating Sertoli cells assemble around clusters of germ cells to form testis cords (see Section 5.3) and become polarized and virtually columnar with cytoplasmic protrusions around PGCs (Combes, Wilhelm, et al., 2009). They form incomplete tight junctions in both mouse and human fetal testes (Gondos, 1981; Nagano & Suzuki, 1976), which will ultimately develop into the blood–testis barrier in the adult testis.

As the supporting cell lineage, Sertoli cells play a central role in the development and function of the mature testis. During development, Sertoli cells orchestrate male-specific processes such as steroidogenic precursor cell differentiation into Leydig cells (Yao, Whoriskey, & Capel, 2002) and endothelial cell migration from the mesonephros (Capel et al., 1999; Martineau et al., 1997). Sertoli cells also protect PGCs from the influence of retinoic acid by expressing CYP26B1, a retinoic acid-degrading enzyme, and therefore prevent entry into meiosis (Bowles et al., 2006). In addition, Sertoli cells produce anti-Müllerian hormone, resulting in the degeneration of the Müllerian duct. Postnatally, their main role is to support and nurture the development of sperm through the stages of spermatogenesis and also function in the production of androgen-binding protein and phagocytosis of degenerating germ cells (Ritzen et al., 1981).

5.2. Leydig cells

The second cell type to differentiate in the developing testis is Leydig cells. These cells were named after Franz Leydig, a German zoologist and anatomist who first described them in 1850 (Leydig, 1850). There are two generations of Leydig cells, fetal and adult Leydig cells. Fetal Leydig cells differentiate in response to signals from Sertoli cells in the interstitium at around 12.5 dpc in mouse and at 8–9 weeks of human development. The origin of these cells has been subject to great debate. Arguably, the most likely precursor cells are mesenchymal cells (Chemes et al., 1985; Moon & Hardy, 1973), although other cellular origins have been hypothesized such as mesonephric cells (Merchant-Larios & Moreno-Mendoza, 1998; Witschi, 1951), coelomic epithelial cells (Karl & Capel, 1998), or macrophages (Clegg & Macmillan, 1965). However, recent work in mice has demonstrated that fetal Leydig cells are derived from precursors in the coelomic epithelium as well as from the mesonephric border (DeFalco, Takahashi, & Capel, 2011). Fully differentiated Leydig cells do not proliferate

(Gondos, Morrison, & Renston, 1977), although the number of Leydig cells steadily increases during development (Codesal, Regadera, Nistal, Regadera-Sejas, & Paniagua, 1990), suggesting that new cells are recruited from existing precursor cells rather than cell division. Leydig cells are the most potent cells in androgen synthesis (Eik-Nes, 1969), which is required for the development of secondary sexual characteristics such as the differentiation of the Wolffian duct into the male reproductive tract and development of the male external genitalia.

Several molecular mechanisms have been described to play a role in fetal Leydig cell specification and differentiation. As is the case for PM cells (see in the succeeding text), the secreted factor desert hedgehog (DHH) and its receptor PTCH1 play an important role in the specification of Leydig cells from mesenchymal precursor cells. *Dhh*-null mice display severe defects in Leydig cell differentiation (Yao et al., 2002). Similarly, mutations in *DHH* in humans are associated with 46,XY partial or complete gonadal dysgenesis (Canto, Soderlund, Reyes, & Mendez, 2004; Canto, Vilchis, Soderlund, Reyes, & Mendez, 2005; Umehara et al., 2000), supporting its role in testis differentiation. In addition, the Notch signaling pathway has also been demonstrated to play a role in Leydig cell generation. Inhibition and constitutive activation of Notch signaling in a mouse model resulted in an increase or decrease in the Leydig cell population, respectively (Tang et al., 2008). A number of other genes have also been implicated in the regulation of mouse fetal Leydig cell differentiation including platelet-derived growth factor receptor α, betaglycan, the microRNA genes *miR-140-3p* and *miR-140-5p*, hepatocyte growth factor, Sertoli cell-expressed androgen receptor, and aristaless-related homeobox gene *Arx* (Brennan, Tilmann, & Capel, 2003; Hazra, Jimenez, Desai, Handelsman, & Allan, 2013; Miyabayashi et al., 2013; Rakoczy et al., 2013; Ricci et al., 2012; Sarraj et al., 2010). However, the contributions of these genes in the development of human fetal Leydig cells have not been determined to date.

5.3. Testis cord formation

Testis cords are the precursor structure to the seminiferous tubules, which are essential for the maturation and export of sperm. Elegant live-cell imaging experiments demonstrated that testis cords are formed over a 24-h period from 11.5 to 12.5 dpc in mouse (Coveney, Cool, Oliver, & Capel, 2008) and at 7–8 weeks of gestation in human (Hanley et al., 2000; Wartenberg, Kinsky, Viebahn, & Schmolke, 1991). Cord formation is initiated by the assembly of Sertoli cell clusters, each of which encloses a small number of germ cells (Fig. 3.4). Concomitantly, these agglomerations of PGCs and Sertoli cells are segregated into regular arches by endothelial cells migrating into the developing testis in stochastically spaced streams from the underlying mesonephros (Combes, Lesieur, et al., 2009; Combes, Wilhelm, et al., 2009; Coveney et al., 2008). These arches or loops develop first in the center of the testis (Bullejos & Koopman, 2001) and are connected at the base next to the mesonephros (Combes, Lesieur, et al., 2009), which will become the rete testis, an anastomosing network of tubules. Inhibition of endothelial cell migration prevents testis cord formation without affecting Sertoli cell differentiation, demonstrating that this migration is necessary for cord formation (Combes, Wilhelm, et al., 2009; Coveney et al., 2008). The migrating endothelial cells also form the testis-specific vasculature with a prominent coelomic vessel on the dorsal side of the testis (Fig. 3.4) and interstitial vascular branches in between the cords (Brennan, Karl, & Capel, 2002).

Newly formed cords at 12.5 dpc in mouse vary in form and diameter (Combes, Lesieur, et al., 2009) but quickly become more distinct and regular. PM cells, the only cell type for which no counterpart appears to exist in the ovary, differentiate from mesenchymal cells and surround the testis cords. In mice, the cords are enclosed by one layer of these long, flat cells (Gardner & Holyoke, 1964), whereas in human, three to four layers exist (Ross & Long, 1966). The differentiation of PM cells is regulated by the secreted factor DHH, which is produced by Sertoli cells (Clark, Garland, & Russell, 2000; Pierucci-Alves, Clark, & Russell, 2001) and acts through its receptor patched 1, PTCH1, expressed on PM and Leydig cells (Bitgood, Shen, & McMahon, 1996; Clark et al., 2000). In addition, the nuclear receptor DAX1, encoded by the gene *Nr0b1*, has been shown to be important for PM cell proliferation and differentiation. Null mutation in mice resulted in a reduced number of PM cells, disrupted basal lamina, and incompletely formed cords (Meeks, Crawford, et al., 2003; Meeks, Weiss, & Jameson, 2003). In humans, *NR0B1* mutations have been associated with gonadal dysgenesis that is independent of the gonadotropin deficiency (Ozisik, Achermann, & Jameson, 2002), similar to the phenotype of the *Nr0b1*-null mice as well as adrenal hypoplasia congenita with associated hypogonadotropic hypogonadism. PM cells, together with Sertoli cells, secrete extracellular matrix proteins to form a basal lamina between the two cell types and will become contractile for the transportation of the sperm (Tripiciano, Filippini, Ballarini, & Palombi, 1998; Tripiciano, Filippini, Giustiniani, & Palombi, 1996).

6. OVARY DIFFERENTIATION

In contrast to testis differentiation, which is marked by the formation of clearly distinguishable testis cords in all species, there is a greater variability in the differentiation of the ovary between different species (Jimenez, 2009). Nevertheless, ovarian development can be characterized by three general processes: (i) oogonia enter prophase of meiosis I to become primary oocytes, (ii) oocytes are surrounded by granulosa cells to form follicles, and (iii) the differentiation of steroid-producing theca cells.

In most mammals, there is, depending on the species, a shorter or longer lag phase between the time of sex determination and the onset of meiosis (Byskov, 1985). In the murine embryonic ovary, the delay between sex determination at around 11.5 dpc and the onset of meiosis at 13.5 dpc is minimal. During this time interval, germ cells are localized in clusters uniformly distributed throughout the ovarian tissue with no clear morphological distinction between a cortex and a medulla, in contrast to the ovary of most vertebrates (Jimenez, 2009). However, a molecular regionalization already exists, with genes such as *Bmp2* and *Lypd6* in the presumptive cortical region and *Wnt4* and *Fst* in the medulla (Chen et al., 2012; Yao et al., 2004). In the human ovary, cords of cells, called primordial sex cords or germ cell cords, are visible from the 6th week of gestation onward, which extend and branch from the basal region into the periphery (Satoh, 1991). These primordial sex cords are contiguous with the mesonephros, but not the coelomic epithelium, and are surrounded by basal lamina (Satoh, 1991). In contrast to primordial sex cords, the so-called primary sex cords are formed by cord-like arrangements of the stratified coelomic epithelium. By the 7th week, pronounced proliferation of germ cells is detected, which results in the enlargement of primordial sex cords. The cords appear fragmented and pushed more toward the periphery by newly formed interstitial tissue at the basis, resulting in the formation of medulla with rete ovarii and the main branches of the ovarian artery and a germ cell-rich cortex by 13 weeks (Pinkerton, Mc, Adams, & Hertig, 1961; Satoh, 1991). Oogonia in the inner cortex start entering meiosis at around week 20, whereas germ cells in the outer cortex continue to proliferate. Clusters of oogonia and oocytes in the mouse and human fetal ovary are connected by intercellular bridges (Pepling & Spradling, 1998; Ruby, Dyer, & Skalko, 1969) and undergo synchronous division (Borum, 1967).

The ovary is connected with the mesonephros through the rete ovarii, which consists of three regions, the extraovarian, the connecting, and the intraovarian retia. The extraovarian rete comprises of five to nine mesonephric tubules, of which the cranial three to five are connected to the Wolffian duct, whereas the caudal tubules are blind ended. These tubules are made of a single layer of cuboidal cells on a basal membrane. They are connected to the ovary through the connecting and intraovarian rete, both of which are cell cords of one to a few cells in thickness with a common basal lamina. Many ovarian somatic cells, especially at the anterior pole, have rete characteristics, suggesting that these cells have migrated in from the mesonephros (Byskov, 1978). The connection to the mesonephros via the rete ovarii has been suggested to be important for the triggering of germ cells to enter meiosis (Byskov, 1975), which was further corroborated by the finding that retinoic acid produced by mesonephric cells is this meiosis-inducing factor (Bowles et al., 2006; Koubova et al., 2006).

The next step during ovarian differentiation, the formation of follicles, does not start until shortly after birth in mouse and from week 21 during human gestation. At these time points, somatic granulosa cells start to break up germ cell clusters and surround single oocytes. An intact basal lamina is formed and encloses this unit, the primordial follicle. The first follicles that form and start growing in mouse and human are close to the medulla of the ovary. The granulosa cells of these central follicles are connected to the intraovarian rete, suggesting that these cells are of mesonephric origin (Byskov & Lintern-Moore, 1973). Recent lineage-tracing experiments in mice showed that these cells differentiate from the supporting cell precursors that in a testis express *Sry* (Mork et al., 2011). In contrast, granulosa cells of follicles within the ovarian cortex that are activated later in life are derived from the ovarian surface epithelium through proliferation and ingression (Mork et al., 2011), supporting the previous hypothesis that granulosa cells differentiate from coelomic epithelial cells (Motta & Makabe, 1982). By week 28 of gestation in human, most primordial follicles have already formed (Pinkerton et al., 1961), and between weeks 28 and 36, follicular development continues, with oocytes growing and granulosa cells increasing in size and number.

Coinciding with the growth of the first follicle, steroidogenic theca cells differentiate from an unknown progenitor population a few days after birth in mouse (Pehlemann & Lombard, 1978) and at midgestation in human (Goldman, Yakovac, & Bongiovanni, 1966; Pinkerton et al., 1961). Until then, there is no *de novo* synthesis of steroid hormones in the developing ovary, although aromatase is expressed earlier, enabling the conversion of extragonadally produced testosterone into estrogen (George & Wilson, 1978).

7. MOLECULAR GENETICS OF OVARIAN DEVELOPMENT

The lack of morphological changes in the fetal ovary is, at least partially, the reason why the molecular genetics of ovarian determination is poorly understood. Recent studies focusing on gene expression changes in XX and XY gonads across the sex-determining period, combined with increasingly complex genetic approaches, point to a model in which at least two independent pathways act in concert to promote ovarian development and repress male differentiation. To date, evidence is lacking for the existence of single ovarian-determining gene with comparable function to *Sry* in males. Rather, the picture that emerges is one of mutual antagonism, whereby the testis- and ovary-promoting factors wrestle for control over their environment and press home their advantage to reinforce lineage-specific differentiation.

The "canonical" wingless-type MMTV integration site family (WNT) signaling pathway is widely used during development to control progenitor cell differentiation and morphogenesis (reviewed in Nelson & Nusse, 2004; Tolwinski & Wieschaus, 2004). This pathway is activated by the interaction of WNT protein ligand with the Frizzled/LRP5/6 receptor complex at the cell surface. This leads to disruption of the β-catenin destruction complex allowing β-catenin to translocate to the nucleus and regulate target gene expression in collaboration with TCF/LEF family transcription factors. Activation of the canonical WNT pathway is specific to XX gonads and, although several WNT ligands are expressed in the developing gonad, WNT4 appears to be the critical player in early ovarian development. *Wnt4* is initially expressed in XX and XY gonads, but from approximately 11.5 dpc, it is upregulated in the differentiating ovary and downregulated in the differentiating testis. *Wnt4*-null XX mice develop gonads that are partially sex-reversed containing testicular features including the production of androgens and testis-like vasculature as well as ovarian features such as oocytes, although these are greatly reduced in number and lack defined cord structures (Heikkila et al., 2005; Vainio, Heikkila, Kispert, Chin, & McMahon, 1999). In humans, three female individuals carrying heterozygous *WNT4* mutations with a similar reproductive tract phenotype to *Wnt4*-null XX mice have been reported, indicating that the role of

WNT4 is broadly conserved between mouse and human (Biason-Lauber et al., 2007; Biason-Lauber, Konrad, Navratil, & Schoenle, 2004; Correa et al., 2012; Philibert et al., 2011, 2008).

While these data show that WNT4 is required for some aspects of ovarian development, transgenic overexpression experiments in XY mice indicate that WNT4 is not sufficient to trigger ovarian development, although development of the testis-specific coelomic vessel is altered (Correa et al., 2012). However, strikingly, expression of stabilized, constitutively active β-catenin in XY gonads results in XY female sex reversal indicating that the β-catenin pathway has potent pro-ovarian and anti-testis activity (Maatouk et al., 2008) (see also cross-repression section later). Complete and partial sex reversal phenotypes also have been identified in humans with duplication of the *WNT4* locus (Jordan et al., 2001).

Recently, analysis of human patients with 46,XX sex reversal (*SRY*-negative) has shown that WNT4 is not the only factor during ovarian differentiation, which functions through the stabilization of β-catenin (Chassot et al., 2008). Molecular genetic analysis of families with a recessive syndrome that includes XX male sex reversal and skin abnormalities revealed a causative mutation in the R-spondin 1 (*RSPO1*) gene (Parma et al., 2006). RSPO1 protein is a secreted molecule that activates the WNT/β-catenin signaling pathway (Chassot et al., 2008). Subsequent studies in mice demonstrated that the deletion of *Rspo1* results in partial XX sex reversal with the presence of ovotestes (Chassot et al., 2008; Tomizuka et al., 2008). The analysis of *Rspo1* expression in gonads of the *Wnt4*-null mouse and *Wnt4* expression in *Rspo1*-null mice indicated that RSPO1 is necessary for robust *Wnt4* expression and is therefore upstream of *Wnt4* during ovarian differentiation (Chassot et al., 2008; Tomizuka et al., 2008).

The second pathway, which appears to function, at least partially, independently of the WNT4/RSPO1 pathway during ovarian development, is marked by the expression of the forkhead transcription factor FOXL2. In contrast to *Wnt4* and *Rspo1*, which both are expressed in XX and XY genital ridges before sex differentiation, *Foxl2* expression is never detected in the XY gonad. Mutation of *FOXL2* in human results in blepharophimosis, ptosis, and epicanthus inversus syndrome (BPES), which is associated with a narrowing of the eye opening (blepharophimosis), droopy eyelids (ptosis), and an upward fold of the skin of the lower eyelid near the inner corner of the eye (epicanthus inversus), and in BPES type I with premature ovarian failure (Crisponi et al., 2001). This phenotype is recapitulated in *Foxl2*-null mice, which display distinctive craniofacial morphology with missing upper eyelids and primary ovarian failure from primordial follicle arrest (Schmidt et al., 2004; Uda et al., 2004). Interestingly, *Foxl2/Wnt4* and *Foxl2/Rspo1* compound mutant mice have a more severe ovarian phenotype compared to the single mutant (Chassot et al., 2008), suggesting both pathways are important for ovarian differentiation.

8. MUTUAL ANTAGONISM

In contrast to environmentally determined sex, genetically determined sex such as in mammals suggests that, once the decision is made, it is irreversible. Surprisingly, a number of studies have shown that this is not the case (Fig. 3.5). Deletion of *Foxl2* postnatally in a conditional null mouse showed rapid upregulation of *Sox9* expression, transdifferentiation

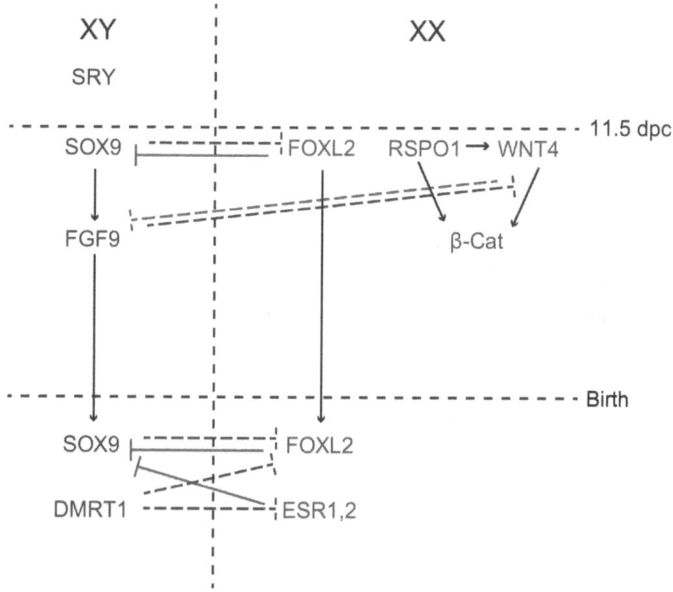

Figure 3.5 Summary of key male-promoting (blue) and female-promoting (pink) factors and their regulatory relationship during embryogenesis and after birth. Mutual cross-repression of the transcription factors SOX9 and FOXL2 serves to canalize testis and ovarian differentiation during embryogenesis and maintain these phenotypes after birth. DMRT1 is also important for postnatal maintenance of the testes through repression of FOXL2 and ESR1/2. Antagonistic interactions between pro-ovarian canonical WNT signaling pathway components (RSPO1, WNT4, and β-catenin) and protestis FGF9 and SOX9 during embryogenesis also reinforce sex-specific gonadogenesis. See text for additional details. (See color plate.)

of granulosa cells into Sertoli cells, and primary female-to-male sex reversal (Uhlenhaut et al., 2009). These data indicate that the ovarian phenotype is not irreversibly set during sex determination but instead must be actively maintained throughout life. Similarly, removal of *Dmrt1*, a gene conserved in its role in male differentiation from *Drosophila* to humans (Raymond, Murphy, O'Sullivan, Bardwell, & Zarkower, 2000) after birth results in the upregulation of *Foxl2* and the downregulation of *Sox9* and therefore transdifferentiation of Sertoli cells into granulosa cells (Matson et al., 2011), suggesting that the gonadal phenotype is maintained by repressing the pathways of the opposite sex. A comparable scenario has been described for the fetal period, when sex is determined, whereby *Sox9* and *Fgf9* repress WNT/β-catenin signaling and *vice versa* (Kim et al., 2006). However, the molecular mechanism of this repression has not been elucidated to date. This mutual antagonism leads to the hypothesis that deletion of both pathways, double knockout of either *Fgf9* and *Wnt4* (Jameson, Lin, & Capel, 2012) or *Sox9* and *Rspo1* (Lavery et al., 2012), should result in neither testicular nor ovarian differentiation. However surprisingly, in both cases in XY gonads testis differentiation occurred, suggesting that *Sox9* is not absolutely necessary for testicular differentiation and the role of *Fgf9* is suppressing the female program rather than supporting the male program. In both cases, the fate of the XX gonad, that is, partial sex reversal, was not changed in comparison to the *Wnt4* and *Rspo1* single knockout, respectively (Jameson et al., 2012; Lavery et al., 2012), suggesting that either these factors are not required to initiate testis-specific vasculature formation and steroidogenesis or other factors such as other FGFs and SOX8/SOX10, respectively, can act redundantly.

9. CONCLUSIONS AND FUTURE PERSPECTIVES

Sex determination is an inherently fascinating process that, when perturbed in humans, can lead to debilitating disorders with severe psychosocial issues. Spawned by the identification of *SRY* over 20 years ago, the field has witnessed the development of increasingly sophisticated models of sex determination and gonadogenesis that include several unexpected findings. It has emerged that SRY has a surprisingly tenuous hold on testis development; indeed, it has been disposed completely by mole voles (Graves, 2002), and its direct target, *SOX9*, encodes the true workhorse of testis development. It has also become clear that ovarian specification and development is an active process and that cross-repression pathways canalize gonad differentiation during embryogenesis and, surprisingly, remain active throughout life. For the future, it seems likely that advances in sequencing technology will drive further understanding of this field. As affordability of exome and whole-genome sequencing increases, molecular analysis of familial and sporadic cases of DSD should reveal additional causative genes for these syndromes and provide further insight into the prevalence and impact of copy number variation and genomic rearrangements. Forward genetic approaches in model species including ENU mutagenesis screens in mice and mapping of spontaneous sex reversal mutants in domestic and livestock species will also likely prove informative. Given the central roles of SOX9 and FOXL2, the genome-wide perspective of transcription factor function afforded by ChIP-seq should reveal insights into hitherto unrecognized functional components of the testicular and ovarian differentiation pathways. This deluge of genetic information will require significant investment into functional studies. The recent advent of genome-editing systems such as TALEN and CRISPR-CAS, which have been utilized in many species including mice, will be crucial for dissecting gene function *in vivo*. Looks like exciting times are ahead!

REFERENCES

Achermann, J. C., Ito, M., Ito, M., Hindmarsh, P. C., & Jameson, J. L. (1999). A mutation in the gene encoding steroidogenic factor-1 causes XY sex reversal and adrenal failure in humans. *Nature Genetics*, *22*(2), 125–126. http://dx.doi.org/10.1038/9629.

Achermann, J. C., Ozisik, G., Ito, M., Orun, U. A., Harmanci, K., Gurakan, B., et al. (2002). Gonadal determination and adrenal development are regulated by the orphan nuclear receptor steroidogenic factor-1, in a dose-dependent manner. *Journal of Clinical Endocrinology and Metabolism*, *87*(4), 1829–1833.

Allen, B. (1904). The embryonic development of the ovary and testis in mammals. *American Journal of Anatomy*, *3*, 88–153.

Anderson, R. A., Fulton, N., Cowan, G., Coutts, S., & Saunders, P. T. (2007). Conserved and divergent patterns of expression of DAZL, VASA and OCT4 in the germ cells of the human fetal ovary and testis. *BMC Developmental Biology*, 7, 136. http://dx.doi.org/10.1186/1471-213X-7-136.

Bagheri-Fam, S., Sim, H., Bernard, P., Jayakody, I., Taketo, M. M., Scherer, G., et al. (2008). Loss of Fgfr2 leads to partial XY sex reversal. *Developmental Biology*, *314*(1), 71–83. http://dx.doi.org/10.1016/j.ydbio.2007.11.010.

Bagheri-Fam, S., Sinclair, A. H., Koopman, P., & Harley, V. R. (2010). Conserved regulatory modules in the Sox9 testis-specific enhancer predict roles for SOX, TCF/LEF, Forkhead, DMRT, and GATA proteins in vertebrate sex determination. *International Journal of Biochemistry & Cell Biology*, *42*(3), 472–477. http://dx.doi.org/10.1016/j.biocel.2009.07.001.

Barrionuevo, F., Bagheri-Fam, S., Klattig, J., Kist, R., Taketo, M. M., Englert, C., et al. (2006). Homozygous inactivation of Sox9 causes complete XY sex reversal in mice. *Biology of Reproduction*, *74*(1), 195–201. http://dx.doi.org/10.1095/biolreprod.105.045930.

Biason-Lauber, A., De Filippo, G., Konrad, D., Scarano, G., Nazzaro, A., & Schoenle, E. J. (2007). WNT4 deficiency—A clinical phenotype distinct from the classic Mayer–Rokitansky–Kuster–Hauser syndrome: A case report. *Human Reproduction*, *22*(1), 224–229. http://dx.doi.org/10.1093/humrep/del360.

Biason-Lauber, A., Konrad, D., Navratil, F., & Schoenle, E. J. (2004). A WNT4 mutation associated with Mullerian-duct regression and virilization in a 46, XX woman. *New England Journal of Medicine*, *351*(8), 792–798. http://dx.doi.org/10.1056/NEJMoa040533.

Birk, O. S., Casiano, D. E., Wassif, C. A., Cogliati, T., Zhao, L., Zhao, Y., et al. (2000). The LIM homeobox gene Lhx9 is essential for mouse gonad formation. *Nature*, *403*(6772), 909–913. http://dx.doi.org/10.1038/35002622.

Bitgood, M. J., Shen, L., & McMahon, A. P. (1996). Sertoli cell signaling by desert hedgehog regulates the male germline. *Current Biology*, *6*, 298–304.

Bogani, D., Siggers, P., Brixey, R., Warr, N., Beddow, S., Edwards, J., et al. (2009). Loss of mitogen-activated protein kinase kinase kinase 4 (MAP3K4) reveals a requirement for MAPK signalling in mouse sex determination. *PLoS Biology*, 7(9), e1000196. http://dx.doi.org/10.1371/journal.pbio.1000196.

Borum, K. (1967). Oogenesis in the mouse. A study of the origin of the mature ova. *Experimental Cell Research*, *45*(1), 39–47.

Bowles, J., Knight, D., Smith, C., Wilhelm, D., Richman, J., Mamiya, S., et al. (2006). Retinoid signaling determines germ cell fate in mice. *Science*, *312*(5773), 596–600. http://dx.doi.org/10.1126/science.1125691.

Bowles, J., Schepers, G., & Koopman, P. (2000). Phylogeny of the SOX family of developmental transcription factors based on sequence and structural indicators. *Developmental Biology*, *227*(2), 239–255. http://dx.doi.org/10.1006/dbio.2000.9883.

Bradford, S. T., Wilhelm, D., Bandiera, R., Vidal, V., Schedl, A., & Koopman, P. (2009). A cell-autonomous role for WT1 in regulating Sry in vivo. *Human Molecular Genetics*, *18*(18), 3429–3438. http://dx.doi.org/10.1093/hmg/ddp283.

Brennan, J., Karl, J., & Capel, B. (2002). Divergent vascular mechanisms downstream of *Sry* establish the arterial system in the XY gonad. *Developmental Biology*, *244*, 418–428.

Brennan, J., Tilmann, C., & Capel, B. (2003). Pdgfr-alpha mediates testis cord organization and fetal Leydig cell development in the XY gonad. *Genes & Development*, *17*(6), 800–810.

Buaas, F. W., Val, P., & Swain, A. (2009). The transcription co-factor CITED2 functions during sex determination and early gonad development. *Human Molecular Genetics*, *18*(16), 2989–3001. http://dx.doi.org/10.1093/hmg/ddp237.

Bullejos, M., & Koopman, P. (2001). Spatially dynamic expression of Sry in mouse genital ridges. *Developmental Dynamics*, *221*(2), 201–205. http://dx.doi.org/10.1002/dvdy.1134.

Burgoyne, P. S., Buehr, M., Koopman, P., Rossant, J., & McLaren, A. (1988). Cell-autonomous action of the testis-determining gene: Sertoli cells are exclusively XY in XX—XY chimaeric mouse testes. *Development*, *102*(2), 443–450.

Byskov, A. G. (1975). The role of the rete ovarii in meiosis and follicle formation in the cat, mink and ferret. *Journal of Reproduction and Fertility*, *45*(2), 201–209.

Byskov, A. G. (1978). The anatomy and ultrastructure of the rete system in the fetal mouse ovary. *Biology of Reproduction*, *19*(4), 720–735.

Byskov, A. G. (1985). Control of meiosis. A summary. *Archives d'Anatomie Microscopique et de Morphologie Expérimentale*, *74*(1), 17–18.

Byskov, A. G. (1986). Differentiation of mammalian embryonic gonad. *Physiological Reviews*, *66*(1), 71–117.

Byskov, A. G., & Lintern-Moore, S. (1973). Follicle formation in the immature mouse ovary: The role of the rete ovarii. *Journal of Anatomy*, *116*(Pt. 2), 207–217.

Canto, P., Soderlund, D., Reyes, E., & Mendez, J. P. (2004). Mutations in the desert hedgehog (DHH) gene in patients with 46, XY complete pure gonadal dysgenesis. *Journal of Clinical Endocrinology and Metabolism, 89*(9), 4480–4483. http://dx.doi.org/10.1210/jc.2004-0863.

Canto, P., Vilchis, F., Soderlund, D., Reyes, E., & Mendez, J. P. (2005). A heterozygous mutation in the desert hedgehog gene in patients with mixed gonadal dysgenesis. *Molecular Human Reproduction, 11*(11), 833–836. http://dx.doi.org/10.1093/molehr/gah216.

Capel, B., Albrecht, K. H., Washburn, L. L., & Eicher, E. M. (1999). Migration of mesonephric cells into the mammalian gonad depends on Sry. *Mechanisms of Development, 84*(1–2), 127–131.

Chaboissier, M. C., Kobayashi, A., Vidal, V. I., Lutzkendorf, S., van de Kant, H. J., Wegner, M., et al. (2004). Functional analysis of Sox8 and Sox9 during sex determination in the mouse. *Development, 131*(9), 1891–1901. http://dx.doi.org/10.1242/dev.01087.

Chassot, A. A., Ranc, F., Gregoire, E. P., Roepers-Gajadien, H. L., Taketo, M. M., Camerino, G., et al. (2008). Activation of beta-catenin signaling by Rspo1 controls differentiation of the mammalian ovary. *Human Molecular Genetics, 17*(9), 1264–1277. http://dx.doi.org/10.1093/hmg/ddn016, [pii] ddn016.

Chemes, H. E., Gottlieb, S. E., Pasqualini, T., Domenichini, E., Rivarola, M. A., & Bergada, C. (1985). Response to acute hCG stimulation and steroidogenic potential of Leydig cell fibroblastic precursors in humans. *Journal of Andrology, 6*(2), 102–112.

Chen, H., Palmer, J. S., Thiagarajan, R. D., Dinger, M. E., Lesieur, E., Chiu, H., et al. (2012). Identification of novel markers of mouse fetal ovary development. *PLoS One, 7*(7), e41683. http://dx.doi.org/10.1371/journal.pone.0041683.

Childs, A. J., Cowan, G., Kinnell, H. L., Anderson, R. A., & Saunders, P. T. (2011). Retinoic acid signalling and the control of meiotic entry in the human fetal gonad. *PLoS One, 6*(6), e20249. http://dx.doi.org/10.1371/journal.pone.0020249.

Clark, J. M., & Eddy, E. M. (1975). Fine structural observations on the origin and associations of primordial germ cells of the mouse. *Developmental Biology, 47*(1), 136–155.

Clark, A. M., Garland, K. K., & Russell, L. D. (2000). Desert hedgehog (Dhh) gene is required in the mouse testis for formation of adult-type Leydig cells and normal development of peritubular cells and seminiferous tubules. *Biology of Reproduction, 63*(6), 1825–1838.

Clegg, E. J., & Macmillan, E. W. (1965). The phagocytic nature of Schiff-positive interstitial cells in the rat testis. *Journal of Endocrinology, 31*, 299–300.

Codesal, J., Regadera, J., Nistal, M., Regadera-Sejas, J., & Paniagua, R. (1990). Involution of human fetal Leydig cells. An immunohistochemical, ultrastructural and quantitative study. *Journal of Anatomy, 172*, 103–114.

Colvin, J. S., Green, R. P., Schmahl, J., Capel, B., & Ornitz, D. M. (2001). Male-to-female sex reversal in mice lacking fibroblast growth factor 9. *Cell, 104*(6), 875–889.

Combes, A. N., Lesieur, E., Harley, V. R., Sinclair, A. H., Little, M. H., Wilhelm, D., et al. (2009). Three-dimensional visualization of testis cord morphogenesis, a novel tubulogenic mechanism in development. *Developmental Dynamics, 238*(5), 1033–1041. http://dx.doi.org/10.1002/dvdy.21925.

Combes, A. N., Wilhelm, D., Davidson, T., Dejana, E., Harley, V., Sinclair, A., et al. (2009). Endothelial cell migration directs testis cord formation. *Developmental Biology, 326*(1), 112–120. http://dx.doi.org/10.1016/j.ydbio.2008.10.040, [pii] S0012-1606(08)01319-5.

Correa, R. V., Domenice, S., Bingham, N. C., Billerbeck, A. E., Rainey, W. E., Parker, K. L., et al. (2004). A microdeletion in the ligand binding domain of human steroidogenic factor 1 causes XY sex reversal without adrenal insufficiency. *Journal of Clinical Endocrinology and Metabolism, 89*(4), 1767–1772.

Correa, S. M., Washburn, L. L., Kahlon, R. S., Musson, M. C., Bouma, G. J., Eicher, E. M., et al. (2012). Sex reversal in C57BL/6J XY mice caused by increased expression of ovarian genes and insufficient activation of the testis determining pathway. *PLoS Genetics*, *8*(4), e1002569. http://dx.doi.org/10.1371/journal.pgen.1002569.

Coveney, D., Cool, J., Oliver, T., & Capel, B. (2008). Four-dimensional analysis of vascularization during primary development of an organ, the gonad. *In: Proceedings of the National Academy of Sciences of the United States of America*, *105*(20), 7212–7217. http://dx.doi.org/10.1073/pnas.0707674105, [pii] 0707674105.

Crisponi, L., Deiana, M., Loi, A., Chiappe, F., Uda, M., Amati, P., et al. (2001). The putative forkhead transcription factor FOXL2 is mutated in blepharophimosis/ptosis/epicanthus inversus syndrome. *Nature Genetics*, *27*(2), 159–166. http://dx.doi.org/10.1038/84781.

De Felici, M. (2009). Primordial germ cell biology at the beginning of the XXI century. *International Journal of Developmental Biology*, *53*(7), 891–894. http://dx.doi.org/10.1387/ijdb.082815mf.

DeFalco, T., Takahashi, S., & Capel, B. (2011). Two distinct origins for Leydig cell progenitors in the fetal testis. *Developmental Biology*, *352*(1), 14–26. http://dx.doi.org/10.1016/j.ydbio.2011.01.011.

Donovan, P. J., Stott, D., Cairns, L. A., Heasman, J., & Wylie, C. C. (1986). Migratory and postmigratory mouse primordial germ cells behave differently in culture. *Cell*, *44*(6), 831–838.

Eik-Nes, K. B. (1969). An effect of isoproterenol on rates of synthesis and secretion of testosterone. *American Journal of Physiology*, *217*(6), 1764–1770.

Fischel, A. (1930). Uber die Entwicklung der Keimdruesen des Menschen. *Anatomy and Embryology (Berlin)*, *92*, 34–72.

Fonseca, A. C., Bonaldi, A., Bertola, D. R., Kim, C. A., Otto, P. A., & Vianna-Morgante, A. M. (2013). The clinical impact of chromosomal rearrangements with breakpoints upstream of the SOX9 gene: Two novel de novo balanced translocations associated with acampomelic campomelic dysplasia. *BMC Medical Genetics*, *14*, 50. http://dx.doi.org/10.1186/1471-2350-14-50.

Foster, J. W., Dominguez-Steglich, M. A., Guioli, S., Kwok, C., Weller, P. A., Stevanovic, M., et al. (1994). Campomelic dysplasia and autosomal sex reversal caused by mutations in an SRY-related gene. *Nature*, *372*(6506), 525–530. http://dx.doi.org/10.1038/372525a0.

Fujimoto, T., Miyayama, Y., & Fuyuta, M. (1977). The origin, migration and fine morphology of human primordial germ cells. *Anatomical Record*, *188*(3), 315–330. http://dx.doi.org/10.1002/ar.1091880305.

Fuss, A. (1912). Ueber die Geschlechtszellen des Menschen und der Saeugetiere. *Archiv für Mikroskopische Anatomie*, *81*, 1–23.

Gardner, P. J., & Holyoke, E. A. (1964). Fine structure of the seminiferous tubule of the Swiss mouse. I. The limiting membrane, Sertoli cell, spermatogonia, and spermatocytes. *Anatomical Record*, *150*, 391–404.

George, F. W., & Wilson, J. D. (1978). Conversion of androgen to estrogen by the human fetal ovary. *Journal of Clinical Endocrinology and Metabolism*, *47*(3), 550–555.

Gierl, M. S., Gruhn, W. H., von Seggern, A., Maltry, N., & Niehrs, C. (2012). GADD45G functions in male sex determination by promoting p38 signaling and Sry expression. *Developmental Cell*, *23*(5), 1032–1042. http://dx.doi.org/10.1016/j.devcel.2012.09.014.

Ginsburg, M., Snow, M. H., & McLaren, A. (1990). Primordial germ cells in the mouse embryo during gastrulation. *Development*, *110*(2), 521–528.

Goldman, A. S., Yakovac, W. C., & Bongiovanni, A. M. (1966). Development of activity of 3 beta-hydroxysteroid dehydrogenase in human fetal tissues and in two anencephalic newborns. *Journal of Clinical Endocrinology and Metabolism*, *26*(1), 14–22.

Gondos, B. (1981). Cellular interrelationships in the human fetal ovary and testis. *Progress in Clinical and Biological Research, 59B*, 373–381.

Gondos, B., Morrison, K. P., & Renston, R. H. (1977). Leydig cell differentiation in the prepubertal rabbit testis. *Biology of Reproduction, 17*(5), 745–748.

Graves, J. A. M. (2002). The rise and fall of *SRY*. *Trends in Genetics, 18*, 259–264.

Gubbay, J., Collignon, J., Koopman, P., Capel, B., Economou, A., Munsterberg, A., et al. (1990). A gene mapping to the sex-determining region of the mouse Y chromosome is a member of a novel family of embryonically expressed genes. *Nature, 346*(6281), 245–250. http://dx.doi.org/10.1038/346245a0.

Hacker, A., Capel, B., Goodfellow, P., & Lovell-Badge, R. (1995). Expression of Sry, the mouse sex determining gene. *Development, 121*(6), 1603–1614.

Hammes, A., Guo, J. K., Lutsch, G., Leheste, J. R., Landrock, D., Ziegler, U., et al. (2001). Two splice variants of the Wilms' tumor 1 gene have distinct functions during sex determination and nephron formation. *Cell, 106*(3), 319–329.

Hanley, N. A., Hagan, D. M., Clement-Jones, M., Ball, S. G., Strachan, T., Salas-Cortes, L., et al. (2000). SRY, SOX9, and DAX1 expression patterns during human sex determination and gonadal development. *Mechanisms of Development, 91*(1–2), 403–407.

Hazra, R., Jimenez, M., Desai, R., Handelsman, D. J., & Allan, C. M. (2013). Sertoli cell androgen receptor expression regulates temporal fetal and adult Leydig cell differentiation, function and population size. *Endocrinology, 154*(9), 3410–3422. http://dx.doi.org/10.1210/en.2012-2273.

Heikkila, M., Prunskaite, R., Naillat, F., Itaranta, P., Vuoristo, J., Leppaluoto, J., et al. (2005). The partial female to male sex reversal in Wnt-4-deficient females involves induced expression of testosterone biosynthetic genes and testosterone production, and depends on androgen action. *Endocrinology, 146*(9), 4016–4023.

Hiramatsu, R., Matoba, S., Kanai-Azuma, M., Tsunekawa, N., Katoh-Fukui, Y., Kurohmaru, M., et al. (2009). A critical time window of Sry action in gonadal sex determination in mice. *Development, 136*(1), 129–138. http://dx.doi.org/10.1242/dev.029587.

Houston, C. S., Opitz, J. M., Spranger, J. W., Macpherson, R. I., Reed, M. H., Gilbert, E. F., et al. (1983). The campomelic syndrome: Review, report of 17 cases, and follow-up on the currently 17-year-old boy first reported by Maroteaux et al. in 1971. *American Journal of Medical Genetics, 15*(1), 3–28. http://dx.doi.org/10.1002/ajmg.1320150103.

Huang, B., Wang, S., Ning, Y., Lamb, A. N., & Bartley, J. (1999). Autosomal XX sex reversal caused by duplication of SOX9. *American Journal of Medical Genetics, 87*(4), 349–353.

Jameson, S. A., Lin, Y. T., & Capel, B. (2012). Testis development requires the repression of Wnt4 by Fgf signaling. *Developmental Biology, 370*(1), 24–32. http://dx.doi.org/10.1016/j.ydbio.2012.06.009.

Jeanes, A., Wilhelm, D., Wilson, M. J., Bowles, J., McClive, P. J., Sinclair, A. H., et al. (2005). Evaluation of candidate markers for the peritubular myoid cell lineage in the developing mouse testis. *Reproduction, 130*(4), 509–516. http://dx.doi.org/10.1530/rep.1.00718.

Jeske, Y. W., Bowles, J., Greenfield, A., & Koopman, P. (1995). Expression of a linear Sry transcript in the mouse genital ridge. *Nature Genetics, 10*(4), 480–482. http://dx.doi.org/10.1038/ng0895-480.

Jimenez, R. (2009). Ovarian organogenesis in mammals: Mice cannot tell us everything. *Sexual Development, 3*(6), 291–301. http://dx.doi.org/10.1159/000280584, [pii] 000280584.

Jirasek, J. E. (1971). Genital ducts and external genitalia: Development and anomalies. *Birth Defects Original Article Series*, 7(6), 131–139.

Jordan, B. K., Mohammed, M., Ching, S. T., Délot, E., Chen, X. N., Dewing, P., et al. (2001). Up-regulation of WNT-4 signaling and dosage-sensitive sex reversal in humans. *American Journal of Human Genetics, 68*, 1102–1109.

Karl, J., & Capel, B. (1998). Sertoli cells of the mouse testis originate from the coelomic epithelium. *Developmental Biology, 203*(2), 323–333. http://dx.doi.org/10.1006/dbio.1998.9068.

Katoh-Fukui, Y., Miyabayashi, K., Komatsu, T., Owaki, A., Baba, T., Shima, Y., et al. (2012). Cbx2, a polycomb group gene, is required for Sry gene expression in mice. *Endocrinology, 153*(2), 913–924. http://dx.doi.org/10.1210/en.2011-1055.

Kent, J., Wheatley, S. C., Andrews, J. E., Sinclair, A. H., & Koopman, P. (1996). A male-specific role for SOX9 in vertebrate sex determination. *Development, 122*(9), 2813–2822.

Kim, Y., Bingham, N., Sekido, R., Parker, K. L., Lovell-Badge, R., & Capel, B. (2007). Fibroblast growth factor receptor 2 regulates proliferation and Sertoli differentiation during male sex determination. *In: Proceedings of the National Academy of Sciences of the United States of America, 104*(42), 16558–16563. http://dx.doi.org/10.1073/pnas.0702581104.

Kim, Y., Kobayashi, A., Sekido, R., DiNapoli, L., Brennan, J., Chaboissier, M. C., et al. (2006). Fgf9 and Wnt4 act as antagonistic signals to regulate mammalian sex determination. *PLoS Biology, 4*(6), e187. http://dx.doi.org/10.1371/journal.pbio.0040187.

Koopman, P., Gubbay, J., Vivian, N., Goodfellow, P., & Lovell-Badge, R. (1991). Male development of chromosomally female mice transgenic for Sry. *Nature, 351*(6322), 117–121. http://dx.doi.org/10.1038/351117a0.

Koubova, J., Menke, D. B., Zhou, Q., Capel, B., Griswold, M. D., & Page, D. C. (2006). Retinoic acid regulates sex-specific timing of meiotic initiation in mice. *In: Proceedings of the National Academy of Sciences of the United States of America, 103*(8), 2474–2479. http://dx.doi.org/10.1073/pnas.0510813103.

Lavery, R., Chassot, A. A., Pauper, E., Gregoire, E. P., Klopfenstein, M., de Rooij, D. G., et al. (2012). Testicular differentiation occurs in absence of R-spondin1 and Sox9 in mouse sex reversals. *PLoS Genetics, 8*(12), e1003170. http://dx.doi.org/10.1371/journal.pgen.1003170.

Lawson, K. A., & Hage, W. J. (1994). Clonal analysis of the origin of primordial germ cells in the mouse. *Ciba Foundation Symposium, 182*, 68–84, discussion 84–91.

Leydig, F. (1850). Zur Anatomie der maennlichen Geschlechtsorgane und Analdruesen der Saeugethiere. *Zeitschrift für wissenschartliche Zoologie, 2*, 1–57.

Luo, X., Ikeda, Y., & Parker, K. L. (1994). A cell-specific nuclear receptor is essential for adrenal and gonadal development and sexual differentiation. *Cell, 77*(4), 481–490.

Maatouk, D. M., DiNapoli, L., Alvers, A., Parker, K. L., Taketo, M. M., & Capel, B. (2008). Stabilization of beta-catenin in XY gonads causes male-to-female sex-reversal. *Human Molecular Genetics, 17*(19), 2949–2955. http://dx.doi.org/10.1093/hmg/ddn193.

Malki, S., Nef, S., Notarnicola, C., Thevenet, L., Gasca, S., Mejean, C., et al. (2005). Prostaglandin D2 induces nuclear import of the sex-determining factor SOX9 via its cAMP-PKA phosphorylation. *EMBO Journal, 24*(10), 1798–1809. http://dx.doi.org/10.1038/sj.emboj.7600660.

Mallet, D., Bretones, P., Michel-Calemard, L., Dijoud, F., David, M., & Morel, Y. (2004). Gonadal dysgenesis without adrenal insufficiency in a 46, XY patient heterozygous for the nonsense C16X mutation: A case of SF1 haploinsufficiency. *Journal of Clinical Endocrinology and Metabolism, 89*(10), 4829–4832. http://dx.doi.org/10.1210/jc.2004-0670.

Martineau, J., Nordqvist, K., Tilmann, C., Lovell-Badge, R., & Capel, B. (1997). Male-specific cell migration into the developing gonad. *Current Biology, 7*(12), 958–968.

Matson, C. K., Murphy, M. W., Sarver, A. L., Griswold, M. D., Bardwell, V. J., & Zarkower, D. (2011). DMRT1 prevents female reprogramming in the postnatal mammalian testis. *Nature, 476*, 101–104.

McCarrey, J. R. (1993). Development of the germ cell. In L. L. Ewing & C. Desjardins (Eds.), *Cellular and molecular biology of the testis* (pp. 58–89). New York: Oxford University Press.
McKay, D. G., Hertig, A. T., Adams, E. C., & Danziger, S. (1953). Histochemical observations on the germ cells of human embryos. *Anatomical Record*, *117*(2), 201–219.
Meeks, J. J., Crawford, S. E., Russell, T. A., Morohashi, K., Weiss, J., & Jameson, J. L. (2003). Dax1 regulates testis cord organization during gonadal differentiation. *Development*, *130*(5), 1029–1036.
Meeks, J. J., Weiss, J., & Jameson, J. L. (2003). Dax1 is required for testis determination. *Nature Genetics*, *34*(1), 32–33.
Merchant-Larios, H., & Moreno-Mendoza, N. (1998). Mesonephric stromal cells differentiate into Leydig cells in the mouse fetal testis. *Experimental Cell Research*, *244*(1), 230–238. http://dx.doi.org/10.1006/excr.1998.4215, [pii] S0014-4827(98)94215-X.
Miyabayashi, K., Katoh-Fukui, Y., Ogawa, H., Baba, T., Shima, Y., Sugiyama, N., et al. (2013). Aristaless related homeobox gene, Arx, is implicated in mouse fetal Leydig cell differentiation possibly through expressing in the progenitor cells. *PLoS One*, *8*(6), e68050. http://dx.doi.org/10.1371/journal.pone.0068050.
Miyamoto, N., Yoshida, M., Kuratani, S., Matsuo, I., & Aizawa, S. (1997). Defects of urogenital development in mice lacking Emx2. *Development*, *124*(9), 1653–1664.
Moniot, B., Declosmenil, F., Barrionuevo, F., Scherer, G., Aritake, K., Malki, S., et al. (2009). The PGD2 pathway, independently of FGF9, amplifies SOX9 activity in Sertoli cells during male sexual differentiation. *Development*, *136*(11), 1813–1821. http://dx.doi.org/10.1242/dev.032631.
Moon, Y. S., & Hardy, M. H. (1973). The early differentiation of the testis and interstitial cells in the fetal pig, and its duplication in organ culture. *The American Journal of Anatomy*, *138*(2), 253–267. http://dx.doi.org/10.1002/aja.1001380209.
Morais da Silva, S., Hacker, A., Harley, V., Goodfellow, P., Swain, A., & Lovell-Badge, R. (1996). Sox9 expression during gonadal development implies a conserved role for the gene in testis differentiation in mammals and birds. *Nature Genetics*, *14*(1), 62–68. http://dx.doi.org/10.1038/ng0996-62.
Mork, L., Maatouk, D. M., McMahon, J. A., Guo, J. J., Zhang, P., McMahon, A. P., et al. (2011). Temporal differences in granulosa cell specification in the ovary reflect distinct follicle fates in mice. *Biology of Reproduction*, *86*(2), 37. http://dx.doi.org/10.1095/biolreprod.111.095208, [pii] biolreprod.111.095208.
Motta, P. M., & Makabe, S. (1982). Development of the ovarian surface and associated germ cells in the human fetus. A correlated study by scanning and transmission electron microscopy. *Cell and Tissue Research*, *226*(3), 493–510.
Nagamine, C. M., Morohashi, K., Carlisle, C., & Chang, D. K. (1999). Sex reversal caused by Mus musculus domesticus Y chromosomes linked to variant expression of the testis-determining gene Sry. *Developmental Biology*, *216*(1), 182–194.
Nagano, T., & Suzuki, F. (1976). The postnatal development of the junctional complexes of the mouse Sertoli cells as revealed by freeze-fracture. *Anatomical Record*, *185*(4), 403–417. http://dx.doi.org/10.1002/ar.1091850403.
Nanda, S., DeFalco, T. J., Loh, S. H., Phochanukul, N., Camara, N., Van Doren, M., et al. (2009). Sox100B, a Drosophila group E Sox-domain gene, is required for somatic testis differentiation. *Sexual Development*, *3*(1), 26–37. http://dx.doi.org/10.1159/000200079.
Nef, S., Verma-Kurvari, S., Merenmies, J., Vassalli, J. D., Efstratiadis, A., Accili, D., et al. (2003). Testis determination requires insulin receptor family function in mice. *Nature*, *426*(6964), 291–295. http://dx.doi.org/10.1038/nature02059.
Nelson, W. J., & Nusse, R. (2004). Convergence of Wnt, beta-catenin, and cadherin pathways. *Science*, *303*(5663), 1483–1487. http://dx.doi.org/10.1126/science.1094291.
Nussbaum, M. (1880). Die Differenzierung des Geschlechts im Tierreich. *Archiv für Mikroskopische Anatomie*, *18*, 1–121.

Orth, J. M. (1982). Proliferation of Sertoli cells in fetal and postnatal rats: A quantitative autoradiographic study. *Anatomical Record*, *203*(4), 485–492. http://dx.doi.org/10.1002/ar.1092030408.

O'Shaughnessy, P. J., Baker, P. J., Monteiro, A., Cassie, S., Bhattacharya, S., & Fowler, P. A. (2007). Developmental changes in human fetal testicular cell numbers and messenger ribonucleic acid levels during the second trimester. *Journal of Clinical Endocrinology and Metabolism*, *92*(12), 4792–4801. http://dx.doi.org/10.1210/jc.2007-1690.

Ozisik, G., Achermann, J. C., & Jameson, J. L. (2002). The role of SF1 in adrenal and reproductive function: Insight from naturally occurring mutations in humans. *Molecular Genetics and Metabolism*, *76*(2), 85–91.

Palmer, S. J., & Burgoyne, P. S. (1991). *In situ* analysis of fetal, prepuberal and adult XX′XY chimaeric mouse testes: Sertoli cells are predominantly, but not exclusively, XY. *Development*, *112*, 265–268.

Parma, P., Radi, O., Vidal, V., Chaboissier, M. C., Dellambra, E., Valentini, S., et al. (2006). R-spondin1 is essential in sex determination, skin differentiation and malignancy. *Nature Genetics*, *38*(11), 1304–1309. http://dx.doi.org/10.1038/ng1907, [pii] ng1907.

Pearlman, A., Loke, J., Le Caignec, C., White, S., Chin, L., Friedman, A., et al. (2010). Mutations in MAP3K1 cause 46, XY disorders of sex development and implicate a common signal transduction pathway in human testis determination. *American Journal of Human Genetics*, *87*(6), 898–904. http://dx.doi.org/10.1016/j.ajhg.2010.11.003.

Pehlemann, F. W., & Lombard, M. N. (1978). Differentiation of ovarian and testicular interstitial cells during embryonic and post-embryonic development in mice. Ultrastructural observations. *Cell and Tissue Research*, *188*(3), 465–480.

Pepling, M. E., & Spradling, A. C. (1998). Female mouse germ cells form synchronously dividing cysts. *Development*, *125*(17), 3323–3328.

Philibert, P., Biason-Lauber, A., Gueorguieva, I., Stuckens, C., Pienkowski, C., Lebon-Labich, B., et al. (2011). Molecular analysis of WNT4 gene in four adolescent girls with mullerian duct abnormality and hyperandrogenism (atypical Mayer–Rokitansky–Kuster–Hauser syndrome). *Fertility and Sterility*, *95*(8), 2683–2686. http://dx.doi.org/10.1016/j.fertnstert.2011.01.152.

Philibert, P., Biason-Lauber, A., Rouzier, R., Pienkowski, C., Paris, F., Konrad, D., et al. (2008). Identification and functional analysis of a new WNT4 gene mutation among 28 adolescent girls with primary amenorrhea and mullerian duct abnormalities: A French collaborative study. *Journal of Clinical Endocrinology and Metabolism*, *93*(3), 895–900. http://dx.doi.org/10.1210/jc.2007-2023.

Pierucci-Alves, F., Clark, A. M., & Russell, L. D. (2001). A developmental study of the Desert hedgehog-null mouse testis. *Biology of Reproduction*, *65*(5), 1392–1402.

Pinkerton, J. H., Mc, Kay Dg, Adams, E. C., & Hertig, A. T. (1961). Development of the human ovary—A study using histochemical techniques. *Obstetrics and Gynecology*, *18*, 152–181.

Pitetti, J. L., Calvel, P., Romero, Y., Conne, B., Truong, V., Papaioannou, M. D., et al. (2013). Insulin and IGF1 receptors are essential for XX and XY gonadal differentiation and adrenal development in mice. *PLoS Genetics*, *9*(1), e1003160. http://dx.doi.org/10.1371/journal.pgen.1003160.

Rakoczy, J., Fernandez-Valverde, S. L., Glazov, E. A., Wainwright, E. N., Sato, T., Takada, S., et al. (2013). MicroRNAs-140-5p/140-3p modulate Leydig cell numbers in the developing mouse testis. *Biology of Reproduction*, *88*(6), 143. http://dx.doi.org/10.1095/biolreprod.113.107607.

Raymond, C. S., Murphy, M. W., O'Sullivan, M. G., Bardwell, V. J., & Zarkower, D. (2000). *Dmrt1*, a gene related to worm and fly sexual regulators, is required for mammalian testis differentiation. *Genes and Development*, *14*, 2587–2595.

Ricci, G., Guglielmo, M. C., Caruso, M., Ferranti, F., Canipari, R., Galdieri, M., et al. (2012). Hepatocyte growth factor is a mouse fetal Leydig cell terminal differentiation factor. *Biology of Reproduction, 87*(6), 146. http://dx.doi.org/10.1095/biolreprod.112.104638.

Ritzen, E. M., Van Damme, M. P., Froysa, B., Reuter, C., De La Torre, B., & Diczfalusy, E. (1981). Identification of estradiol produced by Sertoli cell enriched cultures during incubation with testosterone. *Journal of Steroid Biochemistry, 14*(6), 533–535.

Ross, M. H., & Long, I. R. (1966). Contractile cells in human seminiferous tubules. *Science, 153*(3741), 1271–1273.

Ruby, J. R., Dyer, R. F., & Skalko, R. G. (1969). Continuities between mitochondria and endoplasmic reticulum in the mammalian ovary. *Zeitschrift für Zellforschung und Mikroskopische Anatomie, 97*(1), 30–37.

Runyan, C., Schaible, K., Molyneaux, K., Wang, Z., Levin, L., & Wylie, C. (2006). Steel factor controls midline cell death of primordial germ cells and is essential for their normal proliferation and migration. *Development, 133*(24), 4861–4869. http://dx.doi.org/10.1242/dev.02688.

Sadovsky, Y., Crawford, P. A., Woodson, K. G., Polish, J. A., Clements, M. A., Tourtellotte, L. M., et al. (1995). Mice deficient in the orphan receptor steroidogenic factor 1 lack adrenal glands and gonads but express P450 side-chain-cleavage enzyme in the placenta and have normal embryonic serum levels of corticosteroids. *In: Proceedings of the National Academy of Sciences of the United States of America, 92*(24), 10939–10943.

Sarraj, M. A., Escalona, R. M., Umbers, A., Chua, H. K., Small, C., Griswold, M., et al. (2010). Fetal testis dysgenesis and compromised Leydig cell function in Tgfbr3 (beta glycan) knockout mice. *Biology of Reproduction, 82*(1), 153–162. http://dx.doi.org/10.1095/biolreprod.109.078766, [pii] biolreprod.109.078766.

Satoh, M. (1991). Histogenesis and organogenesis of the gonad in human embryos. *Journal of Anatomy, 177*, 85–107.

Schmahl, J., Eicher, E. M., Washburn, L. L., & Capel, B. (2000). Sry induces cell proliferation in the mouse gonad. *Development, 127*(1), 65–73.

Schmidt, D., Ovitt, C. E., Anlag, K., Fehsenfeld, S., Gredsted, L., Treier, A. C., et al. (2004). The murine winged-helix transcription factor Foxl2 is required for granulosa cell differentiation and ovary maintenance. *Development, 131*(4), 933–942.

Schnabel, C. A., Selleri, L., & Cleary, M. L. (2003). Pbx1 is essential for adrenal development and urogenital differentiation. *Genesis, 37*(3), 123–130. http://dx.doi.org/10.1002/gene.10235.

Sekido, R., Bar, I., Narvaez, V., Penny, G., & Lovell-Badge, R. (2004). SOX9 is up-regulated by the transient expression of SRY specifically in Sertoli cell precursors. *Developmental Biology, 274*(2), 271–279. http://dx.doi.org/10.1016/j.ydbio.2004.07.011.

Sekido, R., & Lovell-Badge, R. (2008). Sex determination involves synergistic action of SRY and SF1 on a specific Sox9 enhancer. *Nature, 453*(7197), 930–934. http://dx.doi.org/10.1038/nature06944.

Sekido, R., & Lovell-Badge, R. (2013). Genetic control of testis development. *Sexual Development*, 7(1–3), 21–32. http://dx.doi.org/10.1159/000342221.

Sertoli, E. (1865). Dell' esistenza di particulari cellule ramificate nei canalicoli seminiferi del testicolo umano. *Morgagni*, 7, 31–40.

Sharpe, R. M., McKinnell, C., Kivlin, C., & Fisher, J. S. (2003). Proliferation and functional maturation of Sertoli cells, and their relevance to disorders of testis function in adulthood. *Reproduction, 125*(6), 769–784.

Spotila, L. D., Spotila, J. R., & Hall, S. E. (1998). Sequence and expression analysis of WT1 and Sox9 in the red-eared slider turtle, Trachemys scripta. *Journal of Experimental Zoology, 281*(5), 417–427.

Stoop, H., Honecker, F., Cools, M., de Krijger, R., Bokemeyer, C., & Looijenga, L. H. (2005). Differentiation and development of human female germ cells during prenatal gonadogenesis: An immunohistochemical study. *Human Reproduction*, *20*(6), 1466–1476. http://dx.doi.org/10.1093/humrep/deh800.

Sutton, E., Hughes, J., White, S., Sekido, R., Tan, J., Arboleda, V., et al. (2011). Identification of SOX3 as an XX male sex reversal gene in mice and humans. *Journal of Clinical Investigation*, *121*(1), 328–341. http://dx.doi.org/10.1172/JCI42580.

Tam, P. P., & Snow, M. H. (1981). Proliferation and migration of primordial germ cells during compensatory growth in mouse embryos. *Journal of Embryology and Experimental Morphology*, *64*, 133–147.

Tang, H., Brennan, J., Karl, J., Hamada, Y., Raetzman, L., & Capel, B. (2008). Notch signaling maintains Leydig progenitor cells in the mouse testis. *Development*, *135*(22), 3745–3753. http://dx.doi.org/10.1242/dev.024786, [pii] dev.024786.

Tevosian, S. G., Albrecht, K. H., Crispino, J. D., Fujiwara, Y., Eicher, E. M., & Orkin, S. H. (2002). Gonadal differentiation, sex determination and normal Sry expression in mice require direct interaction between transcription partners GATA4 and FOG2. *Development*, *129*(19), 4627–4634.

Tolwinski, N. S., & Wieschaus, E. (2004). Rethinking WNT signaling. *Trends in Genetics*, *20*(4), 177–181. http://dx.doi.org/10.1016/j.tig.2004.02.003.

Tomizuka, K., Horikoshi, K., Kitada, R., Sugawara, Y., Iba, Y., Kojima, A., et al. (2008). R-spondin1 plays an essential role in ovarian development through positively regulating Wnt-4 signaling. *Human Molecular Genetics*, *17*(9), 1278–1291. http://dx.doi.org/10.1093/hmg/ddn036, [pii] ddn036.

Tripiciano, A., Filippini, A., Ballarini, F., & Palombi, F. (1998). Contractile response of peritubular myoid cells to prostaglandin F2alpha. *Molecular and Cellular Endocrinology*, *138*(1–2), 143–150.

Tripiciano, A., Filippini, A., Giustiniani, Q., & Palombi, F. (1996). Direct visualization of rat peritubular myoid cell contraction in response to endothelin. *Biology of Reproduction*, *55*(1), 25–31.

Uda, M., Ottolenghi, C., Crisponi, L., Garcia, J. E., Deiana, M., Kimber, W., et al. (2004). Foxl2 disruption causes mouse ovarian failure by pervasive blockage of follicle development. *Human Molecular Genetics*, *13*(11), 1171–1181.

Uhlenhaut, N. H., Jakob, S., Anlag, K., Eisenberger, T., Sekido, R., Kress, J., et al. (2009). Somatic sex reprogramming of adult ovaries to testes by FOXL2 ablation. *Cell*, *139*(6), 1130–1142. http://dx.doi.org/10.1016/j.cell.2009.11.021, [pii] S0092-8674(09)01433-0.

Umehara, F., Tate, G., Itoh, K., Yamaguchi, N., Douchi, T., Mitsuya, T., et al. (2000). A novel mutation of desert hedgehog in a patient with 46, XY partial gonadal dysgenesis accompanied by minifascicular neuropathy. *American Journal of Human Genetics*, *67*(5), 1302–1305. http://dx.doi.org/10.1016/S0002-9297(07)62958-9.

Vainio, S., Heikkila, M., Kispert, A., Chin, N., & McMahon, A. P. (1999). Female development in mammals is regulated by Wnt-4 signalling. *Nature*, *397*, 405–409.

Vidal, V. P., Chaboissier, M. C., de Rooij, D. G., & Schedl, A. (2001 Jul). Sox9 induces testis development in XX transgenic mice. *Nature Genetics*, *28*(3), 216–217.

Wagner, T., Wirth, J., Meyer, J., Zabel, B., Held, M., Zimmer, J., et al. (1994). Autosomal sex reversal and campomelic dysplasia are caused by mutations in and around the SRY-related gene SOX9. *Cell*, *79*(6), 1111–1120.

Waldeyer, W. (1870). *Eierstock und Ei*. Leipzig: Engelmann.

Wang, Q., Lan, Y., Cho, E. S., Maltby, K. M., & Jiang, R. (2005). Odd-skipped related 1 (Odd 1) is an essential regulator of heart and urogenital development. *Developmental Biology*, *288*(2), 582–594. http://dx.doi.org/10.1016/j.ydbio.2005.09.024.

Warr, N., Carre, G. A., Siggers, P., Faleato, J. V., Brixey, R., Pope, M., et al. (2012). Gadd45gamma and Map3k4 interactions regulate mouse testis determination via p38

MAPK-mediated control of Sry expression. *Developmental Cell, 23*(5), 1020–1031. http://dx.doi.org/10.1016/j.devcel.2012.09.016.

Wartenberg, H., Kinsky, I., Viebahn, C., & Schmolke, C. (1991). Fine structural characteristics of testicular cord formation in the developing rabbit gonad. *Journal of Electron Microscopy Technique, 19*(2), 133–157. http://dx.doi.org/10.1002/jemt.1060190203.

Western, P. S., Harry, J. L., Graves, J. A., & Sinclair, A. H. (1999). Temperature-dependent sex determination: Upregulation of SOX9 expression after commitment to male development. *Developmental Dynamics, 214*(3), 171–177. http://dx.doi.org/10.1002/(SICI)1097-0177(199903)214:3<171::AID-AJA1>3.0.CO;2-S.

Wilhelm, D., & Englert, C. (2002). The Wilms tumor suppressor WT1 regulates early gonad development by activation of Sf1. *Genes & Development, 16*(14), 1839–1851. http://dx.doi.org/10.1101/gad.220102.

Wilhelm, D., Hiramatsu, R., Mizusaki, H., Widjaja, L., Combes, A. N., Kanai, Y., et al. (2007). SOX9 regulates prostaglandin D synthase gene transcription in vivo to ensure testis development. *Journal of Biological Chemistry, 282*(14), 10553–10560. http://dx.doi.org/10.1074/jbc.M609578200.

Wilhelm, D., Martinson, F., Bradford, S., Wilson, M. J., Combes, A. N., Beverdam, A., et al. (2005). Sertoli cell differentiation is induced both cell-autonomously and through prostaglandin signaling during mammalian sex determination. *Developmental Biology, 287*(1), 111–124. http://dx.doi.org/10.1016/j.ydbio.2005.08.039.

Witschi, E. (1931). Studies on sex differentiation and sex determination in amphibians. V. Range of the cortex-medulla antagonism in parabiotic twins of Ranidae and Hylidae. *Journal of Experimental Zoology, 58*, 113–145.

Witschi, E. (1948). Migration of germ cells of human embryos from the yolk sac to the primitive gonadal folds. *Contributions to Embryology, Carnegie Institute, 148*, 67–80.

Witschi, E. (1951). Embryogenesis of the adrenal and the reproductive glands. *Recent Progress in Hormone Research, 6*, 1–27.

Yao, H. H., Matzuk, M. M., Jorgez, C. J., Menke, D. B., Page, D. C., Swain, A., et al. (2004). Follistatin operates downstream of Wnt4 in mammalian ovary organogenesis. *Developmental Dynamics, 230*(2), 210–215.

Yao, H. H., Whoriskey, W., & Capel, B. (2002). Desert Hedgehog/Patched 1 signaling specifies fetal Leydig cell fate in testis organogenesis. *Genes & Development, 16*(11), 1433–1440. http://dx.doi.org/10.1101/gad.981202.

Zamboni, L., & Upadhyay, S. (1982). The contribution of the mesonephros to the development of the sheep fetal testis. *The American Journal of Anatomy, 165*(3), 339–356. http://dx.doi.org/10.1002/aja.1001650309.

CHAPTER FOUR

Mechanisms of Thyroid Development and Dysgenesis: An Analysis Based on Developmental Stages and Concurrent Embryonic Anatomy

Mikael Nilsson[*,1], **Henrik Fagman**[†]
[*]Sahlgrenska Cancer Center, Institute of Biomedicine, University of Gothenburg, Göteborg, Sweden
[†]Department of Pathology, Sahlgrenska University Hospital, Göteborg, Sweden
[1]Corresponding author. e-mail address: mikael.nilsson@gu.se

Contents

Abstract

Thyroid dysgenesis is the most common cause of congenital hypothyroidism that affects 1 in 3000 newborns. Although a number of pathogenetic mutations in thyroid developmental genes have been identified, the molecular mechanism of disease is unknown in most cases. This chapter summarizes the current knowledge of normal thyroid development and puts the different developmental stages in perspective, from the time of foregut endoderm patterning to the final shaping of pharyngeal anatomy, for understanding how specific malformations may arise. At the cellular level, we will also discuss fate determination of follicular and C-cell progenitors and their subsequent embryonic growth, migration, and differentiation as the different thyroid primordia evolve and merge to establish the final size and shape of the gland.

Current Topics in Developmental Biology, Volume 106
ISSN 0070-2153
http://dx.doi.org/10.1016/B978-0-12-416021-7.00004-3

1. INTRODUCTION

Thyroid hormone (TH) and other iodinated compounds are found in many species throughout the animal kingdom, but the development of follicular thyroid tissue or an encapsulated thyroid gland is specific for vertebrates, heralding the evolution of a regulated machinery for the biosynthesis, storage, and release of TH under hypothalamic control primarily mediated by thyroid-stimulating hormone (TSH, also named thyrotropin) from the pituitary. Although iodine may be highly enriched by other means in other tissues, for example, as observed in algae, the follicular epithelium provides an efficient uptake mechanism required to accumulate inorganic iodine (iodide, I^-) where iodination of thyroglobulin (TG), the thyroid prohormone, takes place. In fact, the first steps in TH formation are manufactured extracellularly in the follicle lumen secluded from the cell interior (Ekholm, 1981; Ekholm & Wollman, 1975). Mechanisms that initiate follicle formation and maintain the follicle structure are therefore of fundamental interest for the understanding of thyroid development from a functional point of view. Thyroid organogenesis relies in the first place on proper specification of thyroid progenitor cells in the anterior endoderm followed by a timely onset of the differentiation program comprising expression of thyroid-specific genes involved in TH production once morphogenesis is finished. The anatomical sculpturing and positioning of the gland is a multifaceted process that forms the basis of a number of malformations collectively named thyroid dysgenesis that may lead to congenital hypothyroidism (CH), the most common preventable cause of mental retardation in children (Gruters & Krude, 2011), in principal due to shortage of thyroid tissue volume. In contrast to other vertebrates, the mammalian thyroid contains a second endocrine cell type, the parafollicular C cell, that enters the gland by a unique fusion event between the thyroid primordium proper and the ultimobranchial bodies (UB) that arise bilaterally in the most inferior of the pharyngeal pouches. In this chapter, we will briefly summarize the current knowledge on thyroid development with emphasis on mice and human, which display nearly identical thyroid anatomy and also share as far as we know many of the morphogenetic traits and mechanisms of dysgenesis. For comprehensive reviews on other aspects of this topic, readers are referred to Castanet, Marinovic, Polak, and Leger (2010) and Deladoey, Vassart, and Van Vliet (2007) for epidemiology; De Felice and Di Lauro (2004) and Grasberger and Refetoff (2011) for genetics; Fagman and Nilsson (2010) for morphogenesis; and Porazzi, Calebiro, Benato, Tiso, and Persani (2009) for zebrafish.

2. ANATOMY AND PHYLOGENIC ASPECTS OF THYROID DEVELOPMENT

The human thyroid is a bilobed structure with the shape of a butterfly rather than a shield (recall the name of the gland in, e.g., German (shilddrüse) and Swedish (sköldkörtel); refer to the latter), the lobes being connected by a slender isthmus bridging the midline just in front of the proximal trachea (Fig. 4.1A). The left and right lobes extend along either side of the trachea and larynx with close proximity to the carotid arteries passing by laterally; as will be further discussed later, the anatomical relation to these great vessels is probably developmentally important for the positioning of the gland (Alt, Elsalini, et al., 2006; Fagman, Andersson, & Nilsson, 2006; Fagman,

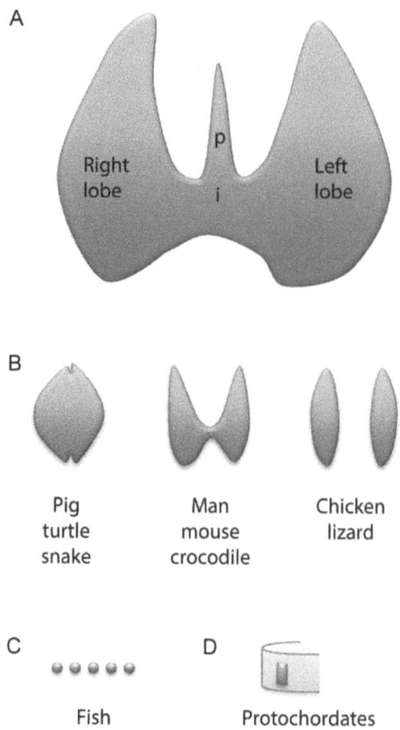

Figure 4.1 Thyroid anatomy. (A) The bilobed human thyroid gland. p, pyramidal lobe; i, isthmus. (B) Shape variation of thyroid gland in (exemplified) different species. (C) Dispersed follicular thyroid in, for example, zebrafish. (D) Thyroid homolog in the wall of invertebrate endostyle. (See color plate.)

Grande, Gritli-Linde, & Nilsson, 2004; Fagman et al., 2007; Opitz et al., 2012). A third, smaller, pyramidal lobe is frequently present in the midline extending from the isthmus in cranial direction with the blind end located anterior to the thyroid cartilage (Park, Kim, Park, Kang, & Kim, 2012). This is a vestigial remnant of the embryonic thyroglossal duct to which the median thyroid primordium is transiently connected in early development after being delaminated from the foregut endoderm.

The thyroid shape varies considerably between species without a distinctive pattern related to major vertebrate groups (Fig. 4.1B). Thus, a bilobed gland with a connecting isthmus is present in most mammals but can also be found in some reptiles, for example, crocodiles. Not only in turtles and snakes but also in pigs, the thyroid is an unpaired gland, ellipsoidal or ovoid in shape, and positioned more or less in the midline. In avian species and many lizards, the thyroid is completely separated into two distinct organs. It is likely that these differences relate to variations in the extent the presumptive lobes will grow bilaterally after the median thyroid primordium bifurcates. Notably, the paired UB contribute to lobe size but are not required for bilobation of the embryonic thyroid. Fish do not develop an encapsulated thyroid but consist of a variable number of loosely connected follicular structures that have a widespread distribution in the midline embedded in the subpharyngeal mesenchyme (Alt, Reibe, et al., 2006; Wendl et al., 2002; Fig. 4.1C).

In protochordates, the predecessor of the vertebrate thyroid comprises a restricted region of the ventral pharyngeal wall that is part of a food-capturing longitudinal groove in the pharyngeal cavity named the endostyle (Hiruta, Mazet, Yasui, Zhang, & Ogasawara, 2005; Kluge, Renault, & Rohr, 2005; Fig. 4.1D). The lamprey represents a unique intermediate species that in the larval stage exhibits an endostyle from which a follicular thyroid develops during metamorphosis, indicating the developmental programs determining these structurally different features coexist in the evolutionary transition from invertebrates to vertebrates. A row of cells in the invertebrate endostyle has iodinating capacity (Fredriksson, Ericson, & Olsson, 1984; Fredriksson, Ofverholm, & Ericson, 1985) and expresses thyroperoxidase (TPO) (Ogasawara, 2000; Ogasawara, Di Lauro, & Satoh, 1999), but the utility of TG as a precursor for TH production is specific for the vertebrate thyroid (Paris, Brunet, Markov, Schubert, & Laudet, 2008; Takagi, Omura, & Go, 1991). The TG gene probably evolved by duplication at the divergence of vertebrates and invertebrates (Mori, Itoh, & Salvaterra, 1987; Takagi et al., 1991). Thus, development of a follicular thyroid and expression of TG are evolutionary linked processes. This notion is consistent with observations that thyroid-like cells of the larval endostyle and thyroid progenitors in the metamorphosed lamprey consist of distinct cell populations in which only the latter express Nkx2-1 (formerly thyroid transcription factor-1 (TTF-1)) known to regulate the vertebrate thyroid (Kluge et al., 2005).

Likewise, the UB, also referred to as the lateral thyroid anlage, appears in vertebrates coinciding with the evolution of the pharyngeal arches and pouches and a more complex skull skeleton (Graham, 2008). In non-mammalian species, these structures never fuse with the thyroid but persist as paired or unpaired ultimobranchial glands, the major function of which is to regulate calcium homeostasis mediated by the peptide hormone calcitonin (CT) in aquatic species including fish and amphibians (Pang, 1971; Robertson, 1971). That the thyroid is the only source of circulating CT in mammals was discovered in 1964 (Foster et al., 1964), and a few years later, the ultimobranchial origin of CT, at the time referred to as thyrocalcitonin (Foster, 1968), was documented (Copp, Cockcroft, & Kueh, 1967; Pearse & Carvalheira, 1967; Tauber, 1967). As indicated from classical quail–chick transplantation experiments (Le Douarin & Le Lievre, 1970; Polak, Pearse, Le Lievre, Fontaine, & Le Douarin, 1974), CT-producing cells in the ultimobranchial gland originate from the neural crest (NC), suggesting that thyroid C cells are neuroectodermal. It is believed that a subpopulation of NC cells (NCCs) entering the inferior pharyngeal arches diverges from the main NC stream and joins the UB before it fuses with the thyroid. The integration of UB in the thyroid, and hence the development of parafollicular C cells, is intriguing considering the fact that CT is physiologically much less important for calcium regulation in higher vertebrates (Hirsch & Baruch, 2003). It is possible that other requirements during evolution of tetrapods for terrestrial life constituted by the emergence of parathyroid glands from the same embryonic tissues that form gills in fish (Zajac & Danks, 2008) eventually made CT production by the ultimobranchial glands redundant. Although there are a few reports on the effects of CT on thyroid function (Ahren, 1989; Isaac, Merceron, Caillens, Raymond, & Ardaillou, 1980) and TSH on C cells (Barasch, Gershon, Nunez, Tamir, & al-Awqati, 1988; Morillo-Bernal et al., 2009), the physiological significance of the intrathyroidal position and intimate seating of this cell type alongside the follicular epithelium in mammals has not be elucidated. Regardless of this, being the cell origin of medullary thyroid cancer (MTC) (Meyer & Abdel-Bari, 1968; Tashjian & Melvin, 1968;

Williams, 1966), detailed studies on the embryonic development of mammalian C cells will potentially reveal novel mechanisms of their propagation that might be recapitulated in tumorigenesis.

3. THYROID ORGANOGENESIS AND SPECTRUM OF DEVELOPMENTAL DEFECTS IN HUMAN

In mammals, the median thyroid anlage in which endoderm progenitors destined to a follicular fate are first recognized forms a placode in the midline of the pharyngeal floor just behind the prospective tongue (Fig. 4.2). Concomitantly, at a more inferior level in the presumptive

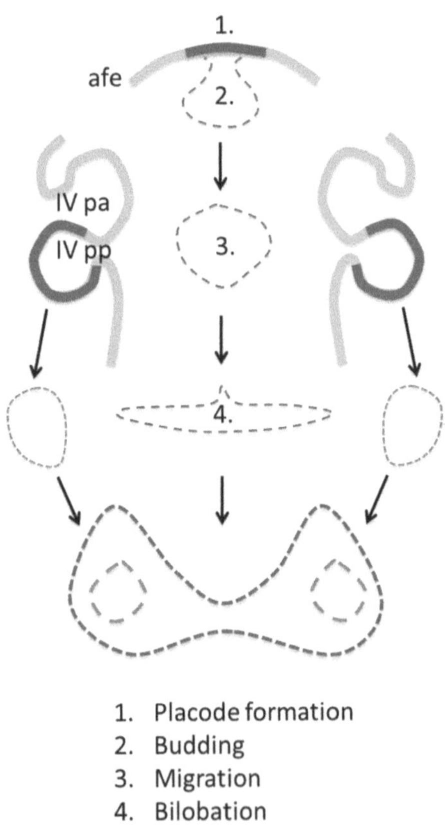

Figure 4.2 Developmental stages of thyroid morphogenesis and organogenesis in mouse. Red (solid or encircled)—midline thyroid anlage and primordium. Green (solid or encircled)—ultimobranchial epithelium/body. afe, anterior foregut endoderm; pa, pharyngeal arch; pp, pharyngeal pouch. (See color plate.)

pharynx, the paired UBs representing the lateral thyroid anlagen are identified as evaginations of the last pharyngeal pouch, numbered fourth and fifth in, respectively, mice and human (Grevellec & Tucker, 2010). After budding from the endoderm, both primordia transiently exist as single structures surrounded by mesoderm and NC-derived ectomesenchyme that is abundant also in this part of the presumptive pharynx (Jiang, Rowitch, Soriano, McMahon, & Sucov, 2000; Kameda, Nishimaki, Chisaka, Iseki, & Sucov, 2007). As surveyed in detail for the embryonic thyroid in mice (Fagman et al., 2006), fusion of primordia depends first on their individual displacements as the median thyroid descends in the midline and the position of the UB is shifted medially relative to other mobile structures, for example, the parathyroid and thymus that are also present in this part of the pharyngeal mesenchyme. Subsequently, lateral elongation of the midline primordium is required to eventually reach and embrace the UB on both sides (Fig. 4.2). Further growth establishes the left and right lobes along with the onset of folliculogenesis. The entire morphogenetic process takes less than 5 days in mouse and several weeks in human (Fagman & Nilsson, 2010).

The developmental path defines spatiotemporally the stages where different thyroid malformations may arise and therefore also sets the scene for putative pathogenetic mechanisms to each of them (Fig. 4.2). From a clinical point of view, athyreosis means lack of a morphologically discernable gland or remnant thyroid tissue as identified by ultrasound or technetium-99m scintigraphy in children with severe CH (Chang, Hong, & Choi, 2009; Clerc et al., 2008). Apart from accidental exposure to high I-131 doses, that is, by inadvertent radioiodine treatment during pregnancy that will destroy the entire fetal thyroid by irradiation damage (Berg, Jacobsson, Nystrom, Gleisner, & Tennvall, 2008), this condition indicates an early developmental defect caused by either agenesis of the anlage or regression of the thyroid bud that was initially normally formed. Hypoplasia of the orthotopic thyroid may be either global, that is, the attenuated gland has a normal shape, or asymmetrical in which one lobe typically is missing, referred to as hemiagenesis (Wu, Wein, & Carter, 2012). A small thyroid may also be ectopically located at any point along the tract for downward migration of the median primordium (Noussios, Anagnostis, Goulis, Lappas, & Natsis, 2011). The presence of a lingual thyroid indicates a complete migration defect. Even more rarely ectopic thyroid tissue can be found in more distant locations in the thoracic or abdominal cavities. Dual ectopy of the thyroid also exists in different sites in the neck or elsewhere. An ectopic thyroid is mostly asymptomatic and may be incidentally discovered by thyroid scintigraphy (Clerc et al., 2008) but may give rise to local symptoms if it is enlarged (Toso, Colombani, Averono, Aluffi, & Pia, 2009) or present as a conspicuous tumor mass of unknown identity subjected to surgery (Hagiuda et al., 2006).

4. THYROID SPECIFICATION AND DETERMINATION: AGENESIS VERSUS ATHYREOSIS

To date, it is not known whether true agenesis of the thyroid exists in human; the reason is it is impossible to distinguish this from athyreosis related to fetal regression of a primordial gland that initially developed properly. Studies on mouse models indicate that this condition probably exists, but as yet, there is no genetic deletion that recapitulates thyroid agenesis without simultaneous gross abnormalities in endoderm and foregut development that are established before the onset of organogenesis.

With markers available today, progenitor cells committed and determined to a thyroid fate cannot be detected before the thyroid placode is established in the pharyngeal floor, which takes place between E8.5 and 9.5 in the mouse embryo. The cells gathered there coexpress Nkx2-1 (Lazzaro, Price, de Felice, & Di Lauro, 1991) and Pax8 (Plachov et al., 1990), which clearly distinguish them from cells in the adjoining endoderm (Fig. 4.3A). Recent seminal work in embryonic stem cells (ESC) indicates that Nkx2-1 and Pax8 are sufficient to determine the follicular thyroid lineage (Antonica et al., 2012). However, additionally, two other transcriptional factors, Foxe1 and Hhex, the expression of which is shared by most if not all cells in the anterior foregut endoderm including thyroid progenitors (Bogue, Ganea, Sturm, Ianucci, & Jacobs, 2000; Zannini et al., 1997), are critical for normal thyroid development. Accordingly, single knockout of these factors in mice produces a severe phenotype, leading to athyreosis (De Felice et al., 1998; Kimura et al., 1996; Mansouri, Chowdhury, & Gruss, 1998; Martinez Barbera et al., 2000). It is noteworthy, however, that in all mutants, the thyroid placode is formed seemingly normally, indicating the emergence and survival of the thyroid bud but not the antecedent specification of thyroid progenitors requires participation of these transcription factors. The clinical relevance of these important discoveries has been confirmed in many studies indicating that a heterozygous mutation of *Nkx2-1*, *Pax8*, or *Foxe1* is sufficient to induce thyroid dysgenesis and overt CH in a subset of patients (for a recent update of the patient cohorts, see Montanelli and Tonacchera (2010).

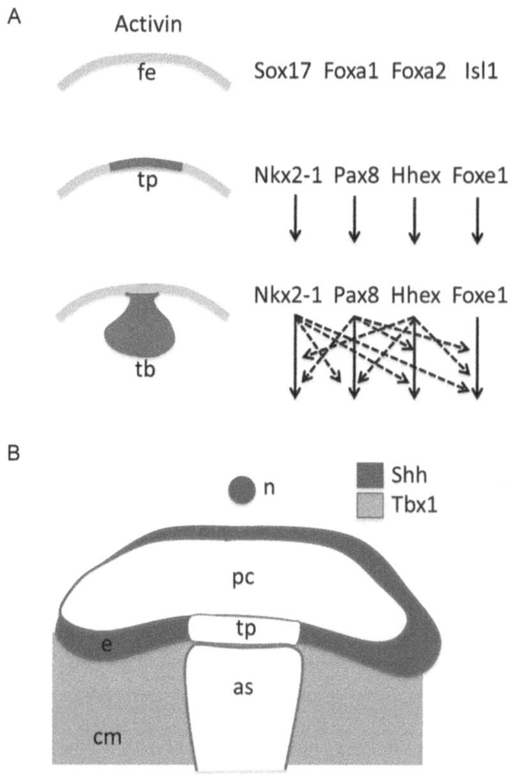

Figure 4.3 Early thyroid development. (A) Transcriptional control of definitive endoderm formation as induced by Activin and regulatory role of Nkx2-1, Pax8, Hhex, and Foxe1 in thyroid progenitor cells. fe, foregut endoderm; tp, thyroid placode; tb, thyroid bud. (B) Expression pattern of sonic hedgehog (Shh) and Tbx1 implicated as non-cell-autonomous factors in early thyroid development. tp, thyroid placode; e, endoderm; pc, pharyngeal cavity; cm, cardiac mesoderm; as, aortic sac; n, notochord. (See color plate.)

The identities of the factor or factors that trigger the onset of Nkx2-1 and Pax8 expression specifically in authentic mouse thyroid progenitor cells *in vivo* are not known, and investigations on ESC have so far not come up with a decisive answer. Most previous studies employing mouse ESC show the requirements of TSH to obtain differentiated thyrocytes (Davies, Latif, Minsky, & Ma, 2011), but TSH is not required for thyroid development *in vivo* (Postiglione et al., 2002). TSH-independent induction of thyroid-specific markers as TSH receptor (TSHR) and sodium–iodide symporter (NIS) is recognized in embryonic bodies treated with activin A (Ma, Latif, & Davies, 2009), but this is likely the result of one of several preprogrammed endoderm lineage traits that are turned on by default during prolonged culture of ESC after being triggered to differentiate into definitive endoderm by activation of the nodal/activin pathway (Sui, Bouwens, & Mfopou, 2013). In a recent report, Nkx2-1+ lung/thyroid progenitors were enriched from activin-exposed endoderm by sequential inhibition of TGF-beta and BMP signaling and stimulation with BMP4 and FGF2 (Longmire et al., 2012), suggesting that lung and thyroid specifications share inductive mechanisms. Interestingly, epigenetic silencing of Oct4 in Nkx2-1+ cells and altered histone modifications of the *Nkx2-1* promoter in Nkx2-1-negative cells accompanied this response. The Nkx2-1+ endoderm progenitor population could also be distinguished from neurogenic progenitors assumed to give rise to Nkx2-1-expressing neurons of the forebrain, indicating cell fate decision was restricted to the lungs and thyroid by this treatment (Longmire et al., 2012). The importance of epigenetic prepatterning of multipotent endoderm to restrict cell fate was recently documented for liver and pancreatic specifications (Xu et al., 2011). ACTIVIN is sufficient to induce a respiratory epithelial fate in human ESC (Li, Eggermont, Vanslembrouck, & Verfaillie, 2013), but whether this in addition comprised the thyroid lineage was not investigated.

An interesting question is whether undifferentiated progenitors destined to lung and thyroid lineage development, respectively, derive from distinct domains of the anterior endoderm or have a common stem cell origin with bipotential features. The Nkx2-1+ thyroid and lung buds are evidently separated from the start of their emergence from the pharyngeal floor, but as will be further discussed, at least, thyroid progenitors may be recruited from outside the proper anlage, that is, the prospective thyroid domain of the foregut endoderm may be larger than is evident from the expression of Nkx2-1 and Pax8, making this possibility not unrealistic for anatomical reasons. However, a necessary factor for the induction of lung development is Wnt2/Wnt2b-mediated activation of canonical beta-catenin signaling, which is required for Nkx2-1 to be expressed in nascent lung progenitor cells, whereas the thyroid is correctly specified in *Wnt2/Wnt2b*-null mutants (Goss et al., 2009, *Developmental Cell*). It is likely therefore that lung and thyroid progenitors although sharing some features of importance for their propagation represent distinct entities with different endoderm origins. Recent transcriptome profiling of the thyroid and lung buds further corroborates that thyroid and lung progenitors, although sharing some gene expression, diverge early in development (Fagman et al., 2011).

Zebrafish disruption of nodal signaling inhibits formation of the thyroid anlage, but this is accompanied by gross defects in endoderm and gut tube formation (Elsalini, von Gartzen, Cramer, & Rohr, 2003), indicating that development was arrested at an earlier stage foregoing thyroid specification. A downstream mediator of the nodal pathway regulating endoderm formation is Sox17 (Zorn & Wells, 2009), a member of the SRY-related HMG box transcription factor family that is first expressed in the nascent endoderm lineage of the inner cell mass (Morris et al., 2010). Sox17 in turn transactivates the forkhead transcription factors Foxa1 and Foxa2 (formerly known as HNF3α and HNF3β, respectively), both of which are ubiquitously expressed in the foregut endoderm and required for its formation (Ang et al., 1993; Monaghan, Kaestner, Grau, & Schutz, 1993). The sequential activation of nodal/Activin–Sox17–Foxa1/2 is thus essential to establish the definitive endoderm and confer competence for subsequent lineage-specific development of endoderm-derived organs (Fig. 4.3A). Recent findings in mouse ESC suggest that the switch from pluripotency to endodermal specification in mouse preimplantation development is driven by Sox17 replacing Sox2 for a partnered regulation with Oct4 of a specific *Sox/Oct* enhancer motif that triggers the endoderm differentiation program (Aksoy et al., 2013). Thus, early defects explain the selective depletion of definitive endoderm recognized in *Sox17*-deficient mice (Kanai-Azuma et al., 2002). The phenotype of this mutant is most severe in mid- and hindgut portions, presumably related to the fact that Sox17 is downregulated during early formation of the foregut endoderm. This may explain why the thyroid bud is properly formed in *Sox17*-null embryos, a shared feature with the liver primordium, although the Hhex domain of the foregut endoderm is much reduced (Kanai-Azuma et al., 2002). Sox17 is thus dispensable for the induction of a thyroid fate. Notably, Sox17 may be reexpressed in later developmental stages and has been shown to participate in morphogenesis and differentiation of endoderm-derived organs as lung (Park, Wells, Zorn, Wert, & Whitsett, 2006) and bile ducts (Uemura et al., 2010, 2013). Whether Sox17 plays a role in thyroid development after the anlage is established in the pharyngeal floor has not been investigated.

Normal liver bud development in the absence of Sox17 most likely depends on the preserved expression of Foxa1 and Foxa2 (Kanai-Azuma et al., 2002), the concerted action of which is critical to establish competence for liver specification in multipotent endoderm progenitor cells (Friedman & Kaestner, 2006; Lee, Friedman, Fulmer, & Kaestner, 2005). Based on the expression pattern of Foxa2 in endoderm-derived organs including the embryonic thyroid (Monaghan et al., 1993), a potential role of Foxa2 in thyroid development has been envisaged (De Felice & Di Lauro, 2004), although there are no experimental studies that confirm this possibility. A first indication that Foxa2 may not be essential for thyroid development was recently provided in a microarray analysis of the gene expression profile in the mouse thyroid bud (Fagman et al., 2011). This showed that thyroid progenitor cells at difference with lung bud cells do not express *Foxa2*. Immunostaining further indicated that Foxa2 is missing in the migrating thyroid (Fagman et al., 2011). This is noteworthy since it highlights an early difference in the transcriptional machinery between thyroid and lung primordial tissues. Previous studies have reported that Foxa2 and Nkx2-1 are coexpressed in the developing lung (Stahlman, Gray, & Whitsett, 1998) and that Foxa2 conjointly with Nkx2-1 regulate lung-specific gene expression (Bohinski, Di Lauro, & Whitsett, 1994). Foxa1 and Foxa2 have also been shown to bind to and regulate the *Nkx2-1* promoter *in vitro* (Minoo et al., 2007). Although Nkx2-1 is expressed in the lungs in *Foxa1/Foxa2* conditional double knockouts (Wan et al., 2005), these observations indicate that the pivotal role Nkx2-1 plays in early lung development (Kimura et al., 1996; Minoo, Su, Drum, Bringas, & Kimura, 1999) is coordinated with and possibly regulated by a Foxa1/2-dependent developmental program. The absence of Foxa2 in the embryonic thyroid suggests that thyroid development differs from both lung and liver morphogenesis on this aspect. Of course, it cannot be excluded that the commitment of the endoderm to a thyroid fate at an earlier stage requires the cooperation of Foxa2 (and/or Foxa1). Due to the early lethality of *Foxa2*-null mutant embryos, this question is presently impossible to elucidate since it will need deletion of *Foxa2* in the prospective thyroid domain of the endoderm for which molecular markers suitable for conditional targeting are currently not available.

In the prospective dorsal pancreatic bud, a natural repression of sonic hedgehog (Shh) induced by notochord-derived factors is required for specification (Hebrok, 2003). Conversely, ectopic expression of Shh in this location of the foregut endoderm promotes a fate switch from the pancreas to the intestinal lineage. Interestingly, the expression of Shh and Nkx2-1/Pax8 is mutually exclusive in the pharyngeal endoderm encompassing the thyroid placode (Fagman et al., 2004; Parlato et al., 2004; Fig. 4.3B), suggesting that silencing of Shh expression might be permissive for the induction of thyroid development as well. However, this possibility is ruled out by observations of a normal-sized thyroid placode in *Shh*-null mice (Fagman et al., 2004).

The mechanisms by which Shh is repressed in pancreatic and thyroid progenitors are probably also different since the thyroid primordium unlike the dorsal pancreas is so distant from the notochord, located on opposite sides of the anterior gut tube (Fig. 4.3B), analogous to the ventral pancreatic anlage that develops independently of Shh. As will be discussed later, Shh acting non-cell-autonomously has a distinctive morphogenetic role in subsequent stages of thyroid organogenesis for which lack of expression in the progenitor cells themselves probably is important.

Recent studies on Xenopus show that exogenous retinoic acid (RA) triggers a lung differentiation program in the anterior endoderm corresponding to the location for the presumptive thyroid anlage and that the expression of Pax2, the Pax8 ortholog in lower vertebrates, is lost by this treatment (Wang et al., 2011). In addition, RA was previously found to abolish the Hhex+ thyroid domain in zebrafish embryos (Stafford & Prince, 2002). Although RA signaling does not seem to play a critical role in mouse thyroid development (Desai, Malpel, Flentke, Smith, & Cardoso, 2004), these observations in lower vertebrates highlight an established regulatory role of endogenous RA in prepatterning of the foregut endoderm into distinct domains that provide positional cues for organogenesis along the alimentary tract (Duester, 2008). The mechanism is apparently adopted for the embryonic thyroid at least in some species.

Findings in various animal models thus collectively support a scenario in which a pool of multipotent endoderm progenitors at a certain time and place are committed to develop along distinct cell lineages depending on the combined permissive actions of more broadly expressed endoderm regulators. Once competence is established, fate determination and further development are promoted by inductive signals secreted from the surrounding tissues, joining a reciprocal interplay between the endoderm and mesoderm that drives organogenesis along the foregut (for a comprehensive overview of this bursting field, see Zorn and Wells (2009)). Still, surprisingly little is known of which factors are necessary to determine the thyroid lineage in the foregut endoderm *in vivo*, in particular in the mouse model. Identification of precardiac mesoderm as a major inductive source for thyroid specification in zebrafish (Wendl et al., 2007) strongly suggests that mammalian thyroid development follows a similar pattern. Another recently discovered similarity between early thyroid development in fish and mouse concerns involvement of the Notch pathway, the disruption of which appears to limit the number of progenitor cells forming the anlage (Carre et al., 2011; Porazzi et al., 2012).

5. EMBRYONIC THYROID GROWTH: HYPOPLASIA AND HEMIAGENESIS

Formation of the thyroid bud implicates that the number of Nkx2-1/Pax8-positive progenitor cells originally present in the placode is gradually increased. The first sign of this is the thickening and pseudostratification of the placode epithelium due to crowding of cells that makes it easily discernable from the adjacent strictly monolayered endoderm (Fagman et al., 2006). In other budding organs as the liver, pancreas, and lungs, the stratified epithelium of the bud is formed by asymmetrical cell division that requires Hhex (Bort, Signore, Tremblay, Martinez Barbera, & Zaret, 2006). Thus, in the absence of Hhex, organogenesis cannot proceed to the budding stage suggesting a cell-autonomous mechanism by which cell fate determination is further established (Bort et al., 2006). However, the thyroid bud forms normally in the absence of Hhex (Parlato et al., 2004), indicating that the developing thyroid does not share this mechanism. In this respect, it is noteworthy that during both placode and bud formation, the thyroid progenitors are distinguished from the endoderm proper and neighboring mesoderm-derived tissues by a very low proliferation rate or even lack of DNA synthesis as monitored by BrdU labeling (Fagman et al., 2006). Thus, at this developmental stage, mitogenic signals stimulating embryonic thyroid growth probably act on another progenitor cell population rather than on those already gathered in the bud.

The hypothesis that the thyroid bud grows by annexation of cells from Nkx2-1/Pax8-negative anterior endoderm is consistent with previous observations in chick embryos (Kinebrew & Hilfer, 2001). The fact that the thyroid placode is normal-sized in null mutants of *Nkx2-1*, *Pax8*, *Hhex*, or *Foxe1* (Parlato et al., 2004) further suggests that intrinsic factors are probably not required for early propagation of the thyroid follicular cell lineage. Interestingly, fate induction and renewal of hepatic progenitor cells by FGF involve different intracellular signaling pathways (Calmont et al., 2006), suggesting these processes are distinct entities that do not necessarily colocalize within the endoderm epithelium. It is also known that the anterior endoderm consists of several distinctive prehepatic domains that contributes to the formation of a single hepatic bud (Tremblay & Zaret, 2005). Growth by annexation of progenitors from outside the actual site of placode formation and budding may thus be a common theme in foregut organogenesis.

When the thyroid primordium progresses through the budding stage, Nkx2-1, Pax8, and Hhex acquire interdependence reflected by loss of expression not seen before in the respective knockouts (Parlato et al., 2004; Fig. 4.3A). Experiments on cell cultures have revealed a hierarchical network of interactions by which this quartet of transcriptional factors cell-autonomously regulate thyroid gene expression (D'Andrea et al., 2006; Nitsch et al., 2010; Puppin et al., 2003, 2004). In addition, Nkx2-1 and Pax8 exert autoregulatory activities that might be of importance to sustain both expression and function (D'Andrea et al., 2006; di Gennaro, Spadaro, Baratta, De Felice, & Di Lauro, 2013). The cardinal role of this regulatory network in early thyroid development has been confirmed in studies on the Nkx2-1, Pax8, and Hhex orthologs in zebrafish (Elsalini et al., 2003; Wendl et al., 2007).

Fewer Nkx2-1 + cells are present in the thyroid primordium in mice deficient in *Isl1* (Westerlund et al., 2008). Isl1 is ubiquitously expressed in the pharyngeal endoderm and required for endoderm cell survival (Cai et al., 2003). However, since adjacent cardiac mesoderm also expresses Isl1 (Cai et al., 2003; Westerlund et al., 2008), it is not possible without targeted deletion experiments to determine from which source, endoderm or mesoderm, Isl1 stimulates the generation of thyroid progenitors (Fig. 4.3A). Nonetheless, pharyngeal mesoderm is a likely source of growth-promoting signals in early thyroid development as recently shown in ablation studies on Tbx1, a key player in global development of the pharyngeal apparatus (Scambler, 2010) and the major candidate gene of 22q11.2 deletion syndromes comprising DiGeorge syndrome in which thyroid dysgenesis and CH are part of a complex phenotype (Stagi et al., 2010). Specifically, targeted deletion of mesodermal *Tbx1* reduced the size of the thyroid primordium, and overexpression of Fgf8 in *Tbx1*-deficient mesoderm was able to rescue the thyroid defect (Lania et al., 2009). This finding is of considerable interest since the actual progenitor cell number forming the anlage may determine final organ size, as originally shown for mouse pancreatic development (Stanger, Tanaka, & Melton, 2007) and more recently suggested for thymus and parathyroid glands emerging from the third pharyngeal pouch (Griffith et al., 2009), thus providing a plausible pathogenetic mechanism that may contribute to the development of thyroid hypoplasia in *Tbx1*-null mice (Fagman et al., 2007). The fact that Nkx2-1-positive cells were reduced in number but not lacking in the targeted *Tbx1* knockout (Lania et al., 2009) indicates that a Tbx1–FGF8 pathway regulates the expansion of the progenitor cell pool rather than constituting an initial inductive signal for thyroid development (Fig. 4.3B).

Although the number of progenitor cells primarily forming the thyroid anlage may be a limiting factor for the final size of the gland, there are now a number of reports indicating that trophic factors are required for propagation of the follicular cell lineage during subsequent stages of morphogenesis. These act both cell-autonomously, most evidently represented by the set of transcription factors expressed from the onset of thyroid lineage development, and non-cell-autonomously comprising yet incompletely characterized morphogens. We will shortly discuss this in relation to the different stages of organogenesis. As already indicated, in the mouse thyroid primordium, cycling cells are not observed before it is already pinched off from the pharyngeal endoderm and descends towards its pretracheal destination (Fagman et al., 2006). Cell proliferation continues thereafter as it stretches bilaterally and beyond, comprising both the fusion with the UBs and the final morphogenetic stage, lobe formation (Fagman et al., 2006). Further enlargement of the gland in the fetal and postnatal periods is the combined effect of addition of new follicles and volume increase of colloid stores. It is evident that disturbed growth regulation at any point during this multifaceted process may result in a reduced thyroid cell mass even if the original number of endoderm progenitors committed to a thyroid fate was normal.

A shared feature of *Nkx2-1*-, *Pax8*-, and *Hhex*-null embryos is the complete regression of the thyroid bud resulting in athyreosis (De Felice et al., 1998; Kimura et al., 1996; Mansouri et al., 1998). It was recently shown that the mouse thyroid bud expresses very high levels of Bcl2 as compared to any neighboring embryonic tissues and that Bcl2 is lost specifically in thyroid progenitor cells in *Pax8*-null mutants (Fagman et al., 2011). Lack of antiapoptotic signals mediated by Bcl2 may thus contribute to the occurrence of apoptotic cells in the Pax8-deficient thyroid. The Bcl2-like gene *bcl2l* is also enriched in the thyroid of zebrafish embryos (Opitz, Maquet, Zoenen, Dadhich, & Costagliola, 2011), suggesting this is a conserved requirement for thyroid progenitor cell survival. In zebrafish, it is Nkx2-1 and Hhex (and Pax2) instead of Pax8 (in mouse) that stimulates Bcl2l expression and ensures survival of the thyroid primordium (Porreca, De Felice, Fagman, Di Lauro, & Sordino, 2012). Whether Nkx2-1 and Hhex regulate Bcl2 or other antiapoptotic factors in mouse thyroid development has not been investigated. However, as the expression of Pax8 in the thyroid bud depends on both (Parlato et al., 2004), it is conceivable that Bcl2 may be affected in Nkx2-1- and Hhex-deficient progenitor cells as well and that this contributes to early regression of the thyroid diverticulum in these mutants. Together, this strongly suggests that embryonic thyroid cell survival in mice requires the concerted action of Nkx2-1, Pax8, and Hhex (Fig. 4.3A). Since these transcription factors in addition are essential for functional differentiation of thyrocytes, it is hypothesized that loss of any one of them may trigger apoptosis as a safeguard mechanism to prevent inappropriate propagation of progenitor cells that lack the ability to differentiate (De Felice & Di Lauro, 2011).

Between E10.5 and 11.5 coinciding with the first signs of proliferation within the primordium (i.e., presence of BrdU-labeled cells), the embryonic thyroid is all surrounded by mesenchyme as it descends towards the site where it one day later will bifurcate (Fagman et al., 2006). In contrast to the important role of mesenchyme-derived signals in pharyngeal arch and pouch development (Graham, 2008; Grevellec & Tucker, 2010), any corresponding effects in the midline are poorly characterized, and the potential influence of mesoderm on thyroid growth at this developmental stage is largely unknown. However, expression of Tbx1 in a population of mesoderm cells close to the migrating thyroid (Fagman et al., 2007) suggests that Tbx1 via FGF8 or other soluble factors might continue stimulating the propagation of thyroid progenitors also after their departure from the endoderm proper (Fig. 4.3B). The hypoplastic thyroid phenotype in *Shh*-null mice is similar to that of *Tbx1* mutant embryos, although Shh is not expressed in the mesoderm or in the thyroid rudiment (Fagman et al., 2004). This suggests that Shh gradients produced by pharyngeal endoderm exert distant effects that directly or indirectly promote the proliferation of thyroid progenitors (Fig. 4.3B).

Between E11.5 and 13.5 starting from a midline position, intense progenitor cell proliferation in left and right directions indicates the onset of the bilobation process (Fig. 4.4). Bilateral growth initially occurs along the course of the third pharyngeal arch artery that connects the cardiac outflow tract with the dorsal aortas (Fagman et al., 2006). It is intriguing that the bifurcating stream of thyroid cells taking this route does not grow randomly around or away from the artery but keep on course to the cranial aspect of the vessel wall without any other apparent structural boundaries. Using the arch artery as guiding track, intrinsic mitogenic signals may be sufficient for further enlargement of the primordium at this stage, the cells being lodged more close to the midline will eventually form the isthmus portion of the gland. At approximately E13, the leading edge cells deviate from the arch artery for a more cranial route passing through a rather loose mesenchyme while approaching the ventromedial aspect of the UB (Fagman et al., 2006; Fig. 4.4). Within the next 24 h, the UB as a whole is engulfed by the

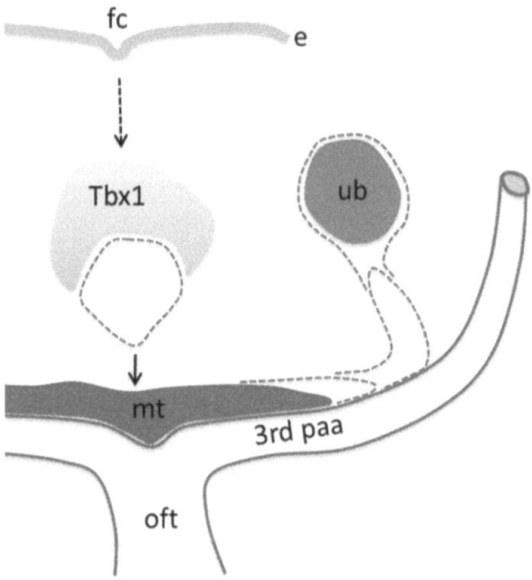

Figure 4.4 Bilobation process. mt, midline thyroid; ub, ultimobranchial body; paa, pharyngeal arch artery; oft, cardiac outflow tract; e, endoderm; fc, foramen cecum. (See color plate.)

expanding median thyroid primordium, the UB surface being covered by proliferating thyroid progenitor cells (Fagman et al., 2006). Based on the anatomical position of the two primordia before fusion, it is difficult to see that the advance of thyroid progenitors towards the UB can take place without specific morphogenetic signals serving as pathfinder. The identity of putative morphogens implicated in embryonic thyroid fusion is unknown.

Thyroid hemiagenesis defined as the absence of one lobe with the other in a normal position is likely caused by a unilateral growth defect manufactured during the bilobation process. Hemiagenesis is a rare clinical entity, and as the amounts of thyroid tissue, similar to after hemithyroidectomy, are mostly sufficient to produce normal TH levels, this condition may be undiagnosed until the thyroid is examined morphologically for other reasons (Chang, Gerscovich, Dublin, & McGahan, 2011; Wu et al., 2012). However, the malformation is of principal interest because it suggests a mechanism by which the midline primordium fails to bifurcate and grow symmetrically. The presence of a fully developed hemi-isthmus adjoining the normal lobe and lack of a corresponding isthmus portion on the affected side in patients (Chang et al., 2011) strongly favor the idea that hemiagenesis is caused by an early asymmetrical growth defect and not by regression of the lobe once formed. Interestingly, the hypoplastic thyroid of *Nkx2-1/Pax8* double-heterozygous-null mice is mostly normal-shaped, but 30% of mutant embryos develop hemiagenesis almost identical to that seen in patients (Amendola et al., 2005). As of all cells in the foregut, only thyroid progenitor cells express both factors (Nkx2-1 but not Pax8 is expressed in the UB and lungs), it is probable that the bilobation defect, unique for this compound mutant, has a cell-autonomous cause. Moreover, as the expression levels of Nkx2-1 and Pax8 were not influenced by one another, this proved for the first time that embryonic thyroid growth depends on gene dosage. Another remarkable finding in this study was the strain specificity indicating that cooperation with other genes is necessary to generate the phenotype (Amendola et al., 2005). A putative candidate gene (*DNAJc17*) encoding a chaperone expressed in the thyroid has been identified (Amendola et al., 2010), but the mechanism by which functional modification of this gene contributes to asymmetrical thyroid growth in $Nkx2\text{-}1^{+/-}/Pax8^{+/-}$ mouse embryos has not been elucidated.

The appearance of a single thyroid lobe in mice has been reported after experimental ablation of a number of genes, for example, *Hoxa3* (Manley & Capecchi, 1995, 1998), *Tbx1* (Fagman et al., 2007; Lania et al., 2009; Liao et al., 2004), and *Shh* (Alt, Elsalini, et al., 2006; Fagman et al., 2004), that are not expressed in the median thyroid primordium, suggesting that hemiagenesis may result from defective extrinsic signals as well. However, a common pattern for these gene deletions is lack of an isthmus, suggesting that the thyroid primordium never enters the bilobation stage but is displaced before this occurs. As will be further commented on later, an alternative mechanism to obtain a hemithyroid phenotype is premature lateralization due to retarded migration of the hypoplastic rudiment. The occurrence of a one-sided thyroid lobe in congenital syndromes with multiple midline anomalies related to defects in laterality formation (Gilbert-Barness, Debich-Spicer, Cohen, & Opitz, 2001) further highlights the importance of extrinsic factors for normal thyroid growth. It is noteworthy that thyroid hemiagenesis in patients is almost exclusively left-sided (Wu et al., 2012) and that the thyroid remnant in *Shh*-deficient mice is mostly located to the right of the midline (Alt, Elsalini, et al., 2006). Although formal experimental proofs are lacking, these observations suggest that thyroid bilobation is, at the very least, influenced by symmetry-breaking signals generated much earlier during embryogenesis (Vandenberg & Levin, 2013).

Shh and FGF8 are required for left–right axis determination in the mouse embryo (Meyers & Martin, 1999). Specifically, Shh prevents left determinants from being expressed on the right and FGF8 determines left-sided asymmetry. How such an asymmetrical information is interpreted by organ primordia and translated to morphology is yet poorly understood (Shiratori & Hamada, 2006). The finding of a predominant left-sidedness of the presumptive carotid arteries to which the single thyroid lobe associates in *Shh*$^{-/-}$ mice (Alt, Elsalini, et al., 2006) suggests that the development of bilateral and essentially symmetrical anatomical structures is paradoxically affected when asymmetrical signals are perturbed or lost. Interestingly, RA signaling promotes bilateral symmetry by a highly conserved mechanism that inhibits the desynchronizing effect of the left–right machinery on somitogenesis (Kawakami, Raya, Raya, Rodriguez-Esteban, & Izpisua Belmonte, 2005; Vermot et al., 2005; Vermot & Pourquie, 2005). Asymmetrical inhibition of FGF8 in the axial mesoderm conveys the buffering role of RA without which symmetry defects of the spine may occur (Vilhais-Neto et al., 2010). Whether a similar mechanism controls the bilateral morphogenesis of internal organs has not been investigated, although it is noteworthy that gut endoderm participates in relaying signals from the node to the mesoderm and the expression there of asymmetrical genes establishes the left–right axis of the body (Saund et al., 2012; Viotti, Niu, Shi, & Hadjantonakis, 2012). The fact that hemithyroid localization to the left or right is entirely stochastic in *Hoxa1*-null mutant mice (Manley & Capecchi, 1995) does not contradict a potential role of asymmetrical signals (or the inhibition thereof) in thyroid development since *Hox* genes provide positional cues to anteroposterior axial identity rather than in left–right patterning.

Reports on congenital causes to general hypoplasia of the orthotopic thyroid are few. As mentioned, mice deficient of one allele of either *Nkx2-1* or *Pax8* exhibit normal thyroid anatomy and function, whereas double-heterozygous mutants display a small gland and manifest CH in the majority of cases (Amendola et al., 2005). This not only confirms that thyroid organogenesis depends on the cooperation of Nkx2-1 and Pax8 also after the budding and bifurcation stages but also provides proof of concept for the possibility that thyroid dysgenesis may be a polygenic disease (Amendola et al., 2005), a notion that gains support from recent observations of ethnic differences in the susceptibility to develop CH with a dysgenic cause (Stoppa-Vaucher, Van Vliet, & Deladoey, 2011). *In vitro* experiments have shown that Nkx2-1 and Pax8 directly interact at the promoter level of target genes (Di Palma et al., 2003). As both transcription factors are required for the development of thyroid follicles from ESC ((Antonica et al., 2012); to be further discussed later), it is tempting to speculate that thyroid hypoplasia in *Nkx2-1*$^{+/-}$/*Pax8*$^{+/-}$ mice depends on defective folliculogenesis. However, the shape of the follicles was reported to be essentially normal although the follicular epithelium was hypertrophic presumably as a result of the high

TSH level reflecting the hypothyroid state. The inability of the mouse thyroid with reduced Nkx2-1 and Pax8 expression to respond to TSH at otherwise goitrogenic concentrations is intriguing and suggests that the follicular cells possess an inborn defect in the growth regulation that cannot be overcome by superstimulation of TSHR (Amendola et al., 2005).

The thyroid gland in *Eya1*-null mice is normal-shaped but small (Xu et al., 2002). The *Eya1* gene encodes a transcriptional coactivator that is ubiquitously expressed in the branchial arches but not in the embryonic thyroid. Accordingly, *Eya1* deficiency primarily leads to severe developmental defects of the thymus and parathyroid glands, but the pharyngeal phenotype also comprises failure of the UB pair to fuse with the midline thyroid primordium (Xu et al., 2002). This indicates first of all that thyroid bilobation does not require interaction with the UBs. However, as suggested by the coincidence of a hypoplastic lobe and an ectopic UB in *Hoxa1* mutants (Manley & Capecchi, 1995), it is likely that the incorporation of UB contributes to the final lobe size of the gland.

Pituitary regulation of the thyroid in mouse embryos commences at the time of functional differentiation, that is, after the entire morphogenetic process is finished. In fact, differentiation of thyroid follicular cells coincides with the differentiation of thyrotropes producing TSH (de Moraes, Vaisman, Conceicao, & Ortiga-Carvalho, 2012). It is therefore not surprising that thyroid organogenesis proceeds normally in mice deficient of *Tsh* or the *Tshr* (Postiglione et al., 2002). However, it is remarkable that further growth of the embryonic thyroid does not require TSH, at difference with the postnatal period and in adult animals in which thyroid enlargement is TSH dependent (Postiglione et al., 2002). In contrast, the fetal human thyroid is highly sensitive to TSH as exemplified by the appearance of congenital goiter when severe iodine deficiency is prevalent (Glinoer, 2007) or in rare cases of fetal hyperthyroidism (Polak et al., 2006). A plausible explanation to the difference is TSH plays a significant role in the intrauterine regulation of thyroid function in human fetuses, but this is likely not required in mice with a much shorter gestation period. Notably, in earlier stages of development, thyroid growth also seems to be autonomous, that is, TSH-independent, in human (Peter, Studer, & Groscurth, 1988).

6. DEFECTIVE THYROID MIGRATION: ECTOPIC GLAND

Lingual thyroid is a rare anomaly that may be discovered in infants during causal investigation of CH, but if the amount of tissue is sufficient to produce normal levels of TH, it is undiagnosed until enlargement for some reason gives local symptoms with complaints of dysphagia, dyspnea, or dysphonia (Chang et al., 2009; Noussios et al., 2011; Toso et al., 2009). Retention of thyroid tissue in this location is caused by a complete migration defect, that is, the descent of the midline primordium does not occur possibly due to failure to disbud properly from the pharyngeal endoderm. Hence, the lack of an orthotopic gland is mandatory when the ectopic thyroid is in a lingual or sublingual position. Etiologically, this condition should be distinguished from thyroglossal cysts caused by incomplete regression of the thyroglossal duct coexisting with a normally located and functioning thyroid. Dual ectopic thyroids present in different locations in the neck also exist, the origins of which have been suggested to represent distinct clones of thyroid progenitor cells that might segregate during morphogenesis due to inborn differences in migrating capacity (Wildi-Runge et al., 2012). However, experimental proof of this possibility is lacking. Divergent migration outside the normal developmental path may explain the rare occurrence of ectopic thyroid tissue, for example, in the wall of the larynx and trachea, inside the submandibular gland, or associated with the pericardium.

Mouse model studies suggest that embryonic thyroid migration is controlled by both intrinsic (cell-autonomous) and extrinsic (non-cell-autonomous) mechanisms. Of the four transcription factors (Hhex, Nkx2-1, Pax8, and Foxe1) that distinguish thyroid progenitor cells at the onset of thyroid morphogenesis, only Foxe1 does not influence the expression of the others, indicating that this factor is located downstream in a regulatory network required for normal thyroid development (De Felice & Di Lauro, 2011; Parlato et al., 2004). In contrast to the phenotype of *Hhex*$^{-/-}$, *Nkx2-1*$^{-/-}$, and *Pax8*$^{-/-}$ embryos, *Foxe1*-deficient thyroid progenitors survive in about 50% of the embryos but fail to migrate after budding leading to a sublingual hypoplastic gland (De Felice et al., 1998). Moreover, knock-in of *Foxe1* under the *Nkx2-1* promoter specifically in the progenitor cells rescues the migration defect (Parlato et al., 2004). Notably, wild-type migrating primordial cells are tightly connected by E-cadherin and stay gathered in a coherent structure (Fagman, Grande, Edsbagge, Semb, & Nilsson, 2003), indicating that they do not undergo epithelial-to-mesenchymal transition (EMT) but rather display the typical features of collective migration. In fact, recent observations on cultured epithelial cells indicate that E-cadherin-mediated cell–cell adhesion coordinates the interactions between front cells at the leading edge and the followers in directional collective migration (Rorth, 2012). It was recently shown that Foxe1 regulates Msx1 and TGF-beta3 in craniofacial development (Venza et al., 2011), but as both these factors are predominantly active in EMT-driven morphogenesis, it is likely that other yet unknown target genes mediate Foxe1-dependent migration of the embryonic thyroid. Findings of a delayed expression of FOXE1 as compared to NKX2-1 and PAX8 in the thyroid of human embryos (Trueba et al., 2005) are consistent with a role of Foxe1 in thyroid migration. On the other hand, Foxe1 is expressed at seemingly normal level in three human lingual thyroids subjected to transcriptome analysis (Abu-Khudir et al., 2010), indicating that pathogenetic mechanisms other than *Foxe1* mutation may be more prevalent in this condition in human.

Collective migration in the embryo cannot take place without interactions with the stromal environment, which provides the substrate for migration and external guidance cues for directed movements. Interestingly, Foxe1 is strongly expressed in thyroid progenitor cells also during bilateral growth before the midline and lateral thyroid primordial fuse (De Felice et al., 1998). This opens up the possibility that Foxe1 may participate also in thyroid bilobation, guided by the arch arteries, which bears strong resemblance to collective migration as well. A direct role of vascular cells determining the distribution of follicles in thyroid morphogenesis has been shown for zebrafish (Alt, Elsalini, et al., 2006; Opitz et al., 2012). In mice, the thyroid bud initially attaches to the aortic sac and follows tightly in the rear of the vessel as it descends into the thoracic cavity (Fagman et al., 2006; Fig. 4.4). With this in mind, it is feasible that altered patterning of the surrounding mesoderm or divergent routes of nearby located embryonic vessels may perturb downward migration of the embryonic thyroid. As already mentioned, the best example of this appears in *Shh*-deficient mice in which the thyroid is severely hypoplastic and lateralized, associated with large vessels that take an abnormal course (Fagman et al., 2004). Three-dimensional reconstruction of *Shh*-null embryos and rescue experiments in corresponding zebrafish mutants have shown that the thyroid primordium after evaginating normally fails to bifurcate and is mislocated due to the asymmetrical development of the carotid arteries (Alt, Elsalini, et al., 2006). Lack of Shh and its receptor patched both in the thyroid and the vessel wall strongly suggests that this phenotype primarily depends on defective Shh signaling from the pharyngeal endoderm or loss of Shh response in other embryonic tissues that indirectly influence thyroid development (Moore-Scott & Manley, 2005).

The severe phenotype of *Shh*-null mouse embryos with general growth retardation and multiple developmental malformations in the pharyngeal region at large makes it difficult to ascertain that thyroid displacement depends on a genuine migration defect in this mutant. Notably, Shh is indispensible for a normal development of the cardiac outflow tract and the pharyngeal arch arteries (Washington Smoak et al., 2005), suggesting an indirect effect by interaction with aberrant vessels. However, mice deficient of *Tbx1*, which is a downstream target of the Shh-patched signaling pathway (Garg et al., 2001), show a unilateral thyroid that associates with the ipsilateral carotid although both carotids exist in the expected positions (Fagman et al., 2007), thus favoring another mechanism. Tracking the thyroid during migration in *Tbx1*$^{-/-}$ embryos revealed that budding was delayed as reflected by maintained connection with the pharyngeal endoderm through a persistent thyroglossal stalk. And once detached, the thyroid was unable to reestablish contact with the aortic sac, which is the normal scenario (Fagman et al., 2006). As Tbx1-positive mesenchyme accumulates in the rear of the descending thyroid in wild-type embryos (Fagman et al., 2007; Fig. 4.4), a plausible hypothesis is that Tbx1 facilitates both regression of the thyroglossal duct and further downward migration of the unleashed primordium. Together, these observations suggest that failed migration of the midline thyroid may phenocopy hemiagenesis by a mechanism before the bilobation stage.

A crucial role of NC-derived mesenchyme for the migration of the thymic rudiment after its separation from the pouch endoderm was recently shown in mice with NC-specific deletion of *ephrinB2* (Foster et al., 2010). Interestingly, this resulted in a cervical thymus without other major anatomical abnormalities in the pharyngeal region. In fact, the migration of *ephrinB2*-deficient NCC to and into the thymus lobes was normal, suggesting that the migration defect was governed by impaired Eph/ephrin interactions between NCC and thymic progenitors (Foster et al., 2010). This study provides the first example of a non-cell-autonomous NC-dependent mechanism that regulates collective migration of organ primordia from the pharyngeal endoderm. A problem with available mouse models with diminished NCC migration is otherwise that the entire pharyngeal apparatus is profoundly affected making it difficult to distinguish direct from indirect effects. For example, in both *Shh*- and *Tbx1*-deficient mice, the diminished NCC population of the pharyngeal arches is accompanied by gross changes in pharyngeal pouch anatomy and malformations of the cardiac outflow tract and the pharyngeal arch arteries (Vitelli, Morishima, Taddei, Lindsay, & Baldini, 2002; Washington Smoak et al., 2005), which, as discussed before, may influence thyroid organogenesis by several means.

Nevertheless, it was originally reported that thyroid dysgenesis comprising a spectrum of malformations (athyreosis, hemiagenesis, or hypoplasia of an orthotopic gland) accompanied thymic aplasia in a chicken model with severely impaired NC development (Bockman & Kirby, 1984). This indicates that NC one way or the other impacts on thyroid development.

As suggested from 3D computer imaging of serially sectioned rat and human embryos (Gasser, 2006), an alternative mechanism to active migration of progenitor cells is that the thyroid primordium is displaced by differential growth of pharyngofacial and cervicothoracic structures accompanying the growth and shape changes of the entire embryo. That migration is a shared feature of all budding organs from the pharynx may be taken as a good argument for the hypothesis. However, although the thyroid, parathyroid, and thymus emerge more or less simultaneously (in mice at E9.5–10.5) and move through the same cervical compartment, the distance of travel from site of origin to the final position varies considerably (Fig. 4.5), indicating that elongation of the neck synchronous with retraction of the heart and central large vessels into the thoracic cavity is likely not the only driving force. In fact, a common migration pathway is shared only by the thymic rudiments and the adjoining pair of parathyroid glands derived from the same pharyngeal pouch, explaining the frequent occurrence of ectopic parathyroid tissue in or close to the mediastinal thymus (Liu et al., 2010). Moreover, although coexistence of ectopic glands has been reported (Ohbuchi et al., 2012; Westbrook, Harsha, & Strenge, 2013), aberrant anatomical locations of the thyroid, parathyroid, and thymus in human are mostly independent entities (Nasseri & Eftekhari, 2010; Noussios et al., 2011; Phitayakorn & McHenry, 2006; Wang, 1976). The extreme anatomy of monotremes, which show both mammalian and reptilian features with a thoracic position of the thyroid gland and a maintained cervical location of the ultimobranchial glands (Haynes, 1999), is another example favoring the idea that the developmental traits of pharyngeal glands follow separate routes of primordial tissue movement.

7. FOLLICULOGENESIS AND THYROID DIFFERENTIATION

As the follicle constitutes the functional unit of the vertebrate thyroid, it is purposive that folliculogenesis and expression of thyroid-specific genes implicated in TH synthesis are coordinated processes. In mice, this starts at E14.5 when thyroid morphogenesis largely is finished (Fagman et al., 2006); thyroid growth thereafter is mostly a matter of increasing the number of

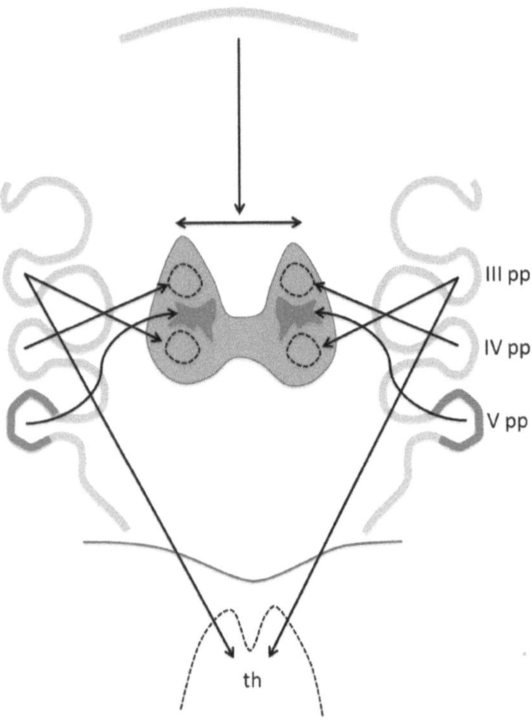

Figure 4.5 Migration pathways of embryonic thyroid, ultimobranchial body, and pharyngeal pouch gland primordia. Parathyroid glands indicated by dashed circles. th, thymus, retrosternal. (See color plate.)

follicles by which the gland obtains its final size. In the presumptive lobes, parenchymal cells that do not yet exhibit a distinctive apical–basal polarity, reflected by the lack of tight junctions, first form columnar cords projecting from center to periphery, septated by connective tissue containing a rudimentary network of microvessels (Fig. 4.6). Subsequently, as the parenchymal cords continue to elongate by cell proliferation predominantly at the tips, they are gradually segmented and converted to diminutive follicles that become invested by capillaries (Fagman et al., 2006). The thyroid lobes thus probably enlarge by addition of new follicles preferentially in the peripheral zone. A conspicuous feature of the young adult mouse thyroid is that the largest follicles are peripherally located subjacent to the enclosing gland capsule, whereas more crowded small-sized follicles mostly occupy the central and medial aspects of the lobes corresponding to the position where the embryonic thyroid once fused with the UB (Fig. 4.6).

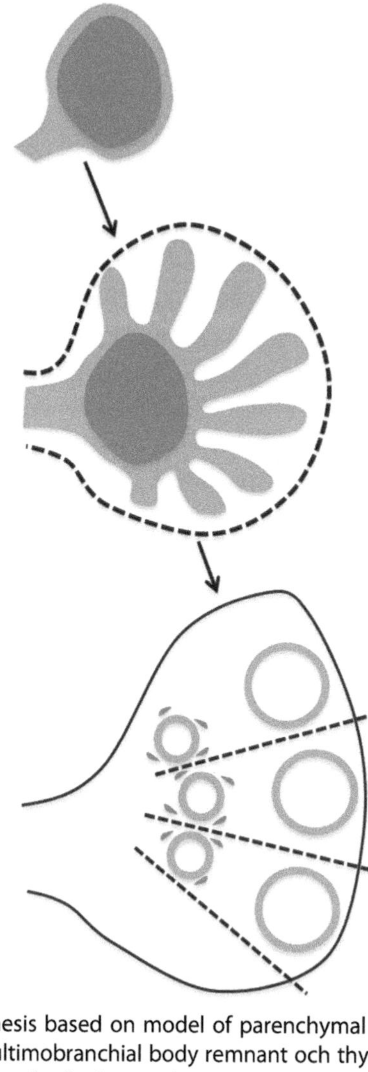

Figure 4.6 Folliculogenesis based on model of parenchymal growth from embryonic thyroid in mouse. The ultimobranchial body remnant och thyroid C cells are indicated with green color. See text for further explanation. (See color plate.)

A plausible explanation to the size difference may simply be that peripheral follicles have more space available for enlargement than those buried deep inside as the gland grows to its mature size. According to this hypothesis, follicular heterogeneity might be inherited from the morphogenetic process rather than reflecting an original multiclonality of thyroid progenitor cells, imprinting the properties of individual follicles. On the other hand, the embryonic patch pattern uncovered in EGFP/BALB/C chimeric mice revealed that individual follicles are variably of monoclonal or polyclonal origin (Ma et al., 2010), suggesting that folliculogenesis and follicular features are influenced by parental genetic differences inborn to the progeny of single ESC. Studer and coworkers highlighted many years ago the basis of follicular heterogeneity in the adult human thyroid and its impact on the pathogenesis of multinodular goiter (Peter, Gerber, Studer, & Smeds, 1985; Peter, Studer, Forster, & Gerber, 1982; Studer, Peter, & Gerber, 1989). One of the most striking observations in these elegant morphological studies was that growth-prone and iodinating follicular epithelial cells represented distinct clones within the same follicle, indicating heterogeneity exists at the single cell level. Whether this difference is genetic, acquired, or a combination thereof is yet an unsolved issue. However, the differential pattern bears strong resemblance with the embryonic patch distribution recognized in X-inactivation studies on the human thyroid (Jovanovic, Delahunt, McIver, Eberhardt, & Grebe, 2003; Novelli et al., 2003). From this, it seems feasible to conclude that differential follicle formation is probably originally based on genetic differences and established during embryonic lobe morphogenesis. As the lobe enlarges, this difference may in turn create micro-environmental niches within the lobe architecture that further augment follicular heterogeneity. To test and validate this hypothesis, *in vivo* imaging of thyroid progenitors with either maternal or paternal features during organogenesis and functional differentiation of the gland will be necessary.

Lumen formation in the embryonic thyroid has been characterized by electron microscopy in different species (recently reviewed in Ref. Rupik, 2011). However, so far, this does not take into account the molecular machinery involved in creating a lumen during epithelial morphogenesis that is now being deciphered (Datta, Bryant, & Mostov, 2011). First elaborated in cell culture, biogenesis of the apical membrane starts with the intracellular formation of intermediate organelles collectively known as the vacuolar apical compartment (VAC) that coalesce to form the lumen (Martin-Belmonte & Mostov, 2008). This process is intimately linked to the establishment of apical–basolateral polarity and the formation of junctional complexes that secludes the lumen. VACs may correspond to intracellular lumens recognized in inactive (thyroxin-treated) adult thyroid follicular cells (Ericson, 1979), but whether a similar mechanism is employed in *de novo* folliculogenesis in the embryonic thyroid has not been investigated. Based on the distribution of ZO-1, a tight junction protein, folliculogenesis in the mouse embryonic thyroid seems to follow a hollowing process, in which the lumen is formed by membrane separation, rather than cavitation generated by apoptosis in the middle of the structure (Fagman et al., 2006).

It has been known for a long time that isolated adult normal thyroid cells are prone to form follicles when cultured in suspension or a gel of extracellular matrix without requirement of TSH stimulation (Chambard, Gabrion, & Mauchamp, 1981; Garbi, Tacchetti, & Wollman, 1986; Nitsch & Wollman, 1980; Toda & Sugihara, 1990; Westermark, Nilsson, Ebendal, & Westermark, 1991). Mouse embryonic thyroid cells appear to have similar features. Hence, a follicular orthotopic gland develops to a normal size in both *Tsh* and *Tshr* knockout mice (Postiglione et al., 2002). This study also showed that cells lacking a functional TSHR express normal levels of TG indicating that thyroid differentiation essentially is TSH-independent. It is noteworthy that knock-in of constitutively active *Tshr* in the *Nkx2-1* locus did not induce precocious folliculogenesis (Postiglione et al., 2002). The inability of thyroid progenitor cells to differentiate before the timely onset of folliculogenesis is of principal interest considering the fact that all known factors implicated in this process and the accompanying functional differentiation of the gland are expressed much earlier in development. It is conceivable to assume that the differentiation program is repressed as long as thyroid progenitors are busy multiplying and migrating, although this hypothesis has not been directly tested experimentally. Speculatively, embryonic growth factors stimulating thyroid cell proliferation before folliculogenesis might exert a dominant negative effect on the cyclic AMP-activated protein kinase A pathway, which constitutes the prime differentiation signal in the thyroid, analogous to the action of peptide growth factors in adult thyroid cells (Roger, van Staveren, Coulonval, Dumont, & Maenhaut, 2010).

A major breakthrough came last year when Costagliola and colleges for the first time were able to generate functional thyroid follicles from mouse ESC by transient overexpression of Nkx2-1 and Pax8 (Antonica et al., 2012). To accomplish folliculogenesis, iodide uptake, and iodination, it was necessary to add TSH. Nonetheless, transplanted to athyroid mice, these *in vitro*-derived follicles successfully restored TH levels and cured the hypothyroid state (Antonica et al., 2012), suggestive of a novel treatment modality in thyroid disease if the technical achievement can be safely transferred to human ESC. Another group recently confirmed the principal findings in this study (R. Ma, Latif, & Davies, 2013).

Maintenance of thyroid follicular structure is also specifically regulated. Mice expressing a mutant *Nkx2-1* allele that generates a hypomorphic thyroid phenotype exhibit major changes in follicle size and shape or even follicular degeneration-associated hypothyroidism (Kusakabe, Kawaguchi, et al., 2006). Similarly, thyroid-specific ablation of *Dicer* required for the biogenesis of microRNAs has no effect on thyroid gland morphogenesis, but at later stages, the follicular architecture is lost leading to overt CH (Frezzetti et al., 2011; Rodriguez et al., 2012). Although the precise mechanisms are not yet revealed, this suggests that a balanced transcriptional activity is required to promote long-lasting thyroid cell survival and functionality of the gland. In this respect, it is interesting to note that the thyroid withstands loss of E-cadherin, the major epithelial adhesion molecule, as demonstrated by a largely preserved histology of the gland after conditional inactivation of this gene from the onset of TG expression (Cali et al., 2007). This finding somewhat contradicts previous observations that TSH both stimulates the expression of E-cadherin (Brabant et al., 1995) and promotes thyroid epithelial integrity directly by stabilizing E-cadherin-mediated adhesion (Larsson, Fagman, & Nilsson, 2004). However, it was recently shown that the developing and adult thyroid in addition expresses cadherin-16 (Ksp-cadherin) under the control of Pax8 (Cali et al., 2012; de Cristofaro et al., 2012), suggesting that partly redundant adhesion mechanisms support maintenance of the epithelial phenotype in normal thyroid cells. Notably, thyroid expression of cadherin-16 is reduced in mutant mice with a dephosphorylated variant of Nkx2-1, leading to a perturbed follicular architecture (Silberschmidt et al., 2011). As Nkx2-1 phosphorylation is not required for thyroid development, this provides the first evidence of a post-translational mechanism by which Nkx2-1 differentially regulate morphogenesis and thyroid function. Microarray analysis showed that three target genes of yet unknown function in thyroid development are specifically upregulated by phosphorylated Nkx2-1 (Silberschmidt et al., 2011). The serine kinase that regulates Nkx2-1 transcriptional activity has not been identified. Other adhesion molecules as R-cadherin that is expressed in mouse thyroid along with the onset of differentiation (Fagman et al., 2003) may also assist in keeping the follicular integrity.

Eph receptors and their cognate ephrin ligands, which also are membrane-bound at the cell surface, exert profound and diverse effects in embryonic morphogenesis by conducting bidirectional (forward and reverse) signaling between adjoining cells (Klein, 2012). Ephs constitute the largest tyrosine receptor family but until recently this was an unexplored field in thyroid development. EphA4, which may bind to both classes (A and B) of the ephrins, is specifically expressed in the follicular cell lineage in all developmental stages and during folliculogenesis (Andersson et al., 2011), suggesting a novel role for Eph/ephrin signaling in thyroid development. *Eph4*-null mutant adult mice have a normal-sized orthotopic gland, are euthyroid, and show only subtle changes in the follicular architecture. However, a histological changes of the thyroid are more pronounced in mutants with a truncated EphA receptor that binds ligand but is unable to convey a forward signal, it is suggested that a more severe phenotype might be compensated for by redundant actions of other coexpressed Ephs (Andersson et al., 2011).

The follicular organization *per se* has been shown to promote TSH-stimulated expression of NIS responsible for iodide uptake *in vitro*, whereas other thyroid-specific genes are less or not at all influenced by 3D culture (Bernier-Valentin, Trouttet-Masson, Rabilloud, Selmi-Ruby, & Rousset, 2006). Accordingly, conditional knockout of *Tshr* does not affect TG but strongly represses the expression of NIS (and TPO) in late-stage mouse embryos (D'Andrea et al., 2006). Interestingly, in the human fetal thyroid, the expression of NIS coincides with the appearance of follicles at gestational week 11, whereas TG is expressed already before this occurs (Szinnai et al., 2007). From an evolutionary viewpoint, no clue is given to why TSH preferentially controls the expression of NIS. Genome analysis suggests that NIS (and TPO) evolved before diversion of the vertebrate lineage, whereas TG is vertebrate-specific emerging concomitantly with a follicular thyroid under neuroendocrine control (Paris et al., 2008; Sower, Freamat, & Kavanaugh, 2009). Nonetheless, the adoptive regulation of NIS by TSH in thyroid follicular cells makes sense considering the fact it is the availability of iodide and not TG that is the limiting factor for iodination and TH production in the normal gland.

Finally, it was recently shown both *in vivo* and in explants of embryonic mouse thyroid that interaction with microvessels invading thyroid primordial tissue greatly influences the follicular organization and possibly also the preceding folliculogenesis (Hick et al., 2013). Interestingly, it is the thyroid progenitor/follicular cells that recruit endothelial cells by a paracrine mode of action involving VEGF. As the effect is evident in the absence of blood supply, it is likely that any endothelium-derived instructive signal(s) does not require normalization of a potential hypoxic state.

8. EMBRYONIC ORIGIN OF THYROID C CELLS: AN UNRESOLVED ISSUE

The parafollicular cells of the thyroid gland possess neuroendocrine features shared by neuroendocrine cells in other organs, for example, the lungs, intestine, prostate, and adrenals. According to our current understanding, thyroid C cells originate from the NC similar to adrenergic chromaffin cells of the adrenal medulla. This notion is primarily based on discoveries in the 1970s by Le Douarin and coworkers who using quail–chick chimeras were able to track the dissemination of NCC to multiple organs and tissues (for an historical overview, see Dupin, Creuzet, & Le Douarin (2006), comprising also CT-producing cells of the ultimobranchial glands (Le Douarin & Le Lievre, 1970; Polak et al., 1974). The fact that MTC, a malignant C-cell-derived thyroid tumor, is caused by germline mutations in the *RET* proto-oncogene (c-ret) that is preferentially expressed in NCC (Pachnis, Mankoo, & Costantini, 1993) and that MTC coexists with pheochromocytoma in patients with multiple endocrine neoplasia type 2 (Moline & Eng, 2011) is consistent with this assumption. However, circumstantial evidence based on a number of observations suggests that mammalian thyroid C cells might have another presumably endoderm origin arguing against the prevailing concept of MTC being a neuro-ectodermal tumor.

The ultimobranchial origin of parafollicular cells, originally suggested from light microscopic observations in dog thyroid (Godwin, 1937), was first experimentally documented in 1967 by the specific uptake of fluorescent amine in embryonic ultimobranchial cells and in the successive stages to the definitive thyroid C cells (Pearse & Carvalheira, 1967). It was subsequently shown using the same labeling technique that cells present in the fourth pharyngeal pouch from which the UB develops exhibited similar amine precursor uptake and decarboxylation (APUD) characteristics (Pearse & Polak, 1971a). Based on findings that such APUD cells were also encountered in the mesenchyme located between the neural tube and the epidermis and extending into the pharyngeal arches, it was postulated that the pouch endoderm was invaded by NCC at an early stage, that is, before the UB buds off and migrates, and that these cells were the genuine C-cell precursors. However, as shown in an accompanying paper by the same authors, amine uptake was not confined to the forth pouch but more broadly distributed in the foregut endoderm including the entire pharynx, suggesting that enteroendocrine cells in general are NC-derived (Pearse & Polak, 1971b). Later, numerous lineage-tracing studies have disqualified an NC origin of gut endocrine cells and proved that endoderm stem cells can differentiate into both exocrine and endocrine phenotypes (May & Kaestner, 2010). Although thyroid C cells belong to the APUD series of neuroendocrine cells, the use of this feature as a marker of embryonic origin is apparently not tenable.

Lineage tracing employing *Wnt1*CRE to stably trace embryonic NCC and their progeny by recombination with a *Rosa26* reporter gene is known to faithfully label the entire pharyngeal arch mesenchyme in mouse embryos (Jiang et al., 2000). As Wnt1 is transiently expressed in the neural plate, dorsal neural tube, and migratory NCC at all axial levels but nowhere else in the embryo (Echelard, Vassileva, & McMahon, 1994), it is likely that all NC-derived cells are marked by this technique, as evidenced by the expected distribution additionally in craniofacial mesenchyme, cardiac outflow tract, peripheral nervous system, adrenal medulla, and melanocytes of the skin (Jiang et al., 2000). Interestingly, although the UB is all surrounded by ectomesenchyme of NC origin, no cells expressing the reporter gene were found to infiltrate the UB at any stage, and moreover, thyroid C cells were unlabeled (Kameda, Nishimaki, Chisaka, Iseki, & Sucov, 2007). This finding thus contradicts previous notions of NCC invading the pharyngeal pouch and suggests that mouse C cells either are not NC-derived or belong to a subpopulation that does not share the typical stem features of NC.

Thyroid C cells express Nkx2-1 (Katoh et al., 2000; Mansouri et al., 1998; Suzuki, Kobayashi, Katoh, Kohn, & Kawaoi, 1998) and Nkx2-1 regulates CT gene expression (Suzuki, Lavaroni, et al., 1998; Suzuki, Katagiri, Ueda, & Tanaka, 2007). This remarkable kinship with thyroid follicular cells applies also to the progenitor stage. Thus, Nkx2-1 is expressed in the UB and no C cells are found in the UB remnant in *Nkx2-1* knockout mice (Kimura et al., 1996). That the UB indeed harbors C-cell progenitors is evidenced by the differentiation of CT-producing C cells in *Pax8*-null mice in which the median thyroid primordium already has regressed (Mansouri et al., 1998). In fact, most cells of the residual UB coexpress Nkx2-1 and CT in this mutant (Mansouri et al., 1998). Interestingly, Nkx2-1 does not seem to play a role in the formation and budding of the UB but is required for its fusion with the thyroid and long-term survival of also the C-cell precursors residing there (Kusakabe, Hoshi, & Kimura, 2006). *Nkx2-1*$^{+/-}$ embryos also display a fusion defect in which the poorly integrated UB forms cystic structures in which Nkx2-1/calcitonin-positive cells are abundant (Kusakabe, Hoshi, & Kimura, 2006). Together, this indicates that the propagation of the C-cell lineage is highly dependent on the transcriptional activity of Nkx2-1 presumably in the UB epithelium.

Retention of C cells in the UB remnant is also observed in embryos with thyroid–UB fusion defects caused by deletion of genes that are not specifically expressed in the thyroid primordia, for example, *Hoxa3* (Manley &

Capecchi, 1998), Eya1 (Xu et al., 2002), and *Hes1* (Carre et al., 2011; Kameda et al., 2013). No C cells are detected in the thyroid of mouse mutants in which the UB is missing related to failure to form the lower pharyngeal arches and pouches, for example, as observed after deletion of *Tbx1* (Fagman et al., 2007). All these studies support the ultimobranchial origin of thyroid C cells.

As yet, there are no reports of C cells in the thyroid of mice lacking UB, indicating that thyroid progenitors from the midline anlage cannot diverge towards the C-cell lineage. However, two papers suggest that the human thyroid might have this plasticity. First, thyroid C cells are not ablated in patients with DiGeorge syndrome (Pueblitz, Weinberg, & Albores-Saavedra, 1993) in which not only thymus and parathyroid fail to develop but also the UB is assumed to be missing due to defective development of all posterior pharyngeal arches and pouches (Liao et al., 2004). More recently, ectopic lingual thyroids located far away from the origin of the UB were found to contain C cells (Abu-Khudir et al., 2010; Vandernoot, Sartelet, Abu-Khudir, Chanoine, & Deladoey, 2012). Although this is an intriguing possibility, it cannot be excluded that the UB yet developed normally and fused with the thyroid in these situations. For example, DiGeorge patients are haploinsufficient of *TBX1*, whereas to reproduce similar malformations in mice, homozygosity of the deleted gene is required (Liao et al., 2004). It is thus possible that the DiGeorge phenotype is milder in humans than mice. However, the reverse plasticity in which the UB adopt features typical of the follicular primordium is evident, for example, Pax8 is ectopically expressed in the UB in $Eya1^{-/-}$ embryos, which might explain the presence of colloid in the UB remnant (Xu et al., 2002). That UB may contribute to both C cells and follicular cells in the human thyroid has previously been highlighted (Williams, Toyn, & Harach, 1989), although it should be noted that follicles generated by the UB epithelium are ultrastructurally distinguished and probably functionally different from follicles derived from the midline anlage at least in rodents (Neve & Wollman, 1971; Wollman & Hilfer, 1978; Wollman & Neve, 1971).

Thyroid C cells share many features of enteroendocrine cells as they are both neuronal and epithelial, the latter evidenced by the expression of E-cadherin (Kameda, Nishimaki, Chisaka, Iseki, & Sucov, 2007). Similar to neurons, neuroendocrine cells require the transcriptional activity of Mash1 to differentiate in neuronal direction (May & Kaestner, 2010). In thyroid development, Mash1 is expressed from E11.5 onward in an increasing number of UB cells, whereas signs of neuronal differentiation coincide with the onset of CT expression when the cells have already infiltrated the thyroid gland (Kameda, Nishimaki, Miura, Jiang, & Guillemot, 2007). Interestingly, in *Mash1*-null mutant mice, the UB develops seemingly normally until the time of fusion with the midline thyroid by which the UB regresses completely by apoptosis and C cells are lacking in the mature thyroid (Kameda, Nishimaki, Miura, Jiang, & Guillemot, 2007). This indicates that Mash1 not only is necessary for C cells to acquire a neuronal phenotype but also acts as a survival factor for both C-cell precursors and the UB epithelium.

RET is expressed not only in NC invading the pharyngeal arches but also in the posterior pharyngeal endoderm (Pachnis et al., 1993). In a mouse model of MEN2A, mutant *RET* induces both MTC and papillary thyroid cancer (PTC) known to arise from thyroid follicular epithelial cells (Reynolds et al., 2001). RET mutation may also give rise to PTC in MTC patients (Melillo et al., 2004). Although these observations may be coincidental, they suggest a more close relationship between C cells and endoderm/endoderm-derived follicular lineage than might be expected if C-cell precursors were solely of NC origin.

In summary, the embryonic origin of thyroid C cells whether it is NC or endoderm or perhaps both remains a controversy. The issue can only be solved by lineage tracing of foregut endoderm progenitors to exclude or verify that mouse C-cell precursors and the epithelial cells of UB are identical.

9. CONCLUDING REMARKS

A wealth of observations based on investigations of vertebrate animal models indicates that appendicular organs derived from the foregut share many developmental features. This comprises the initial formation of a placode composed of endoderm progenitor cells destined to a specific fate followed by budding of the growing primordium, tissue-specific morphogenesis, and eventually terminal differentiation. The entire process seems to be regulated by a limited number of cell-autonomous transcription factors, the combination of which determines organ specificity, and the regionalized activity of morphogens and growth factors produced by the endoderm/organ bud itself or by apposed embryonic tissues such as the notochord and precardiac mesoderm. Budding organs from the anterior foregut, that is, thyroid, parathyroid, and thymus, are distinguished by their unique destiny to disconnect from the ancestral germ layer and move as solitary, condensed assemblies of progenitors to a distant anatomical location by a process that probably involves collective migration. Doing so, these organ primordia pass through a lattice of primitive connective tissue composed of both axial mesoderm and NC-derived ectomesenchyme and close to transient embryonic vessels that appears to greatly influence not only the migration pathways and final position but also the size and shape of these glands. It is therefore not surprising that malformations may arise by a number of spatiotemporally distinct pathogenetic mechanisms that relate to the normal counterparts as the primordium progresses through the different stages of morphogenesis. In this respect, thyroid development is outstanding in complexity since it involves the coordinated fusion of two (actually three) primordia that emerge from different parts of the pharyngeal endoderm.

Distinctive functions of thyroid developmental genes have been elucidated mainly in mouse and zebrafish embryos in which deletion studies reproduce most if not all thyroid malformations as observed in human. However, although CH due to thyroid dysgenesis is one of the most common endocrine disturbances in infants and overall the most frequent curable cause of mental retardation in childhood, pathogenesis of the developmental defect is only rarely attributed to a known gene mutation. Nonetheless, familiar traits indicate a genetic background perhaps best illustrated by the dominant inheritance of the choreoathetosis, CH, and neonatal respiratory distress syndrome (OMIM 610978) caused by monoallelic mutation of *Nkx2-1*, expression of which is restricted to the developing brain, thyroid, and lungs. It is envisaged that further investigation of thyroid phenotypes in mouse mutants of genes expressed primarily not only in thyroid primordia but also in surrounding embryonic tissues impacting on the specification and subsequent morphogenetic growth and migration of the thyroid progenitor cells will be important to uncover yet unknown pathogenetic mechanisms that may explain also the more frequent sporadic cases of thyroid dysgenesis possibly generated by sporadic mutations or epigenetic dysregulation (Deladoey et al., 2007). Instrumental in this search will be transcriptome databases of normal and diseased embryonic thyroid tissues, which now are available for the mouse thyroid bud (Fagman et al., 2011) and the human ectopic lingual thyroid gland (Abu-Khudir et al., 2010). Irrespective of a deepened knowledge on the molecular mechanisms of thyroid morphogenesis and failure thereof, the invention of a protocol to produce functional thyroid follicular cells from pluripotent endoderm, presently established for mouse ESC (Antonica et al., 2012), brings hope for a future novel treatment modality to cure CH with autologous transplantation, eventually making substitution of TH by lifelong medication unnecessary.

ACKNOWLEDGMENTS

MN is supported by grants from the Swedish Research Council and the Swedish Cancer Society. HF is supported by Västra Götalandsregionen under the LUA/ALF agreement, the Assar Gabrielsson Foundation, and Magnus Bergwall Foundation.

REFERENCES

Abu-Khudir, R., Paquette, J., Lefort, A., Libert, F., Chanoine, J. P., Vassart, G., et al. (2010). Transcriptome, methylome and genomic variations analysis of ectopic thyroid glands. *PLoS One*, *5*(10), e13420.

Ahren, B. (1989). Effects of calcitonin, katacalcin, and calcitonin gene-related peptide on basal and TSH-stimulated thyroid hormone secretion in the mouse. *Acta Physiologica Scandinavica*, *135*(2), 133–137.

Aksoy, I., Jauch, R., Chen, J., Dyla, M., Divakar, U., Bogu, G. K., et al. (2013). Oct4 switches partnering from Sox2 to Sox17 to reinterpret the enhancer code and specify endoderm. *The EMBO Journal*, *32*(7), 938–953.

Alt, B., Elsalini, O. A., Schrumpf, P., Haufs, N., Lawson, N. D., Schwabe, G. C., et al. (2006). Arteries define the position of the thyroid gland during its developmental relocalisation. *Development*, *133*(19), 3797–3804.

Alt, B., Reibe, S., Feitosa, N. M., Elsalini, O. A., Wendl, T., & Rohr, K. B. (2006). Analysis of origin and growth of the thyroid gland in zebrafish. *Developmental Dynamics*, *235*(7), 1872–1883.

Amendola, E., De Luca, P., Macchia, P. E., Terracciano, D., Rosica, A., Chiappetta, G., et al. (2005). A mouse model demonstrates a multigenic origin of congenital hypothyroidism. *Endocrinology*, *146*(12), 5038–5047.

Amendola, E., Sanges, R., Galvan, A., Dathan, N., Manenti, G., Ferrandino, G., et al. (2010). A locus on mouse chromosome 2 is involved in susceptibility to congenital hypothyroidism and contains an essential gene expressed in thyroid. *Endocrinology*, *151*(4), 1948–1958.

Andersson, L., Westerlund, J., Liang, S., Carlsson, T., Amendola, E., Fagman, H., et al. (2011). Role of EphA4 receptor signaling in thyroid development: Regulation of folliculogenesis and propagation of the C-cell lineage. *Endocrinology*, *152*(3), 1154–1164.

Ang, S. L., Wierda, A., Wong, D., Stevens, K. A., Cascio, S., Rossant, J., et al. (1993). The formation and maintenance of the definitive endoderm lineage in the mouse: Involvement of HNF3/forkhead proteins. *Development*, *119*(4), 1301–1315.

Antonica, F., Kasprzyk, D. F., Opitz, R., Iacovino, M., Liao, X. H., Dumitrescu, A. M., et al. (2012). Generation of functional thyroid from embryonic stem cells. *Nature*, *491*(7422), 66–71.

Barasch, J., Gershon, M. D., Nunez, E. A., Tamir, H., & al-Awqati, Q. (1988). Thyrotropin induces the acidification of the secretory granules of parafollicular cells by increasing the chloride conductance of the granular membrane. *The Journal of Cell Biology*, *107*(6 Pt 1), 2137–2147.

Berg, G., Jacobsson, L., Nystrom, E., Gleisner, K. S., & Tennvall, J. (2008). Consequences of inadvertent radioiodine treatment of Graves' disease and thyroid cancer in undiagnosed pregnancy. Can we rely on routine pregnancy testing? *Acta Oncologica*, *47*(1), 145–149.

Bernier-Valentin, F., Trouttet-Masson, S., Rabilloud, R., Selmi-Ruby, S., & Rousset, B. (2006). Three-dimensional organization of thyroid cells into follicle structures is a pivotal factor in the control of sodium/iodide symporter expression. *Endocrinology*, *147*(4), 2035–2042.

Bockman, D. E., & Kirby, M. L. (1984). Dependence of thymus development on derivatives of the neural crest. *Science*, *223*(4635), 498–500.

Bogue, C. W., Ganea, G. R., Sturm, E., Ianucci, R., & Jacobs, H. C. (2000). Hex expression suggests a role in the development and function of organs derived from foregut endoderm. *Developmental Dynamics*, *219*(1), 84–89.

Bohinski, R. J., Di Lauro, R., & Whitsett, J. A. (1994). The lung-specific surfactant protein B gene promoter is a target for thyroid transcription factor 1 and hepatocyte nuclear factor 3, indicating common factors for organ-specific gene expression along the foregut axis. *Molecular and Cellular Biology*, *14*(9), 5671–5681.

Bort, R., Signore, M., Tremblay, K., Martinez Barbera, J. P., & Zaret, K. S. (2006). Hex homeobox gene controls the transition of the endoderm to a pseudostratified, cell emergent epithelium for liver bud development. *Developmental Biology*, *290*(1), 44–56.

Brabant, G., Hoang-Vu, C., Behrends, J., Cetin, Y., Potter, E., Dumont, J. E., et al. (1995). Regulation of the cell-cell adhesion protein, E-cadherin, in dog and human thyrocytes in vitro. *Endocrinology*, *136*(7), 3113–3119.

Cai, C. L., Liang, X., Shi, Y., Chu, P. H., Pfaff, S. L., Chen, J., et al. (2003). Isl1 identifies a cardiac progenitor population that proliferates prior to differentiation and contributes a majority of cells to the heart. *Developmental Cell*, *5*(6), 877–889.

Cali, G., Gentile, F., Mogavero, S., Pallante, P., Nitsch, R., Ciancia, G., et al. (2012). CDH16/Ksp-cadherin is expressed in the developing thyroid gland and is strongly down-regulated in thyroid carcinomas. *Endocrinology*, *153*(1), 522–534.

Cali, G., Zannini, M., Rubini, P., Tacchetti, C., D'Andrea, B., Affuso, A., et al. (2007). Conditional inactivation of the E-cadherin gene in thyroid follicular cells affects gland development but does not impair junction formation. *Endocrinology*, *148*(6), 2737–2746.

Calmont, A., Wandzioch, E., Tremblay, K. D., Minowada, G., Kaestner, K. H., Martin, G. R., et al. (2006). An FGF response pathway that mediates hepatic gene induction in embryonic endoderm cells. *Developmental Cell*, *11*(3), 339–348.

Carre, A., Rachdi, L., Tron, E., Richard, B., Castanet, M., Schlumberger, M., et al. (2011). Hes1 is required for appropriate morphogenesis and differentiation during mouse thyroid gland development. *PLoS One*, *6*(2), e16752.

Castanet, M., Marinovic, D., Polak, M., & Leger, J. (2010). Epidemiology of thyroid dysgenesis: The familial component. *Hormone Research in Pædiatrics*, *73*(4), 231–237.

Chambard, M., Gabrion, J., & Mauchamp, J. (1981). Influence of collagen gel on the orientation of epithelial cell polarity: Follicle formation from isolated thyroid cells and from preformed monolayers. *The Journal of Cell Biology*, *91*(1), 157–166.

Chang, J., Gerscovich, E. O., Dublin, A. B., & McGahan, J. P. (2011). Thyroid hemiagenesis: A rare finding. *Journal of Ultrasound in Medicine*, *30*(9), 1309–1310.

Chang, Y. W., Hong, H. S., & Choi, D. L. (2009). Sonography of the pediatric thyroid: A pictorial essay. *Journal of Clinical Ultrasound*, *37*(3), 149–157.

Clerc, J., Monpeyssen, H., Chevalier, A., Amegassi, F., Rodrigue, D., Leger, F. A., et al. (2008). Scintigraphic imaging of paediatric thyroid dysfunction. *Hormone Research*, *70*(1), 1–13.

Copp, D. H., Cockcroft, D. W., & Kueh, Y. (1967). Calcitonin from ultimobranchial glands of dogfish and chickens. *Science*, *158*(3803), 924–925.

D'Andrea, B., Iacone, R., Di Palma, T., Nitsch, R., Baratta, M. G., Nitsch, L., et al. (2006). Functional inactivation of the transcription factor Pax8 through oligomerization chain reaction. *Molecular Endocrinology*, *20*(8), 1810–1824.

Datta, A., Bryant, D. M., & Mostov, K. E. (2011). Molecular regulation of lumen morphogenesis. *Current Biology*, *21*(3), R126–R136.

Davies, T. F., Latif, R., Minsky, N. C., & Ma, R. (2011). Clinical review: The emerging cell biology of thyroid stem cells. *The Journal of Clinical Endocrinology and Metabolism*, *96*(9), 2692–2702.

de Cristofaro, T., Di Palma, T., Fichera, I., Lucci, V., Parrillo, L., De Felice, M., et al. (2012). An essential role for Pax8 in the transcriptional regulation of cadherin-16 in thyroid cells. *Molecular Endocrinology*, *26*(1), 67–78.

De Felice, M., & Di Lauro, R. (2004). Thyroid development and its disorders: Genetics and molecular mechanisms. *Endocrine Reviews, 25*(5), 722–746.

De Felice, M., & Di Lauro, R. (2011). Minireview: Intrinsic and extrinsic factors in thyroid gland development: An update. *Endocrinology, 152*(8), 2948–2956.

De Felice, M., Ovitt, C., Biffali, E., Rodriguez-Mallon, A., Arra, C., Anastassiadis, K., et al. (1998). A mouse model for hereditary thyroid dysgenesis and cleft palate. *Nature Genetics, 19*(4), 395–398.

de Moraes, D. C., Vaisman, M., Conceicao, F. L., & Ortiga-Carvalho, T. M. (2012). Pituitary development: A complex, temporal regulated process dependent on specific transcriptional factors. *The Journal of Endocrinology, 215*(2), 239–245.

Deladoey, J., Vassart, G., & Van Vliet, G. (2007). Possible non-Mendelian mechanisms of thyroid dysgenesis. *Endocrine Development, 10*, 29–42.

Desai, T. J., Malpel, S., Flentke, G. R., Smith, S. M., & Cardoso, W. V. (2004). Retinoic acid selectively regulates Fgf10 expression and maintains cell identity in the prospective lung field of the developing foregut. *Developmental Biology, 273*(2), 402–415.

di Gennaro, A., Spadaro, O., Baratta, M. G., De Felice, M., & Di Lauro, R. (2013). Functional analysis of the murine Pax8 promoter reveals autoregulation and the presence of a novel thyroid-specific DNA-binding activity. *Thyroid, 23*(4), 488–496.

Di Palma, T., Nitsch, R., Mascia, A., Nitsch, L., Di Lauro, R., & Zannini, M. (2003). The paired domain-containing factor Pax8 and the homeodomain-containing factor TTF-1 directly interact and synergistically activate transcription. *The Journal of Biological Chemistry, 278*(5), 3395–3402.

Duester, G. (2008). Retinoic acid synthesis and signaling during early organogenesis. *Cell, 134*(6), 921–931.

Dupin, E., Creuzet, S., & Le Douarin, N. M. (2006). The contribution of the neural crest to the vertebrate body. *Advances in Experimental Medicine and Biology, 589*, 96–119.

Echelard, Y., Vassileva, G., & McMahon, A. P. (1994). Cis-acting regulatory sequences governing Wnt-1 expression in the developing mouse CNS. *Development, 120*(8), 2213–2224.

Ekholm, R. (1981). Iodination of thyroglobulin. An intracellular or extracellular process? *Molecular and Cellular Endocrinology, 24*(2), 141–163.

Ekholm, R., & Wollman, S. H. (1975). Site of iodination in the rat thyroid gland deduced from electron microscopic autoradiographs. *Endocrinology, 97*(6), 1432–1444.

Elsalini, O. A., von Gartzen, J., Cramer, M., & Rohr, K. B. (2003). Zebrafish hhex, nk2.1a, and pax2.1 regulate thyroid growth and differentiation downstream of Nodal-dependent transcription factors. *Developmental Biology, 263*(1), 67–80.

Ericson, L. E. (1979). Intracellular lumens in thyroid follicle cells of thyroxine-treated rats. *Journal of Ultrastructure Research, 69*(2), 297–305.

Fagman, H., Amendola, E., Parrillo, L., Zoppoli, P., Marotta, P., Scarfo, M., et al. (2011). Gene expression profiling at early organogenesis reveals both common and diverse mechanisms in foregut patterning. *Developmental Biology, 359*(2), 163–175.

Fagman, H., Andersson, L., & Nilsson, M. (2006). The developing mouse thyroid: Embryonic vessel contacts and parenchymal growth pattern during specification, budding, migration, and lobulation. *Developmental Dynamics, 235*(2), 444–455.

Fagman, H., Grande, M., Edsbagge, J., Semb, H., & Nilsson, M. (2003). Expression of classical cadherins in thyroid development: Maintenance of an epithelial phenotype throughout organogenesis. *Endocrinology, 144*(8), 3618–3624.

Fagman, H., Grande, M., Gritli-Linde, A., & Nilsson, M. (2004). Genetic deletion of sonic hedgehog causes hemiagenesis and ectopic development of the thyroid in mouse. *The American Journal of Pathology, 164*(5), 1865–1872.

Fagman, H., Liao, J., Westerlund, J., Andersson, L., Morrow, B. E., & Nilsson, M. (2007). The 22q11 deletion syndrome candidate gene Tbx1 determines thyroid size and positioning. *Human Molecular Genetics, 16*(3), 276–285.

Fagman, H., & Nilsson, M. (2010). Morphogenesis of the thyroid gland. *Molecular and Cellular Endocrinology, 323*(1), 35–54.

Foster, G. V. (1968). Calcitonin (thyrocalcitonin). *The New England Journal of Medicine, 279*(7), 349–360.

Foster, G. V., Baghdiantz, A., Kumar, M. A., Slack, E., Soliman, H. A., & Macintyre, I. (1964). Thyroid origin of calcitonin. *Nature, 202*, 1303–1305.

Foster, K. E., Gordon, J., Cardenas, K., Veiga-Fernandes, H., Makinen, T., Grigorieva, E., et al. (2010). EphB-ephrin-B2 interactions are required for thymus migration during organogenesis. In: *Proceedings of the National Academy of Sciences of the United States of America, 107*(30), 13414–13419.

Fredriksson, G., Ericson, L. E., & Olsson, R. (1984). Iodine binding in the endostyle of larval Branchiostoma lanceolatum (Cephalochordata). *General and Comparative Endocrinology, 56*(2), 177–184.

Fredriksson, G., Ofverholm, T., & Ericson, L. E. (1985). Ultrastructural demonstration of iodine binding and peroxidase activity in the endostyle of Oikopleura dioica (Appendicularia). *General and Comparative Endocrinology, 58*(2), 319–327.

Frezzetti, D., Reale, C., Cali, G., Nitsch, L., Fagman, H., Nilsson, O., et al. (2011). The microRNA-processing enzyme Dicer is essential for thyroid function. *PLoS One, 6*(11), e27648.

Friedman, J. R., & Kaestner, K. H. (2006). The Foxa family of transcription factors in development and metabolism. *Cellular and Molecular Life Sciences, 63*(19–20), 2317–2328.

Garbi, C., Tacchetti, C., & Wollman, S. H. (1986). Change of inverted thyroid follicle into a spheroid after embedding in a collagen gel. *Experimental Cell Research, 163*(1), 63–77.

Garg, V., Yamagishi, C., Hu, T., Kathiriya, I. S., Yamagishi, H., & Srivastava, D. (2001). Tbx1, a DiGeorge syndrome candidate gene, is regulated by sonic hedgehog during pharyngeal arch development. *Developmental Biology, 235*(1), 62–73.

Gasser, R. F. (2006). Evidence that some events of mammalian embryogenesis can result from differential growth, making migration unnecessary. *Anatomical Record Part B, New Anatomist, 289*(2), 53–63.

Gilbert-Barness, E., Debich-Spicer, D., Cohen, M. M., Jr., & Opitz, J. M. (2001). Evidence for the "midline" hypothesis in associated defects of laterality formation and multiple midline anomalies. *American Journal of Medical Genetics, 101*(4), 382–387.

Glinoer, D. (2007). Clinical and biological consequences of iodine deficiency during pregnancy. *Endocrine Development, 10*, 62–85.

Godwin, M. C. (1937). Complex IV in the dog with special emphasis on the relation of the ultimobranchial bodies to inter-follicular cells in the post-natal thyroid gland. *The American Journal of Anatomy, 60*, 299–339.

Graham, A. (2008). Deconstructing the pharyngeal metamere. *Journal of Experimental Zoology Part B, Molecular and Developmental Evolution, 310*(4), 336–344.

Grasberger, H., & Refetoff, S. (2011). Genetic causes of congenital hypothyroidism due to dyshormonogenesis. *Current Opinion in Pediatrics, 23*(4), 421–428.

Grevellec, A., & Tucker, A. S. (2010). The pharyngeal pouches and clefts: Development, evolution, structure and derivatives. *Seminars in Cell & Developmental Biology, 21*(3), 325–332.

Griffith, A. V., Cardenas, K., Carter, C., Gordon, J., Iberg, A., Engleka, K., et al. (2009). Increased thymus- and decreased parathyroid-fated organ domains in Splotch mutant embryos. *Developmental Biology, 327*(1), 216–227.

Gruters, A., & Krude, H. (2011). Detection and treatment of congenital hypothyroidism. *Nature Reviews Endocrinology, 8*(2), 104–113.

Goss, A. M., Tian, Y., Tsukiyama, T., Cohen, E. D., Zhou, D., Lu, M. M., et al. (2009). Wnt2/2b and beta-catenin signaling are necessary and sufficient to specify lung progenitors in the foregut. *Developmental Cell, 17*(2), 290–298. http://dx.doi.org/10.1016/j.devcel.2009.06.005.

Hagiuda, J., Kuroda, I., Tsukamoto, T., Ueno, M., Yokota, C., Hirose, T., et al. (2006). Ectopic thyroid in an adrenal mass: A case report. *BMC Urology, 6*, 18.

Haynes, J. I. (1999). Parathyroids and ultimobranchial bodies in monotremes. *The Anatomical Record, 254*, 269–280.

Hebrok, M. (2003). Hedgehog signaling in pancreas development. *Mechanisms of Development, 120*(1), 45–57.

Hick, A. C., Delmarcelle, A. S., Bouquet, M., Klotz, S., Copetti, T., Forez, C., et al. (2013). Reciprocal epithelial:endothelial paracrine interactions during thyroid development govern follicular organization and C-cells differentiation. *Developmental Biology, 381*, 227–240.

Hirsch, P. F., & Baruch, H. (2003). Is calcitonin an important physiological substance? *Endocrine, 21*(3), 201–208.

Hiruta, J., Mazet, F., Yasui, K., Zhang, P., & Ogasawara, M. (2005). Comparative expression analysis of transcription factor genes in the endostyle of invertebrate chordates. *Developmental Dynamics, 233*(3), 1031–1037.

Isaac, R., Merceron, R., Caillens, G., Raymond, J. P., & Ardaillou, R. (1980). Effects of calcitonin on basal and thyrotropin-releasing hormone-stimulated prolactin secretion in man. *The Journal of Clinical Endocrinology and Metabolism, 50*(6), 1011–1015.

Jiang, X., Rowitch, D. H., Soriano, P., McMahon, A. P., & Sucov, H. M. (2000). Fate of the mammalian cardiac neural crest. *Development, 127*(8), 1607–1616.

Jovanovic, L., Delahunt, B., McIver, B., Eberhardt, N. L., & Grebe, S. K. (2003). Thyroid gland clonality revisited: The embryonal patch size of the normal human thyroid gland is very large, suggesting X-chromosome inactivation tumor clonality studies of thyroid tumors have to be interpreted with caution. *The Journal of Clinical Endocrinology and Metabolism, 88*(7), 3284–3291.

Kameda, Y., Nishimaki, T., Chisaka, O., Iseki, S., & Sucov, H. M. (2007). Expression of the epithelial marker E-cadherin by thyroid C cells and their precursors during murine development. *The Journal of Histochemistry and Cytochemistry, 55*(10), 1075–1088.

Kameda, Y., Nishimaki, T., Miura, M., Jiang, S. X., & Guillemot, F. (2007). Mash1 regulates the development of C cells in mouse thyroid glands. *Developmental Dynamics, 236*(1), 262–270.

Kameda, Y., Saitoh, T., Nemoto, N., Katoh, T., Iseki, S., & Fujimura, T. (2013). Hes1 is required for the development of pharyngeal organs and survival of neural crest-derived mesenchymal cells in pharyngeal arches. *Cell and Tissue Research, 353*, 9–25.

Kanai-Azuma, M., Kanai, Y., Gad, J. M., Tajima, Y., Taya, C., Kurohmaru, M., et al. (2002). Depletion of definitive gut endoderm in Sox17-null mutant mice. *Development, 129*(10), 2367–2379.

Katoh, R., Miyagi, E., Nakamura, N., Li, X., Suzuki, K., Kakudo, K., et al. (2000). Expression of thyroid transcription factor-1 (TTF-1) in human C cells and medullary thyroid carcinomas. *Human Pathology, 31*(3), 386–393.

Kawakami, Y., Raya, A., Raya, R. M., Rodriguez-Esteban, C., & Izpisua Belmonte, J. C. (2005). Retinoic acid signalling links left-right asymmetric patterning and bilaterally symmetric somitogenesis in the zebrafish embryo. *Nature, 435*(7039), 165–171.

Kimura, S., Hara, Y., Pineau, T., Fernandez-Salguero, P., Fox, C. H., Ward, J. M., et al. (1996). The T/ebp null mouse: Thyroid-specific enhancer-binding protein is essential for the organogenesis of the thyroid, lung, ventral forebrain, and pituitary. *Genes & Development, 10*(1), 60–69.

Kinebrew, G. M., & Hilfer, S. R. (2001). Cellular dynamics during evagination of the thyroid primordium in the chick embryo. *The Anatomical Record, 264*(2), 146–156.

Klein, R. (2012). Eph/ephrin signalling during development. *Development*, *139*(22), 4105–4109.

Kluge, B., Renault, N., & Rohr, K. B. (2005). Anatomical and molecular reinvestigation of lamprey endostyle development provides new insight into thyroid gland evolution. *Development Genes and Evolution*, *215*(1), 32–40.

Kusakabe, T., Hoshi, N., & Kimura, S. (2006). Origin of the ultimobranchial body cyst: T/ebp/Nkx2.1 expression is required for development and fusion of the ultimobranchial body to the thyroid. *Developmental Dynamics*, *235*(5), 1300–1309.

Kusakabe, T., Kawaguchi, A., Hoshi, N., Kawaguchi, R., Hoshi, S., & Kimura, S. (2006). Thyroid-specific enhancer-binding protein/NKX2.1 is required for the maintenance of ordered architecture and function of the differentiated thyroid. *Molecular Endocrinology*, *20*(8), 1796–1809.

Lania, G., Zhang, Z., Huynh, T., Caprio, C., Moon, A. M., Vitelli, F., et al. (2009). Early thyroid development requires a Tbx1-Fgf8 pathway. *Developmental Biology*, *328*(1), 109–117.

Larsson, F., Fagman, H., & Nilsson, M. (2004). TSH receptor signaling via cyclic AMP inhibits cell surface degradation and internalization of E-cadherin in pig thyroid epithelium. *Cellular and Molecular Life Sciences*, *61*(14), 1834–1842.

Lazzaro, D., Price, M., de Felice, M., & Di Lauro, R. (1991). The transcription factor TTF-1 is expressed at the onset of thyroid and lung morphogenesis and in restricted regions of the foetal brain. *Development*, *113*(4), 1093–1104.

Le Douarin, N., & Le Lievre, C. (1970). Demonstration of neural origin of calcitonin cells of ultimobranchial body of chick embryo. *Comptes rendus hebdomadaires des séances de l'Académie des sciences*, *270*(23), 2857–2860.

Lee, C. S., Friedman, J. R., Fulmer, J. T., & Kaestner, K. H. (2005). The initiation of liver development is dependent on Foxa transcription factors. *Nature*, *435*(7044), 944–947.

Li, Y., Eggermont, K., Vanslembrouck, V., & Verfaillie, C. M. (2013). NKX2-1 activation by SMAD2 signaling after definitive endoderm differentiation in human embryonic stem cell. *Stem Cells and Development*, *22*(9), 1433–1442.

Liao, J., Kochilas, L., Nowotschin, S., Arnold, J. S., Aggarwal, V. S., Epstein, J. A., et al. (2004). Full spectrum of malformations in velo-cardio-facial syndrome/DiGeorge syndrome mouse models by altering Tbx1 dosage. *Human Molecular Genetics*, *13*(15), 1577–1585.

Liu, Z., Farley, A., Chen, L., Kirby, B. J., Kovacs, C. S., Blackburn, C. C., et al. (2010). Thymus-associated parathyroid hormone has two cellular origins with distinct endocrine and immunological functions. *PLoS Genetics*, *6*(12), e1001251.

Longmire, T. A., Ikonomou, L., Hawkins, F., Christodoulou, C., Cao, Y., Jean, J. C., et al. (2012). Efficient derivation of purified lung and thyroid progenitors from embryonic stem cells. *Cell Stem Cell*, *10*(4), 398–411.

Ma, R., Latif, R., & Davies, T. F. (2009). Thyrotropin-independent induction of thyroid endoderm from embryonic stem cells by activin A. *Endocrinology*, *150*(4), 1970–1975.

Ma, R., Latif, R., & Davies, T. F. (2013). Thyroid follicle formation and thyroglobulin expression in multipotent endodermal stem cells. *Thyroid*, *23*(4), 385–391.

Ma, D. F., Sudo, K., Tezuka, H., Kondo, T., Nakazawa, T., Niu, D. F., et al. (2010). Polyclonal origin of hormone-producing cell populations evaluated as a direct in situ demonstration in EGFP/BALB/C chimeric mice. *The Journal of Endocrinology*, *207*(1), 17–25.

Manley, N. R., & Capecchi, M. R. (1995). The role of Hoxa-3 in mouse thymus and thyroid development. *Development*, *121*(7), 1989–2003.

Manley, N. R., & Capecchi, M. R. (1998). Hox group 3 paralogs regulate the development and migration of the thymus, thyroid, and parathyroid glands. *Developmental Biology*, *195*(1), 1–15.

Mansouri, A., Chowdhury, K., & Gruss, P. (1998). Follicular cells of the thyroid gland require Pax8 gene function. *Nature Genetics*, *19*(1), 87–90.

Martin-Belmonte, F., & Mostov, K. (2008). Regulation of cell polarity during epithelial morphogenesis. *Current Opinion in Cell Biology, 20*(2), 227–234.

Martinez Barbera, J. P., Clements, M., Thomas, P., Rodriguez, T., Meloy, D., Kioussis, D., et al. (2000). The homeobox gene Hex is required in definitive endodermal tissues for normal forebrain, liver and thyroid formation. *Development, 127*(11), 2433–2445.

May, C. L., & Kaestner, K. H. (2010). Gut endocrine cell development. *Molecular and Cellular Endocrinology, 323*(1), 70–75.

Melillo, R. M., Cirafici, A. M., De Falco, V., Bellantoni, M., Chiappetta, G., Fusco, A., et al. (2004). The oncogenic activity of RET point mutants for follicular thyroid cells may account for the occurrence of papillary thyroid carcinoma in patients affected by familial medullary thyroid carcinoma. *The American Journal of Pathology, 165*(2), 511–521.

Meyer, J. S., & Abdel-Bari, W. (1968). Granules and thyrocalcitonin-like activity in medullary carcinoma of the thyroid gland. *The New England Journal of Medicine, 278*(10), 523–529.

Meyers, E. N., & Martin, G. R. (1999). Differences in left-right axis pathways in mouse and chick: Functions of FGF8 and SHH. *Science, 285*(5426), 403–406.

Minoo, P., Hu, L., Xing, Y., Zhu, N. L., Chen, H., Li, M., et al. (2007). Physical and functional interactions between homeodomain NKX2.1 and winged helix/forkhead FOXA1 in lung epithelial cells. *Molecular and Cellular Biology, 27*(6), 2155–2165.

Minoo, P., Su, G., Drum, H., Bringas, P., & Kimura, S. (1999). Defects in tracheoesophageal and lung morphogenesis in Nkx2.1(−/−) mouse embryos. *Developmental Biology, 209*(1), 60–71.

Moline, J., & Eng, C. (2011). Multiple endocrine neoplasia type 2: An overview. *Genetics in Medicine, 13*(9), 755–764.

Monaghan, A. P., Kaestner, K. H., Grau, E., & Schutz, G. (1993). Postimplantation expression patterns indicate a role for the mouse forkhead/HNF-3 alpha, beta and gamma genes in determination of the definitive endoderm, chordamesoderm and neuroectoderm. *Development, 119*(3), 567–578.

Montanelli, L., & Tonacchera, M. (2010). Genetics and phenomics of hypothyroidism and thyroid dys- and agenesis due to PAX8 and TTF1 mutations. *Molecular and Cellular Endocrinology, 322*(1–2), 64–71.

Moore-Scott, B. A., & Manley, N. R. (2005). Differential expression of Sonic hedgehog along the anterior-posterior axis regulates patterning of pharyngeal pouch endoderm and pharyngeal endoderm-derived organs. *Developmental Biology, 278*(2), 323–335.

Mori, N., Itoh, N., & Salvaterra, P. M. (1987). Evolutionary origin of cholinergic macromolecules and thyroglobulin. In: *Proceedings of the National Academy of Sciences of the United States of America, 84*(9), 2813–2817.

Morillo-Bernal, J., Fernandez-Santos, J. M., Utrilla, J. C., de Miguel, M., Garcia-Marin, R., & Martin-Lacave, I. (2009). Functional expression of the thyrotropin receptor in C cells: New insights into their involvement in the hypothalamic-pituitary-thyroid axis. *Journal of Anatomy, 215*(2), 150–158.

Morris, S. A., Teo, R. T., Li, H., Robson, P., Glover, D. M., & Zernicka-Goetz, M. (2010). Origin and formation of the first two distinct cell types of the inner cell mass in the mouse embryo. In: *Proceedings of the National Academy of Sciences of the United States of America, 107*(14), 6364–6369.

Nasseri, F., & Eftekhari, F. (2010). Clinical and radiologic review of the normal and abnormal thymus: Pearls and pitfalls. *Radiographics, 30*(2), 413–428.

Neve, P., & Wollman, S. H. (1971). Fine structure of ultimobranchial follicles in the thyroid gland of the rat. *The Anatomical Record, 171*(2), 259–272.

Nitsch, R., Di Dato, V., di Gennaro, A., de Cristofaro, T., Abbondante, S., De Felice, M., et al. (2010). Comparative genomics reveals a functional thyroid-specific element in the far upstream region of the PAX8 gene. *BMC Genomics, 11*, 306.

Nitsch, L., & Wollman, S. H. (1980). Suspension culture of separated follicles consisting of differentiated thyroid epithelial cells. In: *Proceedings of the National Academy of Sciences of the United States of America*, 77(1), 472–476.

Noussios, G., Anagnostis, P., Goulis, D. G., Lappas, D., & Natsis, K. (2011). Ectopic thyroid tissue: Anatomical, clinical, and surgical implications of a rare entity. *European Journal of Endocrinology*, *165*(3), 375–382.

Novelli, M., Cossu, A., Oukrif, D., Quaglia, A., Lakhani, S., Poulsom, R., et al. (2003). X-inactivation patch size in human female tissue confounds the assessment of tumor clonality. In: *Proceedings of the National Academy of Sciences of the United States of America*, *100*(6), 3311–3314.

Ogasawara, M. (2000). Overlapping expression of amphioxus homologs of the thyroid transcription factor-1 gene and thyroid peroxidase gene in the endostyle: Insight into evolution of the thyroid gland. *Development Genes and Evolution*, *210*(5), 231–242.

Ogasawara, M., Di Lauro, R., & Satoh, N. (1999). Ascidian homologs of mammalian thyroid peroxidase genes are expressed in the thyroid-equivalent region of the endostyle. *The Journal of Experimental Zoology*, *285*(2), 158–169.

Ohbuchi, T., Mori, T., Tabata, T., Hanaguri, M., Hisaoka, M., Hashida, K., et al. (2012). Coexistence of pyriform sinus fistula, ectopic lingual thyroid, and ectopic cervical thymus. *Auris, Nasus, Larynx*, *39*(6), 634–637.

Opitz, R., Maquet, E., Huisken, J., Antonica, F., Trubiroha, A., Pottier, G., et al. (2012). Transgenic zebrafish illuminate the dynamics of thyroid morphogenesis and its relationship to cardiovascular development. *Developmental Biology*, *372*(2), 203–216.

Opitz, R., Maquet, E., Zoenen, M., Dadhich, R., & Costagliola, S. (2011). TSH receptor function is required for normal thyroid differentiation in zebrafish. *Molecular Endocrinology*, *25*(9), 1579–1599.

Pachnis, V., Mankoo, B., & Costantini, F. (1993). Expression of the c-ret proto-oncogene during mouse embryogenesis. *Development*, *119*(4), 1005–1017.

Pang, P. K. (1971). Calcitonin and ultimobranchial glands in fishes. *The Journal of Experimental Zoology*, *178*(1), 89–99.

Paris, M., Brunet, F., Markov, G. V., Schubert, M., & Laudet, V. (2008). The amphioxus genome enlightens the evolution of the thyroid hormone signaling pathway. *Development Genes and Evolution*, *218*(11–12), 667–680.

Park, J. Y., Kim, D. W., Park, J. S., Kang, T., & Kim, Y. W. (2012). The prevalence and features of thyroid pyramidal lobes as assessed by computed tomography. *Thyroid*, *22*(2), 173–177.

Park, K. S., Wells, J. M., Zorn, A. M., Wert, S. E., & Whitsett, J. A. (2006). Sox17 influences the differentiation of respiratory epithelial cells. *Developmental Biology*, *294*(1), 192–202.

Parlato, R., Rosica, A., Rodriguez-Mallon, A., Affuso, A., Postiglione, M. P., Arra, C., et al. (2004). An integrated regulatory network controlling survival and migration in thyroid organogenesis. *Developmental Biology*, *276*(2), 464–475.

Pearse, A. G., & Carvalheira, A. F. (1967). Cytochemical evidence for an ultimobranchial origin of rodent thyroid C cells. *Nature*, *214*(5091), 929–930.

Pearse, A. G., & Polak, J. M. (1971a). Cytochemical evidence for the neural crest origin of mammalian ultimobranchial C cells. *Histochemie*, *27*(2), 96–102.

Pearse, A. G., & Polak, J. M. (1971b). Neural crest origin of the endocrine polypeptide (APUD) cells of the gastrointestinal tract and pancreas. *Gut*, *12*(10), 783–788.

Peter, H. J., Gerber, H., Studer, H., & Smeds, S. (1985). Pathogenesis of heterogeneity in human multinodular goiter. A study on growth and function of thyroid tissue transplanted onto nude mice. *The Journal of Clinical Investigation*, *76*(5), 1992–2002.

Peter, H. J., Studer, H., Forster, R., & Gerber, H. (1982). The pathogenesis of "hot" and "cold" follicles in multinodular goiters. *The Journal of Clinical Endocrinology and Metabolism*, *55*(5), 941–946.

Peter, H. J., Studer, H., & Groscurth, P. (1988). Autonomous growth, but not autonomous function, in embryonic human thyroids: A clue to understanding autonomous goiter growth? *The Journal of Clinical Endocrinology and Metabolism*, *66*(5), 968–973.

Phitayakorn, R., & McHenry, C. R. (2006). Incidence and location of ectopic abnormal parathyroid glands. *American Journal of Surgery*, *191*(3), 418–423.

Plachov, D., Chowdhury, K., Walther, C., Simon, D., Guenet, J. L., & Gruss, P. (1990). Pax8, a murine paired box gene expressed in the developing excretory system and thyroid gland. *Development*, *110*(2), 643–651.

Polak, M., Legac, I., Vuillard, E., Guibourdenche, J., Castanet, M., & Luton, D. (2006). Congenital hyperthyroidism: The fetus as a patient. *Hormone Research*, *65*(5), 235–242.

Polak, J. M., Pearse, A. G., Le Lievre, C., Fontaine, J., & Le Douarin, N. M. (1974). Immunocytochemical confirmation of the neural crest origin of avian calcitonin-producing cells. *Histochemistry*, *40*(3), 209–214.

Porazzi, P., Calebiro, D., Benato, F., Tiso, N., & Persani, L. (2009). Thyroid gland development and function in the zebrafish model. *Molecular and Cellular Endocrinology*, *312*(1–2), 14–23.

Porazzi, P., Marelli, F., Benato, F., de Filippis, T., Calebiro, D., Argenton, F., et al. (2012). Disruption of global and JAGGED1-mediated notch signaling affect thyroid morphogenesis in the zebrafish. *Endocrinology*, *153*, 5645–5658.

Porreca, I., De Felice, E., Fagman, H., Di Lauro, R., & Sordino, P. (2012). Zebrafish bcl2l is a survival factor in thyroid development. *Developmental Biology*, *366*(2), 142–152.

Postiglione, M. P., Parlato, R., Rodriguez-Mallon, A., Rosica, A., Mithbaokar, P., Maresca, M., et al. (2002). Role of the thyroid-stimulating hormone receptor signaling in development and differentiation of the thyroid gland. In: *Proceedings of the National Academy of Sciences of the United States of America*, *99*(24), 15462–15467.

Pueblitz, S., Weinberg, A. G., & Albores-Saavedra, J. (1993). Thyroid C cells in the DiGeorge anomaly: A quantitative study. *Pediatric Pathology*, *13*(4), 463–473.

Puppin, C., D'Elia, A. V., Pellizzari, L., Russo, D., Arturi, F., Presta, I., et al. (2003). Thyroid-specific transcription factors control Hex promoter activity. *Nucleic Acids Research*, *31*(7), 1845–1852.

Puppin, C., Presta, I., D'Elia, A. V., Tell, G., Arturi, F., Russo, D., et al. (2004). Functional interaction among thyroid-specific transcription factors: Pax8 regulates the activity of Hex promoter. *Molecular and Cellular Endocrinology*, *214*(1–2), 117–125.

Reynolds, L., Jones, K., Winton, D. J., Cranston, A., Houghton, C., Howard, L., et al. (2001). C-cell and thyroid epithelial tumours and altered follicular development in transgenic mice expressing the long isoform of MEN 2A RET. *Oncogene*, *20*(30), 3986–3994.

Robertson, D. R. (1971). Endocrinology of amphibian ultimobranchial glands. *The Journal of Experimental Zoology*, *178*(1), 101–104.

Rodriguez, W., Jin, L., Janssens, V., Pierreux, C., Hick, A. C., Urizar, E., et al. (2012). Deletion of the RNaseIII enzyme dicer in thyroid follicular cells causes hypothyroidism with signs of neoplastic alterations. *PLoS One*, 7(1), e29929.

Roger, P. P., van Staveren, W. C., Coulonval, K., Dumont, J. E., & Maenhaut, C. (2010). Signal transduction in the human thyrocyte and its perversion in thyroid tumors. *Molecular and Cellular Endocrinology*, *321*(1), 3–19.

Rorth, P. (2012). Fellow travellers: Emergent properties of collective cell migration. *EMBO Reports*, *13*(11), 984–991.

Rupik, W. (2011). Structural and ultrastructural differentiation of the thyroid gland during embryogenesis in the grass snake Natrix natrix L. (Lepidosauria, Serpentes). *Zoology (Jena, Germany)*, *114*(5), 284–297.

Saund, R. S., Kanai-Azuma, M., Kanai, Y., Kim, I., Lucero, M. T., & Saijoh, Y. (2012). Gut endoderm is involved in the transfer of left-right asymmetry from the node to the lateral plate mesoderm in the mouse embryo. *Development*, *139*(13), 2426–2435.

Scambler, P. J. (2010). 22q11 deletion syndrome: A role for TBX1 in pharyngeal and cardiovascular development. *Pediatric Cardiology*, *31*(3), 378–390.

Shiratori, H., & Hamada, H. (2006). The left-right axis in the mouse: From origin to morphology. *Development*, *133*(11), 2095–2104.

Silberschmidt, D., Rodriguez-Mallon, A., Mithboakar, P., Cali, G., Amendola, E., Sanges, R., et al. (2011). In vivo role of different domains and of phosphorylation in the transcription factor Nkx2-1. *BMC Developmental Biology*, *11*, 9.

Sower, S. A., Freamat, M., & Kavanaugh, S. I. (2009). The origins of the vertebrate hypothalamic-pituitary-gonadal (HPG) and hypothalamic-pituitary-thyroid (HPT) endocrine systems: New insights from lampreys. *General and Comparative Endocrinology*, *161*(1), 20–29.

Stafford, D., & Prince, V. E. (2002). Retinoic acid signaling is required for a critical early step in zebrafish pancreatic development. *Current Biology*, *12*(14), 1215–1220.

Stagi, S., Lapi, E., Gambineri, E., Salti, R., Genuardi, M., Colarusso, G., et al. (2010). Thyroid function and morphology in subjects with microdeletion of chromosome 22q11 (del(22)(q11)). *Clinical Endocrinology*, *72*(6), 839–844.

Stahlman, M. T., Gray, M. E., & Whitsett, J. A. (1998). Temporal-spatial distribution of hepatocyte nuclear factor-3beta in developing human lung and other foregut derivatives. *The Journal of Histochemistry and Cytochemistry*, *46*(8), 955–962.

Stanger, B. Z., Tanaka, A. J., & Melton, D. A. (2007). Organ size is limited by the number of embryonic progenitor cells in the pancreas but not the liver. *Nature*, *445*(7130), 886–891.

Stoppa-Vaucher, S., Van Vliet, G., & Deladoey, J. (2011). Variation by ethnicity in the prevalence of congenital hypothyroidism due to thyroid dysgenesis. *Thyroid*, *21*(1), 13–18.

Studer, H., Peter, H. J., & Gerber, H. (1989). Natural heterogeneity of thyroid cells: The basis for understanding thyroid function and nodular goiter growth. *Endocrine Reviews*, *10*(2), 125–135.

Sui, L., Bouwens, L., & Mfopou, J. K. (2013). Signaling pathways during maintenance and definitive endoderm differentiation of embryonic stem cells. *The International Journal of Developmental Biology*, *57*(1), 1–12.

Suzuki, M., Katagiri, N., Ueda, M., & Tanaka, S. (2007). Functional analysis of Nkx2.1 and Pax9 for calcitonin gene transcription. *General and Comparative Endocrinology*, *152*(2–3), 259–266.

Suzuki, K., Kobayashi, Y., Katoh, R., Kohn, L. D., & Kawaoi, A. (1998). Identification of thyroid transcription factor-1 in C cells and parathyroid cells. *Endocrinology*, *139*(6), 3014–3017.

Suzuki, K., Lavaroni, S., Mori, A., Okajima, F., Kimura, S., Katoh, R., et al. (1998). Thyroid transcription factor 1 is calcium modulated and coordinately regulates genes involved in calcium homeostasis in C cells. *Molecular and Cellular Biology*, *18*(12), 7410–7422.

Szinnai, G., Lacroix, L., Carre, A., Guimiot, F., Talbot, M., Martinovic, J., et al. (2007). Sodium/iodide symporter (NIS) gene expression is the limiting step for the onset of thyroid function in the human fetus. *The Journal of Clinical Endocrinology and Metabolism*, *92*(1), 70–76.

Takagi, Y., Omura, T., & Go, M. (1991). Evolutionary origin of thyroglobulin by duplication of esterase gene. *FEBS Letters*, *282*(1), 17–22.

Tashjian, A. H., Jr., & Melvin, E. W. (1968). Medullary carcinoma of the thyroid gland. Studies of thyrocalcitonin in plasma and tumor extracts. *The New England Journal of Medicine*, *279*(6), 279–283.

Tauber, S. D. (1967). The ultimobranchial origin of thyrocalcitonin. In: *Proceedings of the National Academy of Sciences of the United States of America*, *58*(4), 1684–1687.

Toda, S., & Sugihara, H. (1990). Reconstruction of thyroid follicles from isolated porcine follicle cells in three-dimensional collagen gel culture. *Endocrinology*, *126*(4), 2027–2034.

Toso, A., Colombani, F., Averono, G., Aluffi, P., & Pia, F. (2009). Lingual thyroid causing dysphagia and dyspnoea. Case reports and review of the literature. *Acta Otorhinolaryngologica Italica, 29*(4), 213–217.

Tremblay, K. D., & Zaret, K. S. (2005). Distinct populations of endoderm cells converge to generate the embryonic liver bud and ventral foregut tissues. *Developmental Biology, 280*(1), 87–99.

Trueba, S. S., Auge, J., Mattei, G., Etchevers, H., Martinovic, J., Czernichow, P., et al. (2005). PAX8, TITF1, and FOXE1 gene expression patterns during human development: New insights into human thyroid development and thyroid dysgenesis-associated malformations. *The Journal of Clinical Endocrinology and Metabolism, 90*(1), 455–462.

Uemura, M., Hara, K., Shitara, H., Ishii, R., Tsunekawa, N., Miura, Y., et al. (2010). Expression and function of mouse Sox17 gene in the specification of gallbladder/bile-duct progenitors during early foregut morphogenesis. *Biochemical and Biophysical Research Communications, 391*(1), 357–363.

Uemura, M., Ozawa, A., Nagata, T., Kurasawa, K., Tsunekawa, N., Nobuhisa, I., et al. (2013). Sox17 haploinsufficiency results in perinatal biliary atresia and hepatitis in C57BL/6 background mice. *Development, 140*(3), 639–648.

Vandenberg, L. N., & Levin, M. (2013). A unified model for left-right asymmetry? Comparison and synthesis of molecular models of embryonic laterality. *Developmental Biology, 379*, 1–15.

Vandernoot, I., Sartelet, H., Abu-Khudir, R., Chanoine, J. P., & Deladoey, J. (2012). Evidence for calcitonin-producing cells in human lingual thyroids. *The Journal of Clinical Endocrinology and Metabolism, 97*(3), 951–956.

Venza, I., Visalli, M., Parrillo, L., De Felice, M., Teti, D., & Venza, M. (2011). MSX1 and TGF-beta3 are novel target genes functionally regulated by FOXE1. *Human Molecular Genetics, 20*(5), 1016–1025.

Vermot, J., Gallego Llamas, J., Fraulob, V., Niederreither, K., Chambon, P., & Dolle, P. (2005). Retinoic acid controls the bilateral symmetry of somite formation in the mouse embryo. *Science, 308*(5721), 563–566.

Vermot, J., & Pourquie, O. (2005). Retinoic acid coordinates somitogenesis and left-right patterning in vertebrate embryos. *Nature, 435*(7039), 215–220.

Vilhais-Neto, G. C., Maruhashi, M., Smith, K. T., Vasseur-Cognet, M., Peterson, A. S., Workman, J. L., et al. (2010). Rere controls retinoic acid signalling and somite bilateral symmetry. *Nature, 463*(7283), 953–957.

Viotti, M., Niu, L., Shi, S.-H., & Hadjantonakis, A.-K. (2012). Role of the gut endoderm in relying left-right patterning in mice. *PLoS Biology, 10*(3), e1001276.

Vitelli, F., Morishima, M., Taddei, I., Lindsay, E. A., & Baldini, A. (2002). Tbx1 mutation causes multiple cardiovascular defects and disrupts neural crest and cranial nerve migratory pathways. *Human Molecular Genetics, 11*(8), 915–922.

Wan, H., Dingle, S., Xu, Y., Besnard, V., Kaestner, K. H., Ang, S. L., et al. (2005). Compensatory roles of Foxa1 and Foxa2 during lung morphogenesis. *The Journal of Biological Chemistry, 280*(14), 13809–13816.

Wang, C. (1976). The anatomic basis of parathyroid surgery. *Annals of Surgery, 183*(3), 271–275.

Wang, J. H., Deimling, S. J., D'Alessandro, N. E., Zhao, L., Possmayer, F., & Drysdale, T. A. (2011). Retinoic acid is a key regulatory switch determining the difference between lung and thyroid fates in Xenopus laevis. *BMC Developmental Biology, 11*, 75.

Washington Smoak, I., Byrd, N. A., Abu-Issa, R., Goddeeris, M. M., Anderson, R., Morris, J., et al. (2005). Sonic hedgehog is required for cardiac outflow tract and neural crest cell development. *Developmental Biology, 283*(2), 357–372.

Wendl, T., Adzic, D., Schoenebeck, J. J., Scholpp, S., Brand, M., Yelon, D., et al. (2007). Early developmental specification of the thyroid gland depends on han-expressing surrounding tissue and on FGF signals. *Development, 134*(15), 2871–2879.

Wendl, T., Lun, K., Mione, M., Favor, J., Brand, M., Wilson, S. W., et al. (2002). Pax2.1 is required for the development of thyroid follicles in zebrafish. *Development, 129*(15), 3751–3760.

Westbrook, B. J., Harsha, W. J., & Strenge, K. (2013). Original report of bilateral carotid body tumors with 2 rare concomitant anatomic findings, an ectopic parathyroid gland and cervical thymus, with literature review. *Head & Neck, 35*(3), E65–E68.

Westerlund, J., Andersson, L., Carlsson, T., Zoppoli, P., Fagman, H., & Nilsson, M. (2008). Expression of Islet1 in thyroid development related to budding, migration, and fusion of primordia. *Developmental Dynamics, 237*(12), 3820–3829.

Westermark, K., Nilsson, M., Ebendal, T., & Westermark, B. (1991). Thyrocyte migration and histiotypic follicle regeneration are promoted by epidermal growth factor in primary culture of thyroid follicles in collagen gel. *Endocrinology, 129*(4), 2180–2186.

Wildi-Runge, S., Stoppa-Vaucher, S., Lambert, R., Turpin, S., Van Vliet, G., & Deladoey, J. (2012). A high prevalence of dual thyroid ectopy in congenital hypothyroidism: Evidence for insufficient signaling gradients during embryonic thyroid migration or for the polyclonal nature of the thyroid gland? *The Journal of Clinical Endocrinology and Metabolism, 97*(6), E978–E981.

Williams, E. D. (1966). Histogenesis of medullary carcinoma of the thyroid. *Journal of Clinical Pathology, 19*(2), 114–118.

Williams, E. D., Toyn, C. E., & Harach, H. R. (1989). The ultimobranchial gland and congenital thyroid abnormalities in man. *The Journal of Pathology, 159*(2), 135–141.

Wollman, S. H., & Hilfer, S. R. (1978). Embryologic origin of the various epithelial cell types in the second kind of thyroid follicle in the C3H mouse. *The Anatomical Record, 191*(1), 111–121.

Wollman, S. H., & Neve, P. (1971). Ultimobranchial follicles in the thyroid glands of rats and mice. *Recent Progress in Hormone Research, 27*, 213–234.

Wu, Y. H., Wein, R. O., & Carter, B. (2012). Thyroid hemiagenesis: A case series and review of the literature. *American Journal of Otolaryngology, 33*(3), 299–302.

Xu, C. R., Cole, P. A., Meyers, D. J., Kormish, J., Dent, S., & Zaret, K. S. (2011). Chromatin "prepattern" and histone modifiers in a fate choice for liver and pancreas. *Science, 332*(6032), 963–966.

Xu, P. X., Zheng, W., Laclef, C., Maire, P., Maas, R. L., Peters, H., et al. (2002). Eya1 is required for the morphogenesis of mammalian thymus, parathyroid and thyroid. *Development, 129*(13), 3033–3044.

Zajac, J. D., & Danks, J. A. (2008). The development of the parathyroid gland: From fish to human. *Current Opinion in Nephrology and Hypertension, 17*(4), 353–356.

Zannini, M., Avantaggiato, V., Biffali, E., Arnone, M. I., Sato, K., Pischetola, M., et al. (1997). TTF-2, a new forkhead protein, shows a temporal expression in the developing thyroid which is consistent with a role in controlling the onset of differentiation. *The EMBO Journal, 16*(11), 3185–3197.

Zorn, A. M., & Wells, J. M. (2009). Vertebrate endoderm development and organ formation. *Annual Review of Cell and Developmental Biology, 25*, 221–251.

CHAPTER FIVE

The Pineal Gland from Development to Function

Dora Sapède[*,†], **Elise Cau**[*,†,1]
[*]Université de Toulouse, UPS, Centre de Biologie du Développement (CBD), Toulouse, France
[†]CNRS, CBD UMR 5547, Toulouse, France
[1]Corresponding author: e-mail address: elise.cau@univ-tlse3.fr

Contents

Current Topics in Developmental Biology, Volume 106
ISSN 0070-2153
http://dx.doi.org/10.1016/B978-0-12-416021-7.00005-5

Abstract

The pineal gland is a small neuroendocrine organ whose main and most conserved function is the nighttime secretion of melatonin. In lower vertebrates, the pineal gland is directly photosensitive. In contrast, in higher vertebrates, the direct photosensitivity of the pineal gland had been lost. Rather, the action of this gland as a relay between environmental light conditions and body functions involves reception of light information by the retina. In parallel to this sensory regression, the pineal gland (and its accessory organs) appears to have lost several functions in relation to light and temperature, which are important in lower vertebrate species. In humans, the functions of the pineal gland overlap with the functions of melatonin. They are extremely widespread and include general effects both on cell protection and on more precise functions, such as sleep and immunity. Recently, the role of melatonin has received a considerable amount of attention due to increased cancer risk in shift workers and the discovery that patients suffering from neurodegenerative diseases, autism, or depression exhibit abnormal melatonin rhythms.

1. FUNCTION OF THE PINEAL GLAND

1.1. A conserved role in regulating circadian rhythms

Most living things (plants, animals, fungi, and cyanobacteria) adapt their activities according to the day–night cycle, via the so-called circadian clock. In humans, a number of parameters cycle during the day such as temperature, heart rate, concentration of cortisol in the plasma, and the production of red blood cells. In addition, the circadian clock controls a number of behaviors such as the sleep–wake cycles, feeding, and cognition rhythms. In animals, an important output of circadian activity is the rhythmic secretion of melatonin, which is secreted at night and absent during the day. Nocturnal secretion of melatonin is present in all species analyzed so far but is interpreted differently depending on whether the animal is nocturnal or diurnal.

Schematically, the establishment of a rhythmic secretion of melatonin involves (1) the production of a 24 h endogenous rhythm and (2) a photosensitive activity to phase this rhythm with the environment. In lower vertebrates, these two functions are under the control of a single organ: the pineal gland (which is also called epiphysis cerebri). Indeed, this small vesicle, which originates in the dorsal diencephalon, has the ability to both secrete melatonin and directly sense light. In contrast, in higher vertebrates, these two functions have been specifically separated. Indeed, while the pineal gland remains the main site of melatonin production, the light

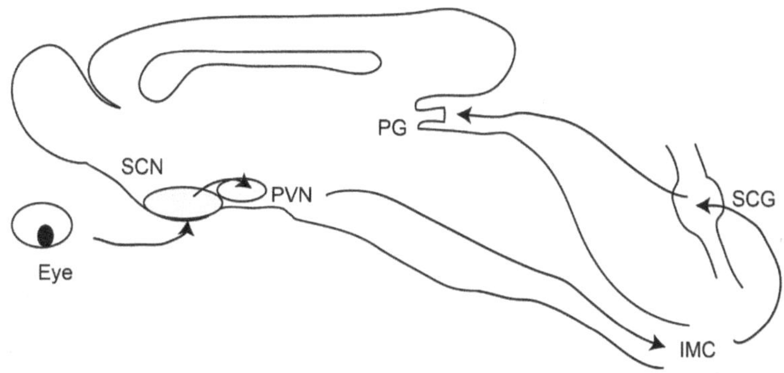

Figure 5.1 *Schematized drawing of the melatonin production circuit in mammals*: Circulation of neuronal information is indicated by an arrow. Dorsal is up and anterior is left. IMC, intermediolateral column of the spinal cord; PG, pineal gland; SCG, superior cervical ganglia; SCN, suprachiasmatic nucleus; PVN, paraventricular nucleus.

information is now sensed by the retina and travels through a complex route involving the suprachiasmatic nucleus (SCN) and the superior cervical ganglia (SCG) (Fig. 5.1), ultimately modulating pineal activity.

Interestingly, melatonin is found in most if not all invertebrates and vertebrates (Hardeland & Poeggeler, 2003; Reiter, 1991) and in very primitive unicellular aerobic organisms such as algae (Poeggeler, 1993). In mammals, several extrapineal nonendocrine sites of melatonin synthesis have been identified (Blask, Dauchy, Sauer, Krause, & Brainard, 2003; Bubenik, 2002; Pandi-Perumal et al., 2008; Slominski, Tobin, Zmijewski, Wortsman, & Paus, 2008). However, pinealectomy eliminates all detectable levels of melatonin in plasma, urine, saliva, and cerebrospinal fluid, pointing to a predominant role for the pineal gland in melatonin production (Lewy, Tetsuo, Markey, Goodwin, & Kopin, 1980) and suggesting that the extrapineal melatonin most probably reflects local production for autocrine or paracrine functions (Slominski et al., 2008; Tan et al., 2003).

How does the pineal gland generate a rhythm of melatonin secretion? Melatonin is produced from serotonin (see Fig. 5.2). The rate-limiting enzyme in this process is AANAT (for arylalkylamine *N*-acetyltransferase), which shows a greater activity at night. Two different strategies are employed to achieve this circadian regulation: Rodents, birds, and fish show a transcriptional control of *aanat*, while in sheep, cows, rhesus monkeys, and humans, transcription of *aanat* occurs uniformly throughout day and night phases but protein degradation is high during the day. Although rodents and

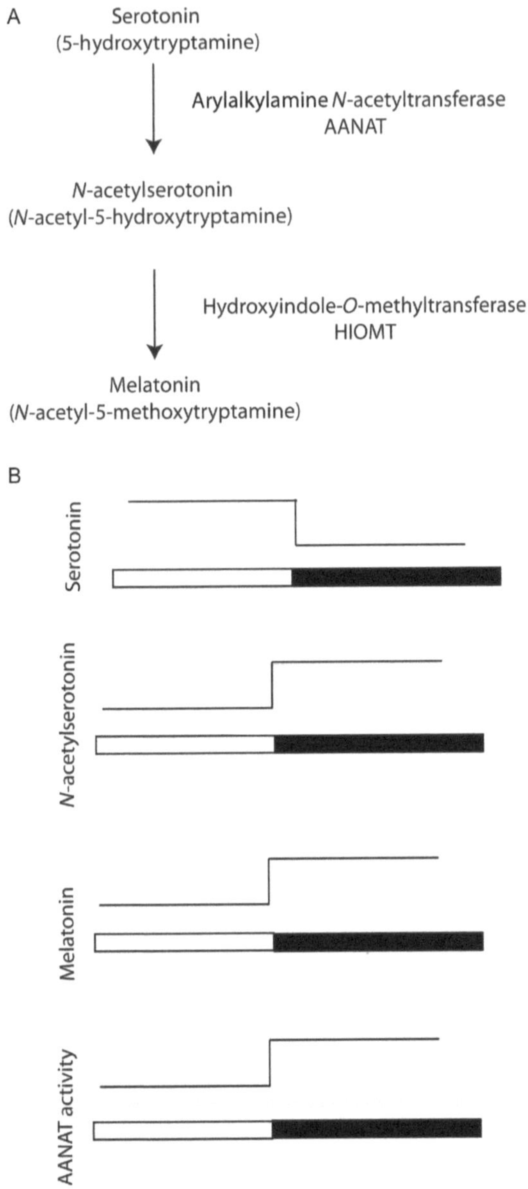

Figure 5.2 *Schematized metabolic pathway for melatonin*: (A) Melatonin is synthesized from serotonin by two enzymes: AANAT and HIOMT; (B) Relative levels of serotonin, *N*-acetylserotonin, and melatonin during light (white bar) and dark phases (black bar).

fish both rely on an *aanat* transcriptional control to achieve a rhythm, the mechanisms that are involved in these two groups of species are rather different. Zebrafish have two *aanat* genes, among which only *aanat2* is expressed in the pineal gland. In this species, the rhythm of *aanat2* transcription is thought to be autonomous to the pineal gland and to be a direct consequence of control by elements of the core circadian clock. Specifically, a heterodimer of two bHLH PAS transcription factors, BMAL and CLOCK, acts on an E-box located in *aanat* regulatory sequences (Appelbaum, Anzulovich, Baler, & Gothilf, 2005). Expression of genes encoding for BMAL and CLOCK shows circadian-dependent transcriptional cycle in a number of tissues including the pineal gland. Transcription of these genes peaks at the beginning of the night (Cermakian, Whitmore, Foulkes, & Sassone-Corsi, 2000; Whitmore, Foulkes, Strahle, & Sassone-Corsi, 1998). BMAL/CLOCK heterodimers both regulate their own transcription and activate the transcription of two groups of inhibitory elements the Period (Per) and Cry (Cryptochrome) genes. Heterodimers of Per and Cry, in turn, inhibit the transcriptional activity of Bmal/clock heterodimers. Therefore, these four groups of players (Bmal, Clock, Per, and Cry) drive a circadian clock analogous to what has been originally described in *Drosophila*. However, the presence of multiple copies of *bmal*, *clock*, *per*, and *cry* genes complicates the story. For instance, while transcription of the six different *cry* genes cycles during the day, they show slightly different cycles ranging from a peak in the morning (*cry3*) to a peak during the evening (*cry2a* and *cry2b*). Moreover, only Cry1a, Cry1b, Cry2a, and Cry2b have the ability to repress Clock/Bmal activity. Similarly, the three Bmal and the three Clock factors can form various combinations of heterodimers, which exhibit different efficiencies and sensitivity to Cry inhibition (see Vatine, Vallone, Gothilf, and Foulkes (2011) for a review). How does the Bmal/Clock/Per/Cry transcriptional loop integrate light information?

Light directly induces *per2* expression (Vatine et al., 2009). In fact, a 430 bp element located in the 5′ upstream regulatory sequences of the *per2* gene confers both light and clock regulation when placed upstream of an EGFP reporter gene and injected in zebrafish embryos. The pattern of expression driven by this element is ubiquitous but enhanced in the pineal gland, especially at embryonic stages. This regulatory region was narrowed down to a 67 bp element using a transfection-based assay in PAC2 cells; it contains both an E-box, which drives responsiveness to Clock/Bmal, and a D-box, which binds the thyrotroph embryonic factor (TEF), a bZIP transcription factor of the PAR subfamily. Expression of TEF is ubiquitous and enhanced in the pineal gland and shows a strong induction by light. In addition, Vatine et al. (2009) showed that TEF is responsible for *per2* induction by light. What drives the light responsiveness of *tef* expression and in particular what are the molecules that directly sense light to control this expression are questions that remain to be answered.

The control of *aanat* expression is very similar in birds and fish. In the chick pineal gland, transcription of the clock genes *cPer2*, *cCry1*, and c*Cry2* oscillates. Interestingly, cE4BP4, a bZIP transcription factor of the PAR subfamily, was found to bind the promoter of *cPer2* and to inhibit its transcription, which suggests a conserved role for PAR transcription factors in mediating the effects of light on transcription of *per* genes (Doi, Nakajima, Okano, & Fukada, 2001).

In rodents, *aanat* transcriptional oscillations are not driven autonomously within the pineal gland but occur as a result of nocturnal release of noradrenaline (NA) from the SCG. In these species, light activates atypical photoreceptors called ipRGCs (for intrinsically photosensitive retinal ganglion cells). IpRGCs are projection neurons (PNs) expressing a particular opsin belonging to the family of melanopsins, which are closely related to the rhabdomeric type of opsins. These cells project to the SCN, which functions as a self-sustained oscillator exhibiting oscillations in *bmal*, *clock*, and *per* expressions. The information given by the SCN travels through the paraventricular nuclei, the intermediolateral column of the spinal cord, and ultimately to the SCG (Fig. 5.1), which activates the pineal gland. The nocturnal release of NA in the pineal gland triggers cyclic AMP (cAMP) accumulation. Accumulation of cAMP leads to the activation of protein kinase A II, which phosphorylates the transcription factor CREB, thereby leading to its accumulation in the nucleus. Phospho-CREB directly activates transcription from the *aanat* promoter. The cyclic pattern of *aanat* expression is caused by the upregulation of ICER, a transcription factor of the CREB/CREM family, which accumulates during the night, as a result of NA activation. As ICER is able to displace pCREB from its binding site but inhibits its targets, the ratio of pCREB/ICER products, which oscillates during the day, is proposed to drive the oscillation in *aanat* expression (see Karolczak, Korf, and Stehle (2005) for a review). Finally, oscillations in the expressions of *clock*, *bmal*, and *per* genes were also described in the rodent pineal gland. These observations suggest that, even in rodents, the pineal gland acts as a circadian clock; however, the current effect of this pineal clock on melatonin production remains to be addressed.

In the last group of species (sheep, cows, rhesus monkeys, and human), AANAT protein stability varies between night and day, while *aanat* transcription is stable. These species also rely on the ipRGC → SCG circuit, which releases NA to the pineal gland (Maronde & Stehle, 2007; Stehle et al., 2011). Intracellular accumulation of cAMP induces AANAT phosphorylation by a cAMP-dependent kinase; once phosphorylated, AANAT binds to the dimeric protein 14-3-3 that can function to protect AANAT from degradation and increases its affinity for its substrate (Ganguly et al., 2001, 2005). Therefore, both AANAT stability and activities are increased at night.

Despite the use of very different mechanisms, in all animal species, the day–night cycle translates in nocturnal activity of ANAAT and therefore nocturnal secretion of melatonin. The physiological effects of melatonin and the detrimental effects of its misregulation are discussed at the end of this chapter.

1.2. Stereotypical nonvisual responses to light in anamniotes

While the best conserved function of the pineal gland is its involvement in circadian rhythm through melatonin secretion, the pineal gland exhibits additional functions in anamniotes. The first set of functions is the mediation of stereotypical motor responses in response to light. Experiments in *Xenopus* tadpoles show that the integrity of the pineal gland is required to elicit swimming upon dimming of the light (Foster & Roberts, 1982; Jamieson & Roberts, 2000). A similar function for the pineal gland has been described in another teleost fish, the cavefish *Astyanax mexicanus* whose larvae tend to swim upward upon light dimming, as do *Xenopus* tadpoles. This stereotypical response strongly diminishes upon surgical ablation of the pineal gland but not after eye removal (Yoshizawa & Jeffery, 2008). Recent data suggest that the pineal gland might not play similar functions in zebrafish. Zebrafish larvae show a number of stereotypical responses to light: They tend to swim and orient their body axis towards a light stimulus; they also show higher locomotor activity after loss of illumination and a quick diving to the bottom of the tank followed by a period where they swim upward. Surprisingly, genetic ablation of the pineal gland using a Tg(*tph2:NfsB-mcherry*) transgenic line does not affect these behaviors. In contrast, specific deep brain photoreceptors located in the preoptic area of the hypothalamus are important for these stereotypical responses (Fernandes et al., 2012).

The zebrafish pineal gland is also involved in a mechanism of background adaptation. Indeed, zebrafish embryos raised on black/gray or white background adapt to the color of the background. This mechanism involves the graded dispersion of pigment granules (the melanophores) within the fish pigment cells (the melanocytes). On a white background, embryos show aggregated melanophores, which make them look lighter. The dispersion/aggregation balance is regulated by two peptides: α-MSH (melanocyte-stimulating hormone), which is produced by the pituitary, and the MCH (melanocyte-concentrating hormone) secreted by the lateral tuberal nuclei (NLT) of the hypothalamus. Exposure to a black background specifically upregulates the secretion of α-MSH, while raising zebrafish on a white background promotes MCH secretion. The pineal gland has been thought to be involved in this process for many years as shining light or darkening the pineal area appeared to be sufficient to drive the adaptive response (Breder & Rasquin, 1950). Interestingly, the zebrafish pineal gland produces an antagonist of the melanocortin receptor (MC1R), the agouti-related protein (AgRP2). Embryos in which *agrp2* activity is reduced using a morpholino or in which pineal neurogenesis is impaired (*floating head* mutant (*flh*)) show a permanent dark color even in a white context. This lack in melanophore contraction correlates with low levels of pro-MCH and pro-MCH-like genes (*pmch* and *pmchl*) in the hypothalamus. Altogether, these results suggest that through the secretion of AgRP2, the pineal gland controls the levels of MCH produced from the NLT and thus melanophore contraction (Zhang et al., 2010). In addition, eye removal seems to produce a similar defect of white background adaptation and lack of upregulation of *pmch* and *pmchl* (Neuhauss et al., 1999; Shiraki, Kojima, & Fukada, 2010; Zhang et al., 2010). It remains to be addressed how background controls the secretion/activity of AgRP2 and to unravel the role of the eyes in this process.

1.3. Function of the pineal complex during thermic regulation in reptiles

In lizards, the pineal territory produces an accessory organ called the parietal eye (see Section 3.3). Experiments in which the parietal eye was surgically removed in *Anolis carolinensis* lizards suggest that this accessory organ is not involved in melatonin production (Underwood & Calaban, 1987). In contrast, the parietal eye plays a role during thermic regulation. Indeed, lizards are poikilothermic animals that adapt their body temperature to the temperature of their environment. When placed in a temperature gradient, they choose a high body temperature during the light phase and a lower body temperature during the dark phase. Upon parietalectomy, lizards of the *Sceloporus magister* species retain a light–dark cycle of body temperature but consistently choose a higher body temperature than control (Engbretson & Hutchison, 1976). Along the same line, analysis of the geographic distribution of lizard species with or without parietal eyes shows that species bearing this accessory organ populate regions of both low and high latitudes, while species that do not have a parietal eye are only found at low latitudes. Therefore, the parietal eye has been proposed to enable lizard species to adapt to the harder climates and more variable photoperiods encountered in the higher latitudes (Gundy, Ralph, & Wurst, 1975). Finally, interactions between the pineal gland and the parietal eye might exist as removal of the pineal gland in *Crotaphytus collaris* induces the opposite phenotype compared to what has been described for parietalectomy in *S. magister*, namely, a choice of lower body temperatures (Firth, Ralph, & Boardman, 1980). The observation that *Drosophila* rhodopsin is capable of sensing thermic differences in a light-independent manner (Shen et al., 2011) suggests that the parietal eye opsins could be implicated in sensing temperature independently of light. However, experiments in which the parietal eye function had been investigated using shielding with foil rather than ablation definitively implicate a light-detection role (Phillips & Howes, 1988).

2. CELLULAR COMPOSITION OF THE PINEAL GLAND

Associated with the evolution of the pineal gland from a photosensitive neuroendocrine organ in nonmammalian vertebrates to a dedicated neuroendocrine structure in mammals are dramatic changes in cell composition.

2.1. The pinealocyte: from neuroendocrine photoreceptor to neuroendocrine cell

The so-called pinealocyte is thought to be the main cell type present in the pineal gland. In fact, this generic term encompasses several distinct cell types, which are classically grouped in three categories:

- A typical photoreceptor (PhR)
- A modified photoreceptor (mPhR), which is often called a rudimentary photoreceptor in the literature
- A nonphotosensory pinealocyte (the pinealocyte sensu stricto PSS)

In lampreys, fish, and amphibians, most pinealocytes are true PhRs bearing a proper outer segment and the entire molecular cascade required to sense light. Hyperpolarizing electrophysiological responses of the pineal gland following illumination were for instance described in the goldfish and trout (Marchiafava & Kusmic, 1993; Meissl, Nakamura, & Thiele, 1986). Birds and reptiles also exhibit photosensitive pinealocytes, but their outer segments are less well developed than in lampreys, fish, and amphibians, and as such, they are considered to be mPhRs. In contrast, adult mammalian pinealocytes do not bear outer segments nor seem to be photosensitive. However, a number of genes involved in light detection are known to be expressed in the rat pineal gland from P0 to P110. Transcripts for rhodopsin, blue cone opsin (but not red opsin), cone transducin, rod transducin, cone phosphodiesterase, rod phosphodiesterase, recoverin, and cone arrestin were indeed observed. These observations led to the proposition that the rat pinealocyte could be photosensitive at least at early stages of postnatal development (Blackshaw & Snyder, 1997). Could this potential early photosensitivity play a physiological role? Fifteen- to twenty-day-old rats (P15–P20) show a circadian rhythm even when sympathectomized, in contrast to what occurs in the adult. This result suggests that their rhythm does not depend on the classical retina → SCN → SCG pathway but might rely instead on a direct photosensitivity from the pineal gland (Machado, Wragg, & Machado, 1969). What triggers the change between the early and the late rat pineal gland? Experiments in which rat pineals are put in culture suggest that, until P5, pineal cells can differentiate as rodlike PhRs, although definitive evidence that these cells are photosensitive is lacking. After P5, this ability is completely lost and interestingly the ability to generate PhR-like cells is strongly inhibited by the addition of NA to the culture medium. Therefore, the establishment of a sympathetic connection could explain the changes between the early and late rat pineal gland (Araki, 1992).

Finally, several studies report the expression of phototransduction molecules in the human pineal gland. For instance, Huang, Klein, and Korf (1992) described expression of rod opsin in a small fraction of adult human pinealocytes. Expression of S-arrestin (Perentes, Rubinstein, Herman, & Donoso, 1986) and recoverin (Korf, Korf, Schachenmayr, Chader, & Wiggert, 1992) was also described. The significance of these observations remains unknown.

Two distinct theories have been proposed to explain the ontogeny of pinealocytes. Collin and Oksche were the first to propose that the mammalian pinealocyte has evolved from an ancestral PhR, which is still found in the pineal gland of lower vertebrates (see Ekstrom and Meissl (2003) and Mano and Fukada (2007) for reviews). In favor of this theory of sensory regression, three arguments can be developed:

1. mPhRs have been described in the pineal of a few mammal species. Indeed, a careful ultrastuctural examination of the pinealocytes of the

young mole suggests the existence of three different categories of pinealocytes (I–III). The type I and type II are mPhRs, while the type III most closely resembles the classical pinealocyte. Type I and type II pinealocytes are only rarely observed in the adult mole (Pevet, 1980; Pevet & Collin, 1976). Pinealocytes bearing the morphology of mPhRs were also observed in the adult noctule bat (Pevet, Kappers, & Voute, 1977).

2. The observation that the early rat pinealocytes express phototransduction molecules and that these expressions decrease with age (see the preceding text) is another strong argument for the theory of sensory regression.
3. Finally, sensory regression has directly been observed in the pineal gland of the viviparous lizard *Lacerta vivipara*. Indeed, in this species, the reduced outer segment of these rudimentary photoreceptors, which compose the pineal gland, is a secondary event as the number of membranous disks that compose these outer segments progressively diminishes after birth (Meiniel, 1976).

More recently, Ekstrom and Meissl (2003) had developed a different view in which the PSSs have not directly evolved from a PhR; but instead the pineal stem cells are able to produce the three categories of pinealocytes. The proportion of stem cells that would choose one of the three fates would then vary between species and possibly evolve depending on the age of the animal. This theory is elegantly put forward by the authors: "It is not the mature photoreceptors that evolve; it is the developmental processes that shape them." The main argument for this theory is the observation that the three subtypes of pinealocytes coexist in a very ancient species such as the lamprey (Meiniel, 1980; Vigh-Teichmann, Vigh, & Wirtz, 1989). It also relies on a different interpretation of the experiment made by Araki in 1992. Indeed, an alternative explanation of this observation is that the rat neonatal pineal gland contains a stem cell that would generate either photosensitive or classical pinealocytes, depending on the stage or the culture condition. However, the stem cell capable of generating the three pinealocyte subtypes remains to be identified. In addition, if this theory is valid, one would expect to discover genes that when mutated modify the proportions between the different types of pinealocytes.

As mentioned previously, the pineal glands of several species contain morphologically distinct pinealocytes. Is there also variability in the pinealocyte population at the molecular level? In the following, we will focus on PhRs and mPhRs as these are the cell populations where diversity has been investigated most thoroughly.

Expression of the opsin photopigments characterizes the photoreceptors, which express them as they determine both the wavelength and the intensity of the light that is detected. Since a number of different opsins have been described in birds, fish, and reptiles (see the succeeding text), one of the key questions now is to understand whether distinct opsins characterize distinct PhR subtypes within a given species. This question also lends itself to understanding whether distinct specific PhR subtypes participate in distinct physiological functions in particular in fish and amphibians where the pineal gland exhibits several distinct functions.

Along this line, the requirement for different wavelengths to suppress melatonin production from perfused zebrafish pineal glands has been analyzed. These experiments suggest that at least two photopigments might participate in melatonin suppression: one photopigment that is sensitive to 408–512 nm (with an optimum at 500 nm) and one photopigment with a sensitivity ranging from 560 to 608 nm (optimum at 570 nm) (Ziv, Tovin, Strasser, & Gothilf, 2007). A similar requirement for two photopigments was inferred from experiments performed in the trout and in the pike (Falcon & Meissl, 1981; Kusmic, Passarelli, & Gualtieri, 1993). Interestingly, exorhodopsin and rhodopsin, two rod-type opsins sharing around 74% identity, have both been shown to be expressed in the zebrafish pineal gland (Mano, Kojima, & Fukada, 1999). Based on their homologies and several predicted structural features, the λ_{max} of exorhodopsin is predicted to be close to 492–503 nm, very close to the λ_{max} of rhodopsin (500 nm). Therefore, exorhodopsin and rhodopsin could contribute to the first peak of melatonin suppression. The red opsin, a classical cone opsin, which is also expressed in the zebrafish pineal gland (Mano et al., 1999), is a good candidate for the second peak with a λ_{max} at 570 nm (Krauss & Neumeyer, 2003). Finally, using RT PCR, two additional opsins were shown to be expressed in the pineal gland: parapinopsin b and parietopsin (Shiraki et al., 2010). Parietopsin is a green-sensitive opsin, which was first described in the lizard parietal eye (Su et al., 2006). It not only is structurally closer to the vertebrate visual opsins but also shares a few structural features of the so-called rhabdomeric opsins that are more prominent in invertebrates (Sakai et al., 2012). Parapinopsin is a UV-sensitive bistable photopigment that was first described to be expressed in the pineal and parapineal glands of the catfish (Blackshaw & Snyder, 1997). It is also expressed in the pineal and parapineal glands of lampreys. Interestingly, the pineal gland of the lamprey shows a remarkable regionalization with a dorsal part, which expresses rhodopsin and a ventral part expressing parapinopsin (Koyanagi et al., 2004). Finally, while rhodopsin and red cone opsin are the two only retinal opsins found in the zebrafish pineal gland, some other teleosts show expression of cone-UV and cone-green opsins in their pineal glands (Forsell, Ekstrom, Flamarique, & Holmqvist, 2001), possibly in relation with ecological differences between these species.

Pinopsin is a blue-sensitive opsin, which is a chimera of rod and cone opsins. Expression of parapinopsin, pinopsin, and parietopsin has been reported in the reptile pineal complex (Frigato, Vallone, Bertolucci, & Foulkes, 2006; Wada, Kawano-Yamashita, Koyanagi, & Terakita, 2012).

The bird pineal gland contains two types of pinealocytes: mPhRs and PSS. Expression of pinopsin, rhodopsin, and iodopsin (the chicken red cone opsin) has been described in the mPhRs, although the extent of coexpression between these three opsins remains to be addressed. The PSSs only seem to be able to express pinopsin (see Ekstrom and Meissl (2003) for a review). A melanopsin gene (*opn4*) has been shown to be expressed in the chick pineal gland and, interestingly, its mRNA cycles with a peak during the night (Bailey & Cassone, 2005; Chaurasia et al., 2005; Takanaka, Okano, Iigo, & Fukada, 1998).

While expression of several opsins was reported in the pineal glands of reptiles, fish, and birds, a considerable amount of work remains to understand whether these opsins characterize distinct types of PhR and to understand their specific functions. Interestingly, however, the expression of a number of opsin genes has been described to respond to light and/or exhibit a circadian rhythm, suggesting a possible involvement in the nocturnal secretion of melatonin. The best described case is *pinopsin* for which a light-responsive element (TGGCACGTGGGGTTCCTC) containing a central E-box was identified (Takanaka et al., 1998; Takanaka, Okano, Yamamoto, & Fukada, 2002).

2.2. The PN: an option found in the most primitive species

The true PhRs of the fish, bird, and reptile pineal gland have been shown to project onto secondary afferent neurons called projection neurons (PNs) that are functionally analogous to the retinal ganglion cells. Interestingly, the regression of the outer segment in birds and reptiles is associated with the diminution in the number of PNs. Intracellular recordings have shown that PNs are always active in the dark and that their activity is reduced upon light stimulation (Ekstrom & Meissl, 1988). This reduction of activity is thought to be a result of photon sensing by PhRs, which directly projects to PN. However, the observation that the chick pineal gland expresses melanopsin (Bailey & Cassone, 2005; Chaurasia et al., 2005) has raised the possibility that some PNs might be intrinsically photosensitive and could function as the ipRGC cells do. Finally, intracellular recording has led to the suggestion that the rainbow trout pineal gland might also contain interneurons (Ekstrom & Meissl, 1988); however, so far, these observations have never been corroborated in another species nor have those cells been characterized at the molecular level.

Whether PNs exist in mammals is much less clear. PNs have been described in the young ferret (Vigh & Vigh-Teichmann, 1993). Finally, enkephalin-positive neurons have been described in the human pineal gland (Moller, 1997; Moore & Sibony, 1988), but whether they represent PNs or specialized pinealocytes remains to be determined. An argument supporting the existence of PNs is the existence of a nerve coming out of the pineal gland. This *nervus pinealis* has been described at fetal stages and seems to connect the pineal gland to the posterior commissure. It has been described in several species including humans and disappears during fetal life (Moller, 1979). Whether this nerve is formed by PNs projections is not clear. However, together with the observation of an early photosensitivity in the rat pineal gland, the presence of this nerve in early life supports the existence of a shift in pineal function between fetal–early postnatal and adult stages.

2.3. Glial cells: a recent addition to the pineal gland

In mammals, two types of glial cells are observed (see Moller & Baeres, 2002).

- An astrocytic type, which is also called interstitial cell expressing GFAP
- A phagocyte or microglial type, which most likely bears antigen-presenting activity

Glial cells have been described in the lizard parietal eye (Engbretson & Linser, 1991) but not in other species of lower vertebrates, suggesting that, during evolution, these cells are a recent addition to the pineal gland.

3. ORGANOGENESIS OF THE PINEAL GLAND, PRODUCTION OF ACCESSORY ORGANS

As explained earlier, the pineal gland is highly variable not only in cell-type composition but also in morphology (see the succeeding text). Surprisingly, the first steps of its formation are very conserved among species. Indeed, the pineal organ develops as an evagination of the neuroepithelium

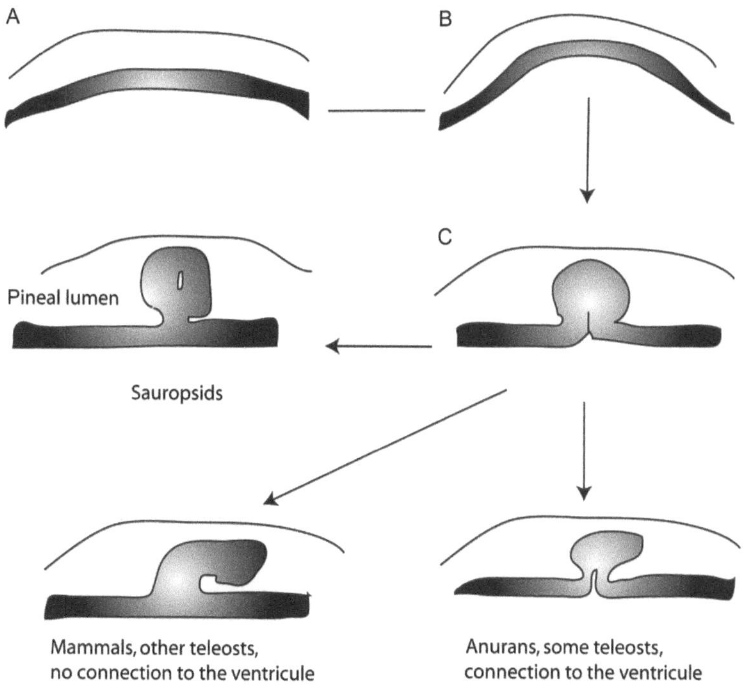

Figure 5.3 *Pineal gland development, convergence at early steps and divergence in adult forms*: Schematized frontal sections of the pineal territory during ontogeny. Dorsal is up. The pineal gland originates from dorsal diencephalic neuroepithelium (in gray, A, B), which evaginates (C). In the adult, a great diversity of morphologies is observed, in particular regarding the presence of a pineal lumen and the maintenance of a connection with the third ventricle (see text for details).

of the dorsal diencephalon (see Fig. 5.3A and B) above the future third ventricle. In a second step, this pineal evagination forms a little vesicle (Fig. 5.3C). Depending on the species, this vesicle retains a connection with the third ventricle or loses it (see Fig. 5.3). For instance, anurans and specific species of teleosts show a continuum between the pineal lumen and the third ventricle, while mammals and other teleost species have a compact pineal gland. In some species of sauropsids, a pineal lumen persists but does not show any continuity with the third ventricle. In addition, in these species, there is a wide range of pineal gland morphologies (see Quay (1965) for a review). There is also a wide range of structures observed for the bird pineal gland. Although variable in shape, the mammalian pineal glands always present a lobular structure (see Oksche, 1965).

Finally, in lower vertebrates, the "pineal anlagen" also produce accessory organs (called parapineal, parietal eye, and frontal organs; see the succeeding text) that are not present in birds or mammals (see Concha and Wilson (2001), Ekstrom and Meissl (2003), and Vigh et al. (2002) for reviews).

3.1. The parapineal organ

Parapineal organs have been described in the teleosts and lampreys. The lamprey parapineal gland is located beneath the pineal gland and occupies a medial position (Fig. 5.4). In zebrafish, the parapineal gland was shown to arise from the anterior third section of the pineal anlagen. Cells from this area migrate first to the left, then ventrally, and therefore end up on the ventral left side of the pineal gland (Fig. 5.4; see also Section 4; Concha et al., 2003; Gamse, Thisse, Thisse, & Halpern, 2003). Both pineal and parapineal glands are located beneath the skull. In addition, the zebrafish parapineal gland seems to be molecularly distinct from the pineal gland, as two

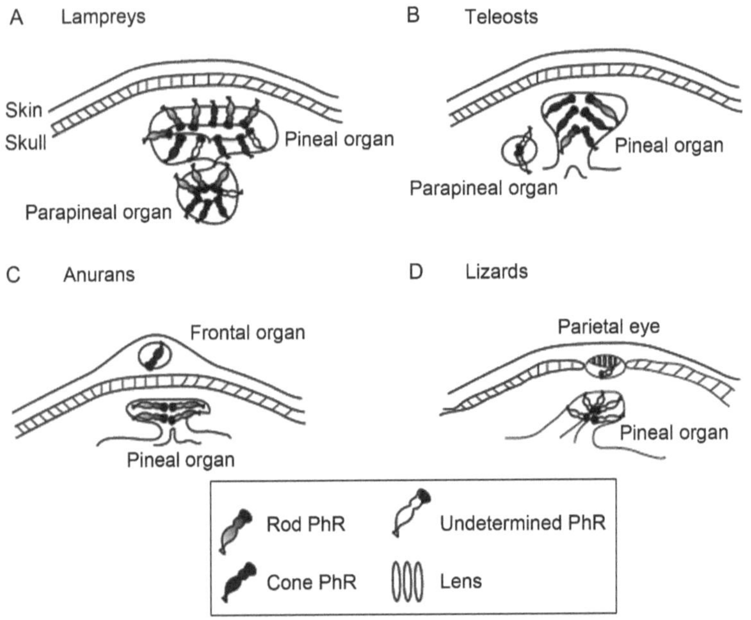

Figure 5.4 *The pineal and its accessory organs*: Schematized pineal territory showing the main pineal organ together with accessory organs (parapineal, frontal eye, and parietal eye) in various species. Frontal sections, dorsal up. Accessory organs can be found in either intracranial or extracranial positions. The type of PhRs found in these organs is indicated. In some cases, it is not clear whether the PhRs belong to the rod or the cone subfamily either due to the lack of available data and differences between species belonging to the same class or because the opsins expressed are not clearly affiliated to the rod or the cone family (i.e., parapinopsin and parietopsin). See text for more details.

transcription factors: *gfi-1.2* (*growth factor-independent*) and *sox1*a are expressed in the parapineal precursors but not in the pineal gland (Clanton, Hope, & Gamse, 2013; Dufourcq, Rastegar, Strahle, & Blader, 2004). The parapineal gland of channel catfish expresses a specific opsin, *parapinopsin*, but a few pineal cells also express this opsin (Blackshaw & Snyder, 1997). Similarly, *parapinopsin* expression is observed in the pineal gland of zebrafish (our unpublished observations). It thus remains to be addressed whether pineal and parapineal glands are devoted to sense different light wavelengths or if they share some light-detection activities. An even more general question that begs for an answer is the specific function of the parapineal gland.

3.2. The frontal organ in amphibians (also called the frontal eye)

Anurans show a frontal organ in addition to their pineal glands, while urodeles do not. In species where it has been described, the frontal organ is extracranial (located between the skull and the skin), while the pineal organ is intracranial. In amphibians, there is an apparent specialization of the pineal gland and its accessory organ as the pineal gland contains more rod-type PhRs, while the frontal organ contains more cones than rods (see Vigh et al. (2002) for a review). In contrast to the parapineal organ, the frontal organ has been suggested to arise from the caudal part of the pineal territory and would thus not be a parapineal homologue (see Concha and Wilson (2001) for a review).

3.3. The parietal eye in reptiles

The parietal eye of lizards originates from the anterior part of the pineal territory, making it a bona fide homologue of the parapineal gland. In addition, both the parapineal gland and the parietal eye project to the left habenula. The structure of the parietal eye strikingly resembles an eye with a dorsal lens and a ventral retina. This parietal eye contains all the cell types described in the pineal gland: PhRs, PNs, and glia (Engbretson & Anderson, 1990; Engbretson & Linser, 1991). Interestingly, the lizard parietal eye expresses *parapinopsin* and *parietopsin*, which are not found in their pineal gland (Wada et al., 2012). Along the same line, the parietal eye plays a unique role during thermic regulation in reptiles (see Section 1.3), suggesting a great extent of functional specialization for the main and accessory pineal organs.

4. DEVELOPMENT OF THE PINEAL GLAND

Zebrafish is a model of choice for genetic and pharmacological studies. Moreover, the zebrafish pineal gland has received a considerable amount of attention, if compared to the rodent pineal gland, owing to pioneering studies performed by Steve Wilson who identified the nerve coming out of the pineal gland as one of the earliest neuronal tractus formed in the zebrafish brain (Wilson & Easter, 1991; Wilson, Ross, Parrett, & Easter, 1990).

In zebrafish, the first sign of pineal specification is the expression of the *floating head* (*flh*) transcription factor. This homeodomain transcription is expressed specifically in the pineal anlagen (Masai et al., 1997). The Wnt and BMP pathways have been shown to act as cues that determine both the position and the size of the *flh* expression domain along the anteroposterior and dorsoventral axis, respectively (Barth et al., 1999; Masai et al., 1997). Loss of *flh* activity in the *flh* mutant leads to an almost complete loss of pineal neurons and to strongly reduced expression of two bHLH factors *ash1a* and *ngn1* (Cau & Wilson, 2003; Masai et al., 1997). The use of both morpholinos and mutants, which abolish *ash1a* and *ngn1* functions, revealed that these two genes are redundantly required for neuronal production (Aurelie Quillien, Patrick Blader and EC, unpublished results; Cau & Wilson, 2003). Since reduction of either *ash1a*, *ngn1*, or *flh* activity affects both the PhR and the PN compartments of the pineal gland, evidence is lacking that any of these genes function in the specification of PhR and PN identities. In contrast, activation of the BMP pathway has been shown to be both necessary and sufficient for the specification of the PhRs. On the other hand, Notch acts as an inhibitor of PN identity. Interestingly, the two pathways cross talk as BMP activity is required for efficient activation of Notch targets (Cau & Blader, 2009; Cau, Quillien, & Blader, 2008; Quillien et al., 2011). Altogether, these results suggest a model whereby establishment of a PhR fate relies on two events: the activation of a PhR program by the BMP pathway and the inhibition of a PN fate through Notch activation. In this model, the cross talk between Notch and BMP would ensure that activation of the proper PhR program and prevention of the inappropriate (PN) program would occur in the same cells. Several questions remain unanswered however; in particular, the mechanisms that drive PN specification remain elusive. Even on the PhR side, further work is needed to identify BMP targets in this system and understand how these targets act to drive PhR specification. Interestingly, *otx5* is expressed downstream of BMP, and while initial expression can be detected in a few PNs, it is specifically expressed in the PhRs at 48 h postfertilization (hpf, Quillien et al., 2011). *otx5* is crucial for the expression of several cycling genes (*aanat2*, *rev-erb-α*, and *irbp*) but not for genes that remain constantly expressed during the day cycle (*rx2*, *crx*, and *flh*) (Appelbaum et al., 2005; Appelbaum & Gothilf, 2006; Gamse et al., 2002). Further mechanistic studies have established that *otx5* directly regulates *aanat2* and *rev-erb-a* through the canonical photoreceptor-conserved element (PCE: TAATC) and the noncanonical bipartite pineal expression-related element (PERE GANNCTTA and TAAA) binding sites, respectively (Appelbaum et al., 2005; Appelbaum & Gothilf, 2006; Nishio et al., 2008). These results identify *otx5* as a major regulator of "circadian identity" in the PhR developmental program. Pierce, Noche, Ponomareva, Chang, and Liang (2008) had obtained results, which suggest that *otx5* might regulate other aspects of PhR identity. Indeed, reducing *otx5* activity using a specific morpholino strongly reduces *exorhod* expression. In addition, reducing *exorhod* activity in turn diminishes *aanat2* expression. Therefore, *otx5* could have both a direct effect on *aanat2* transcription and simultaneously a more general role on PhR specification (Pierce et al., 2008). Moreover, analysis of *exorhod* gene regulatory sequences suggests that *otx5* could directly regulate this gene's expression as CRX/OTX-binding sites were found in a 147 bp element located in the 5′ upstream region of the *exorhodopsin* gene, which is crucial for pineal expression. Finally, a small 12 bp element named PIPE (for pineal expression-promoting element: TGACCCCAATCT) also seems crucial for driving pineal-specific expression as it can generate artificial pineal expression when introduced into a construct containing *rhodopsin* regulatory sequences (Asaoka, Mano, Kojima, & Fukada, 2002). The transcription factors that bind to this PIPE element remain to be discovered.

Another intriguing question is: how does transcription of *aanat2*, *rev-erb-α*, and *irbp* cycle given that *otx5* transcription does not cycle? Different strategies must be employed as the three genes do not cycle in phase since *aanat2* expression peaks at night and *irbp* peaks during the day (Gamse et al., 2002). As described in the Section 1.3, this question has been partly solved in the case of *aanat2*. Indeed, a 257 bp element located in the 3′-flanking region of the *aanat2* gene is necessary for pineal-specific expression. This PRDM (pineal-restrictive downstream module) element contains three PCEs that mediate OTX5 effect on *aanat2* transcription (see the preceding text) and an E-box that is thought to bind a heterodimer of BMAL/CLOCK. Mutation in any of these four sites strongly impairs expression driven by the

PRDM element. Experiments in cell culture show a synergy of OTX5 and BMAL/CLOCK, therefore raising the possibility that a physical interaction exists between OTX5 and BMAL/CLOCK heterodimer. Moreover, *in vivo*, *otx5* and *bmal/clock* genes are required for efficient transcription driven by the PRDM element (Appelbaum et al., 2005; Appelbaum & Gothilf, 2006; Appelbaum et al., 2004).

Bsx1 is another candidate that could act downstream of BMP to specify a PhR fate. Experiments performed in Xenopus show that *bsx1* transcripts can be found in pineal PhRs during the day (but not at night). This expression seems to be regulated by light rather than circadian activity as the rhythm is lost when embryos are kept in the dark. Reduction of *bsx1* activity leads to a decreased PhR number and to an increased proliferation, while overexpression of *bsx1* generates the inverse phenotype (D'Autilia, Broccoli, Barsacchi, & Andreazzoli, 2010). These results suggest that *bsx1* could act to promote PhR differentiation at the expense of proliferation in a PhR fate-restricted progenitor.

The zebrafish pineal gland anlagen also produce a parapineal organ, which is usually located to the left side of the main, medial, and pineal organ. The cell-type composition of this small organ (around 9–12 cells) remains rather elusive. Experiments in which caged fluorescein was photoactivated in embryos bearing a Tg(*flh:egfp*) transgene show that parapineal precursors are located within the anterior third of the "pineal" anlagen at 24 hpf. In addition, these experiments show that the left-sided parapineal organ originates from both left and right sides of the "pineal" territory (Concha et al., 2003). By 28 hpf, parapineal cell precursors begin to migrate to the left of the pineal organ, forming a small "rosette-like" structure (Roussigne, Bianco, Wilson, & Blader, 2009). Both *tbx2b* and FGF (fibroblast growth factor) signaling have been shown to be important for parapineal specification. In a *tbx2b* mutant *from beyond (fby)*, parapineal precursors expressing *gfi-1.2* (*growth factor-independent*) can be identified albeit in smaller numbers, but their migration is impaired (Dufourcq et al., 2004; Snelson, Santhakumar, Halpern, & Gamse, 2008). On the other hand, reduction of FGF activity also results in a decreased number of *gfi1.2+* cells and in impaired parapineal migration, but experiments in which both *tbx2b* and FGF activities are affected show a cumulative phenotype, suggesting that *tbx2b* and FGF act independently of each other (Clanton, Hope, & Gamse, 2013). Moreover, the regulation of cell numbers and the control of parapineal migration are most likely independent events, as reduction of the number of parapineal precursors using laser ablation does not affect the migration of the remaining cells (Snelson et al., 2008). In addition, reduction of FGF activity induces an increased production of arrestin 3a (Arr3a)+pineal photoreceptors, suggesting that FGFs operate to control a binary parapineal versus Arr3a+PhR fate choice. In contrast, reducing *tbx2b* activity does not affect the numbers of Arr3a+PhRs (Clanton, Hope, & Gamse, 2013). Finally, FGF signaling is also required at a subsequent step to promote parapineal migration (Regan, Concha, Roussigne, Russell, & Wilson, 2009).

A pineal-specific homeobox transcription factor has not been identified yet in mammals, to the best of our knowledge. In particular, the well-characterized *floating head* homologue gene *Not1* is not found in the mouse diencephalon (Plouhinec et al., 2004). Pax6 has been proposed to be crucial for pineal specification based on the absence of a pineal gland in *sey/sey* (*small eye*) mutant mice; however, defects in Pax6-deficient embryos encompass several areas of the brain from the cortex anteriorly to the pretectum posteriorly, reflecting the wide expression of this gene in the anterior brain and its general patterning activity (Estivill-Torrus, Vitalis, Fernandez-Llebrez, & Price, 2001; Stoykova, Fritsch, Walther, & Gruss, 1996). A similar agenesis of the mouse pineal gland has been described in mouse in which *otx2* function is impaired in the territory of *crx* expression (Nishida et al., 2003). It is not clear, however, whether this defect reflects a patterning, determination, specification, or differentiation problem. Interestingly, this same study also shows that in absence of late *otx2* expression in the retina, less photoreceptors are produced, while an excess number of amacrine cells are generated, suggesting a photoreceptor→amacrine fate transformation. Conversely, overexpression of *otx2* is sufficient to force a photoreceptor fate. One of the key effector of *otx2* in this process appears to be *crx,* which is orthologous and functionally related to *otx5* (Gamse et al., 2002; Nishida et al., 2003; Plouhinec et al., 2003). The pineal gland of *crx*−/− mice seems morphologically normal, which suggests that *otx2* has additional targets. However, *crx*−/− mice exhibit photoentrainment defects. Indeed, when measured on an exercise wheel, the nocturnal activity of *crx*−/− animals is more fragmented and less nocturnal activity is detected as compared to wild-type. Moreover, *crx*−/− animals need more time to adapt their activities to a 4-h shift in the light–dark cycle. Finally, loss of *crx* function affects expression of *aanat* and two opsins (*rhodopsin* and *blue cone opsin*) in the pineal gland, while other genes such as *cone arrestin* are not affected. Altogether, these results suggest a role for *crx*, downstream of *otx2*, to regulate gene expression in the pineal gland (Furukawa, Morrow, Li, Davis, & Cepko, 1999).

5. PHYSIOLOGICAL ROLES OF MELATONIN

As already mentioned earlier, the best characterized pineal function is the cyclic production and release of melatonin in a daily pattern, with a peak at night in both diurnal and nocturnal organisms. The cyclic nocturnal melatonin production in the pineal gland therefore acts as the "chemical expression of darkness," which provides target organs with information about the duration of the night (Reiter, 1991). Other indolamines have been detected in the pineal gland of mammals, such as 5-methoxytryptophol (Wilson, Lynch, & Ozaki, 1978), which is also secreted in a circadian fashion by the pineal gland (Zawilska, Vivien-Roels, Skene, Pevet, & Nowak, 2000), although with an opposite phase to that of melatonin. Its biosynthesis and physiological activity have been so far poorly investigated, in marked contrast with the function of melatonin.

5.1. The many pathways of melatonin

In mammals, two melatonin membrane-bound receptor subtypes, MT1 and MT2, have been identified (Dubocovich, Masana, & Benloucif, 1999; Reppert et al., 1995; Reppert, Weaver, & Ebisawa, 1994). They belong to a specific group of GPCR, as they possess an NRY motif in the second intracellular loop rather than the usual DRY/ERY (Reppert, Weaver, & Godson, 1996). A third receptor had been identified and called MT3 as it was first assumed to be a new mammalian GPCR melatonin receptor but appeared to correspond to the enzyme quinone reductase II (QR2, Nosjean et al., 2000). Although this enzyme has well-described antioxidative properties, a function shared with melatonin (see the succeeding text), the potential action of melatonin binding to QR2 is currently not known.

Membrane melatonin receptor subtypes are expressed in a wide variety of tissues including the retina, SCN, pars tuberalis (PT), cerebral and peripheral arterial system, kidney, pancreas, adrenal gland, mammary tissue, gonads, and immune system, suggesting widespread activity for this hormone (Liu et al., 1997; Reppert et al., 1994). Membrane-bound melatonin receptors employ a variety of signal transduction pathways. Both MT1 and MT2 are coupled to different heterotrimeric Gi proteins containing α, β, and γ subunits, which mediate adenylate cyclase inhibition followed by a decrease in cAMP and PKA activity. The decrease in PKA activity further leads to a decreased phosphorylation of the transcription factor CREB (Reppert et al., 1995).

The lipophilic structure of melatonin allows it to freely pass through the cell membranes (Costa, Lopes, & Lamy-Freund, 1995), and as such, it has been detected in the nucleus of several cells located in the brain and various peripheral organs such as the pineal gland, gut, and kidney (Menendez-Pelaez & Reiter, 1993). Melatonin has further been shown to bind and activate members of the retinoic acid-related orphan receptors family, which belongs to the superfamily of steroid receptors (ROR/RZR): RORα, RZRα, and RZRβ (Becker-Andre et al., 1994; Wiesenberg, Missbach, Kahlen, Schrader, & Carlberg, 1995). Finally, melatonin has been reported to act as a very potent radical scavenger *in vitro*, independently of the presence of membrane or nuclear receptor (Edgar, Dement, & Fuller, 1993).

5.2. The role of melatonin in entraining circadian rhythms

As mentioned previously, the rhythmic secretion of melatonin by the pineal gland is the best described output of the circadian clock. Melatonin then provides time information to the whole organism including on peripheral oscillators (Hardeland, Madrid, Tan, & Reiter, 2012).

Interestingly, melatonin is itself capable of entraining circadian rhythms. As mentioned earlier, in mammals, the main circadian pacemaker is located in the SCN. Secretion of the pineal hormone melatonin not only acts as an output of the SCN activity but also impacts on the SCN, through melatonin membrane-bound receptors located in this nucleus. Exogenous melatonin administration induces two different effects on the SCN: an acute suppression of neuronal firing and a phase-shifting effect (entrainment) (Liu et al., 1997 and references therein). Inhibition of neuronal firing in the SCN is abolished in MT1 receptor knockout mice, whereas a phase-shifting response to melatonin persists in these mutants. As MT2 membrane-bound receptor mRNA has been detected in the mouse SCN, and the phase-shifting response depends on Gi proteins, a role for MT2 in the entrainment of the SCN clock has been proposed. The use of a pharmacological blocker of MT2 supports the notion of a contribution of this receptor subtype in the phase-shifting effect of melatonin in mouse. Therefore, the two different effects of acute melatonin administration on the SCN employ different receptors in the mouse. What is the mechanism underlying melatonin modulation of SCN activity? One possibility would be that melatonin regulates the oscillation of clock genes; however, such an effect has not been revealed (Poirel et al., 2003).

5.3. The role of melatonin during seasonal rhythms

In addition to its role in regulating the circadian system, melatonin also provides seasonal information in animals that react to seasonal variations in the length of the day (the so-called photoperiodic animals). The duration of the melatonin signal depends on the duration of the dark period, and consequently, winter is associated with a longer duration of melatonin secretion. Therefore, in photoperiodic animals, the endocrine message of the pineal gland gives information about changes in day length through the year. In that manner, it regulates seasonal functions such as color and growth of pelage, body weight variations, and reproductive activity. For instance, many rodents are long-day breeding animals, meaning that they are sexually active during summer and sexually depressed during winter, when nights are longer and melatonin secretion is at its maximal duration. In these animals, it has been shown that the reproductive quiescent period is prevented by ablation of the pineal gland, leading to the initial association of the melatonin with an antigonadotropic function. Moreover, it is now established that variations in the length of nocturnal melatonin secretion function as a chemical "calendar," which provides the organism with a notion of the season of the year. In addition, a specific duration of melatonin secretion, or the equivalent specific photoperiod, is necessary and sufficient to induce a particular reproductive response in seasonal breeders. Then, the message is interpreted differentially depending on the physiology of the species, allowing them to adapt their reproductive activity to the best conditions for the birth of their progeny (Weaver, Stehle, Stopa, & Reppert, 1993).

In the Syrian hamster, administration of exogenous melatonin during the day causes gonadal atrophy even when the experiment is performed during long days. Subsequently, treated animals eventually became unresponsive to melatonin injections and the gonads regenerate (Reiter, 1980). This has led to the assumption that the gonadal involution of long-day breeders during winter months is related to a desensitization of melatonin receptors, which depends on the early history of melatonin secretion episodes.

Melatonin affects reproductive activity by regulating hypothalamo–pituitary axis functions. Melatonin might modulate the feedback of steroids on the hypothalamic cells secreting the gonadotropin-releasing hormone. The seasonally controlled regulation of prolactin secretion by melatonin in the PT of the anterior pituitary is better documented. Several clock genes are expressed in the PT (Bmal1, Clock, Per1, Per2, Cry1, and Cry2) with a proper circadian rhythm independent of the SCN clock (Lincoln, Messager,

Andersson, & Hazlerigg, 2002). Melatonin, secreted at night under the control of the SCN, is able to influence Per1 expression in the PT, which thus translates the "signal of darkness" directly to the peripheral oscillator in the PT.

Although this function in regulating seasonal rhythm appears to be of questionable relevance in humans, which are not photoperiodic organisms, it is interesting to note that the symptoms of some mood disorders (seasonal affective disorders) also follow a seasonal cycle.

5.4. Regulation of sleep patterns

The relationship between the pineal hormone and sleep was identified by Aaron Lerner, the researcher who initially isolated melatonin from bovine pineal extracts. Indeed, he and his coworker reported that administering high pharmacological doses of melatonin to human subjects induces sleepiness (Nordlund & Lerner, 1977). This initial finding was confirmed by several other studies using either pharmacological or physiological doses of melatonin and which concluded that melatonin administered during the day is able to increase sleepiness and to decrease the latency of sleep onset (Zhdanova et al., 1995). However, the effects of daytime administrations are time-dependent, and melatonin administration at night, corresponding with high endogenous levels, gave no consistent results on the promotion of sleep (Tzischinsky & Lavie, 1994; Wyatt, Dijk, Ritz-de Cecco, Ronda, & Czeisler, 2006).

Regulation of sleep is a complex process influenced by two separate components: first, a circadian component that corresponds to SCN activity driving the circadian rhythm of sleep–wake and, second, a homeostatic component that independently dictates the "need" for sleep depending on the duration and amount of sleep during the previous rest episode (Borbely, 1982). Studies in SCN-lesioned primates revealed that the SCN is probably responsible for the high level of wakefulness encountered in the early evening (Edgar et al., 1993). This period is followed by a rapid transition to a phase of high propensity for sleep, which coincides with—and is tightly linked to—the time of the nocturnal rise in melatonin secretion (Lavie, 1997; Tzischinsky & Lavie, 1994).

Although the role of melatonin in sleep regulation is still unclear, one of the most commonly accepted hypothesis is that melatonin facilitates sleep through an inhibitory action on the SCN, which promotes the wake state during the early evening (Lavie, 1997). However, melatonin inhibition of multiunit activity in the SCN also occurs in nocturnal animals in which the melatonin peak coincides with the period of maximal activity (Lavie, 1997).

Although melatonin is often described primarily as a hypnotic, due to its facilitating effect on sleep, it is important to note that melatonin cyclic secretion is independent of sleep–wake rhythms (Morris, Lack, & Barrett, 1990). Moreover, diurnal and nocturnal animals both have elevated levels of melatonin at night, which corresponds to their phase of activity, suggesting that melatonin acts as a hypnotic only in diurnal species in which sleep is appropriate during the dark period (Mendelson, Gillin, Dawson, Lewy, & Wyatt, 1980).

Core body temperature is maximal in the early evening and minimal in the early morning. As the nocturnal peak of melatonin secretion coincides with the trough in core body temperature (Hughes & Badia, 1997), it has been suggested that sleep regulation by melatonin may involve an effect on thermoregulation (Reid, Van den Heuvel, & Dawson, 1996) possibly through a vasodilatory effect on peripheral vasculature (Murphy & Campbell, 1996).

5.5. Cell protection

Free radicals are normal by-products of oxygen metabolism. Some of these chemical species are highly toxic for lipids, proteins, and DNA. Accumulation of damaged cell components (enhanced production) and/or decreased ability to eliminate the hydroxyl radical contributes to aging and age-related pathologies (Harman, 1993, 1994), and its capacity to damage DNA increases the risk to develop cancer (Ames, Shigenaga, & Hagen, 1993).

Melatonin has been reported to act as a very potent radical scavenger *in vitro* (Tan et al., 1993). Moreover, after radical scavenging, the melatonin product is not toxic by itself, but on the contrary is able to neutralize a second free radical.

In vivo, exogenous melatonin administration in rats limits the oxidative damage induced by the chemical carcinogen safrole (Tan et al., 1993) that is supposed to act via the production of toxic free radicals in the cell nucleus. When the carcinogen was given at night, during the phase of elevated endogenous levels of melatonin, DNA damage was 20% lower than in rats treated with the carcinogen only during the day. Finally, the authors showed that pinealectomy enhances the DNA damage induced by safrole as compared with the normal daytime conditions. Altogether, these results demonstrated that physiological levels of melatonin are able to counteract oxidative damage induced by a chemical carcinogen (Tan et al., 1994).

Moreover, glutamate transmission generates free radicals in the postsynaptic cells (Dykens, Stern, & Trenkner, 1987). SCN input from the retinohypothalamic tract (RHT) is glutamatergic and melatonin may act in these neurons as an antioxidant to protect them from oxidative degeneration. Finally, melatonin was also reported to be protective against ionizing radiations (Blickenstaff, Reddy, & Witt, 1994).

5.6. Impact of melatonin on nervous system development

Melatonin receptors are highly expressed in embryos including in zebrafish embryos, which has led to question the role of this hormone during development. Melatonin administration has a general effect on cell proliferation and accelerates the development of zebrafish embryos. Opposite effects are obtained upon treatment with melatonin receptor antagonists, which suggests that melatonin promotes cell division (Danilova, Krupnik, Sugden, & Zhdanova, 2004). In contrast, blocking melatonin receptor activity leads to a delayed neurogenesis in the habenulae in a manner that suggests that melatonin is required for cell cycle exit of neural progenitors (de Borsetti et al., 2011). In addition, the fact that the habenulae are located in close vicinity to the pineal gland suggests that this regulation could involve a short-range direct effect of melatonin. Despite these promising pioneering studies, the study of the developmental role of melatonin is still in its infancy.

5.7. Impact of melatonin on memory formation

Memory is a complex process involving distinct phases: acquisition, consolidation, and retrieval. Animal models can be tested for their performance during these different steps using specific behavioral paradigms. Chaudhury and Colwell (2002) had shown that the circadian clock influences the three different levels of memory formation. Indeed, performance in the fear-conditioned paradigm (association of a tone with an electric footshock) is higher if the training is performed during the day, suggesting that the circadian rhythm might influence memory acquisition. Recall is also higher when tested during the day and long-term extinction of memory occurs quicker if the animal was trained at night. Thus, surprisingly, mice, which are nocturnal animals, learn better during the day, meaning at their inactive phase. What remains to be clarified is whether performance in other tests using positive conditioning would show the same day–night pattern. Finally, in this study, melatonin does not fully account for the effects of circadian rhythms on memory, and the comparison between two strains of mouse, one with and one without rhythm of melatonin secretion, suggests that while this hormone most likely controls the rapidity at which memory is acquired, it does not control the consolidations and retrieval phases (Chaudhury & Colwell, 2002).

Using an active avoidance conditioning test, Rawashdeh et al. demonstrate that, just as mice, fish learn better at day and that melatonin has adverse effect on mnemonic performance, while pinealectomy improves it. However, in contrast to what is observed in mice, melatonin seems to affect memory consolidation (Rawashdeh, de Borsetti, Roman, & Cahill, 2007).

How does melatonin inhibit memory formation? In rodents, this hormone had been shown to directly act on the dentate gyrus neurons of the hippocampus. Indeed, melatonin inhibits the modifications of synaptic strength, which constitute the basic cellular mechanism for memory (the so-called long-term potentiation) (see Rawashdeh and Maronde (2012) for a review). The topic of the adverse effect of melatonin on mnemonic performance is obviously crucial as melatonin is being administered to patients and used by certain populations to alleviate jet lag effects. Surprisingly, the data examining this topic on human subjects are scarce and mostly concern people falling into specific categories such as the sick or aged population (Rawashdeh & Maronde, 2012).

5.8. Stimulation of immune function

Early studies in this field showed that surgical pinealectomy causes an arrest in thymic cell proliferation and depresses the immune system (Csaba & Barath, 1975). Removal of the pineal gland also results in a reduced number of circulating immune cells, such as lymphocytes and leukocytes (Beskonakli et al., 2001). In several species, this function follows daily and seasonal fluctuations. Indeed, this function is enhanced in short days and correlates with the length of melatonin secretory episodes (Nelson & Demas, 1996).

6. PATHOLOGICAL DYSREGULATION OF MELATONIN SYNTHESIS/SIGNALING

6.1. Absence or dysphasing of rhythmic melatonin signaling

Several accidental and developmental conditions have been described that affect the rhythmic melatonin production.

Accidental severing of the spinal cord can induce an absence of melatonin production depending on the level at which the injury occurs. Such injuries most likely affect the sympathetic innervation of the pineal gland (see Zeitzer, Ayas, Shea, Brown, & Czeisler, 2000 and references therein).

Craniopharyngiomas are benign tumorous malformations occurring during embryonic life that can affect the hypothalamic and pituitary regions. While they do not impact on life prognostic, these embryonic malformations lead to health alterations including daytime sleepiness. Interestingly, patients exhibiting craniopharyngiomas display decreased melatonin secretion at night, most likely as a result of SCN impairment. This abnormal secretion of the pineal hormone could account for the alterations in sleep pattern (Muller, Handwerker, Wollny, Faldum, & Sorensen, 2002).

Smith–Magenis syndrome is a congenital disorder with characteristic behavioral and developmental abnormalities and a completely inverted rhythmic melatonin secretion, which rises during the day (De Leersnyder, Claustrat, Munnich, & Verloes, 2006). Patients often reported sleep difficulties, which can be alleviated by artificially reproducing the normal amplitude melatonin rhythm. This can be achieved using a combination of melatonin administration in the evening to compensate for the low endogenous levels with β-adrenergic blockers during the day (Carpizo et al., 2006; De Leersnyder et al., 2006). The main cause of Smith–Magenis syndrome is an haploinsufficiency of the retinoic acid-induced 1 gene (RAI1). Studies in a mouse model for this syndrome reveal that RAI1 regulates expression of circadian genes in the hypothalamus and in particular that RAI1 acts as a positive regulator of *clock* (Williams, Zies, Mullegama, Grotewiel, & Elsea, 2012).

6.2. Polymorphisms impairing normal melatonin biosynthesis and signaling

Low levels of circulating melatonin have been consistently reported in patients with autism spectrum disorders (ASD) (Kulman et al., 2000; Nir et al., 1995; Tordjman, Anderson, Pichard, Charbuy, & Touitou, 2005) although the relationship between pineal hormone deficiency and the possible susceptibility to ASD remains elusive. Recently, it has been found that the pseudoautosomal region 1 of the sex chromosomes is deleted in several cases of ASD. Interestingly, this chromosomal region contains the gene encoding acetylserotonin-*O*-methyltransferase (ASMT), the last enzyme in the synthesis of melatonin. The ASMT gene has been finally associated with susceptibility for ASD (Melke et al., 2008).

Low levels of melatonin production and secretion are not the only causes of defective melatonin signaling. For instance, numerous convergent data point to an association of the human polymorphisms in MT2 receptor subtype and an increased risk to develop type 2 diabetes (Lyssenko et al., 2009; Staiger et al., 2008). This is in line with the changes in insulin secretion resulting from melatonin receptor mutants in mouse (Peschke et al., 2002), although in this species, MT1 is the principal mediator of this effect on pancreatic β-cells. It nevertheless appears that melatonin functional signaling is needed to prevent the development of type 2 diabetes.

Depression is also associated with polymorphisms in several different receptors (RORα and MT1) and enzymes of the melatonin synthesis pathway (AANAT and ASMT). Although a fully detailed report of these genetic variants would exceed the scope of this chapter, it is interesting to note that polymorphisms in human melatonin receptors have been associated with numerous other pathologies, providing evidence for a potential involvement of the pineal hormone in physiological processes that have been so far poorly explored, such as Parkinson disease (QR2), rheumatoid arthritis and idiopathic scoliosis (MT2), polycystic ovarian syndrome (MT1), or macular degeneration (RORβ) (Hardeland et al., 2012).

6.3. Melatonin deficiency in aging and neurodegenerative diseases

Decreased nocturnal levels of circulating melatonin have been consistently detected during normal aging (Reiter & Richardson, 1992; Srinivasan et al., 2005), leading to fragmented sleep–wake patterns (Carskadon, Brown, & Dement, 1982).

Calcification of the pineal gland starts very early in life, probably at birth, and since this phenomenon increases with age, it has been proposed to be the origin of the age-related decline in melatonin production. Whether this is indeed the case is however not clear (Kunz et al., 1999; Schmid, 1993). In contrast, disruption of the RHT that transmits the photic sensory input to the SCN or neuronal loss in this nucleus and altered rhythmic activity of SCN neurons (see Hofman (2000) for a review) lead to a reduced amplitude in the melatonin rhythmic secretion or in dysphasing melatonin secretion compared to the day–night alternance.

Impaired afferent noradrenergic innervation to the pineal gland is another cause of age-related decline in melatonin production (Jengeleski, Powers, O'Connor, & Price, 1989). Finally, highly diverse pathological or environmental conditions specific to the elderly could also indirectly impact the melatonin rhythmic secretion such as the reduced levels of illumination due to a more sedentary way of life (Mishima, Okawa, Shimizu, & Hishikawa, 2001), ocular mydriase, and cataract, which both impair light detection by ipRGCs.

Individuals with neurodegenerative disorders such as Parkinson's disease (PD) or Alzheimer's disease (AD) show a stronger diminution of melatonin rhythms compared to controls of the same age (Bordet et al., 2003; Skene et al., 1990). Are the effects of AD and PD on melatonin secretion only a simple consequence or can they contribute to the etiology of these neurodegenerative disorders?

Melatonin levels in the cerebrospinal fluid are reduced and rhythmic day–night fluctuations flattened even in preclinical stages of AD pathogenesis, in subjects who do not already display cognitive dysfunctions. It has thus been proposed that a concomitant deregulation of SCN activity and/or associated sympathetic innervation to the pineal gland occurs during the early steps of AD and results in defective melatonin secretion. Finally, light therapy is known to alleviate behavioral and cognitive symptoms and sleep difficulties in AD patients (McCurry et al., 2011). Melatonin administration also improves cognitive functions and decreases agitation, probably via its ability to reduce both oxidative stress and neuronal overexcitation (reviewed in Pappolla et al., 2000; Reiter, 2002). Indeed, mitochondrial dysfunction is characteristic of numerous neurodegenerative pathologies.

PD is characterized by a degeneration of the nigrostriatal dopaminergic system. Among several possible causes, oxidative stress and inflammation are the most commonly admitted to explain dopaminergic neural loss in PD (Tansey, McCoy, & Frank-Cannon, 2007; Zhang et al., 1999). Melatonin has been reported to counteract lipid peroxidation in the brain and to prevent neuronal loss in the nigrostriatal pathway in a PD model (Acuna-Castroviejo, Coto-Montes, Gaia Monti, Ortiz, & Reiter, 1997; Antolin et al., 2002). Expression of MT1 and MT2 melatonin receptors is reduced in the brain of PD patients in particular in the substantia nigra, probably as a consequence of neuronal loss (Adi et al., 2010). This result indicates that melatonin signaling impairment might be involved in the physiopathology of PD progression. Indeed, as explained earlier, both the pathway stimulating melatonin secretion and the ability of the brain targets to respond to the hormone are affected in PD patients. This situation leads to a general deficit in the antioxidative and neuroprotective effects of melatonin, which would normally dampen PD symptoms and as a consequence potentially accelerates the degenerative process.

6.4. Melatonin suppression by light at night and cancer risk

Light at night, and more precisely in the blue wavelength range, acutely suppresses pineal melatonin production and secretion (Brainard et al., 2001; Lewy et al., 1980; Thapan, Arendt, & Skene, 2001).

The increasing use of electric light by humans has extended our activity well beyond the natural day. It has been postulated that lighting until late at night or at inappropriate times could impact negatively on melatonin production and have deleterious consequences for health. This is particularly relevant for night-shift workers who experience repeated and prolonged periods of melatonin deficiencies that are not compensated by following rest during the day due to dysphasing between the internal and the "social clock."

Based on the fact that light-induced melatonin suppression increases estrogen production, studies were conducted in women shift worker to test whether the risk for hormone-dependent cancer would increase in these populations. The best documented example is breast cancer for which remarkable differences have been consistently found in women depending on their exposition to light at night. Several studies correlated the low levels of urinary 6-sulphatoxymelatonin before diagnostic of breast cancer (Schernhammer et al., 2008) and found more than 50% elevation in the risk to develop breast cancer in nurses exposed to rotating shift work (Hansen, 2001; Schernhammer et al., 2008; Schernhammer, Kroenke, Laden, & Hankinson, 2006; Schernhammer et al., 2001). Other studies have confirmed a link between nonday shift work and risk of breast cancer in women (Tynes, Hannevik, Andersen, Vistnes, & Haldorsen, 1996).

On the contrary, several studies found that totally blind women have a strongly reduced breast cancer risk possibly because they retain high levels of melatonin throughout the day (Flynn-Evans, Stevens, Tabandeh, Schernhammer, & Lockley, 2009; Kliukiene, Tynes, & Andersen, 2001). Similar observations reported an elevated risk of prostate cancer in men exposed to light at night during shift work (Sigurdardottir et al., 2012).

Shift work increases the risk to develop cancer, and in agreement with these observations, constant light exposure has effect on cancer formation in animal models. For instance, rats exposed to constant bright light (a condition that completely abolishes the nocturnal rise in melatonin secretion) starting 1 week before tumor cell implantation develop tumors in a more rapid fashion when compared to rats exposed to light–dark cycles. In addition, the tumor growth rate is increased upon constant light exposure (Blask et al., 2003; Dauchy, Blask, Sauer, Brainard, & Krause, 1999). What is the molecular mechanism behind the effect of melatonin/light on cancer?

One such mechanism involves the regulation of linoleic acid (LA) metabolism and uptake. LA is the most common omega-6 polyunsaturated fatty acid present in food and it has been associated with a stimulatory effect on the growth of established tumors (Sauer, Dauchy, & Blask, 1997). In the absence of melatonin, the levels of LA uptake and metabolization by tumors are increased, which results in an accelerated tumor growth. Similarly, administration of different light intensities that result in a dose-dependent reduction in melatonin signaling produces a dose-related stimulation of tumor growth, LA uptake, and metabolism for rat hepatic tumors (Dauchy, Sauer, Blask, & Vaughan, 1997).

A second pathway for the oncostatic actions of melatonin is the induction of the expression of the tumor-suppressor protein p53 (Mediavilla, Cos, & Sanchez-Barcelo, 1999). This protein is usually present in normal cells. Permanent and rapid degradation via the ubiquitin proteasome pathway maintains p53 at low levels in healthy cells (Haupt et al., 1997). Upon induction of DNA damage, the p53 pathway is activated, thereby leading to growth inhibition, or apoptosis (Oren, 2003; Unger et al., 1999). Melatonin promotes accumulation of p53 and its phosphorylation on Ser15, without any induction of DNA damage. This induction leads to a reduction in cell proliferation in both normal and transformed cells (Santoro, Marani, Blandino, Muti, & Strano, 2012).

A third pathway is the regulation of clock genes. Indeed, cell division follows a circadian rhythm. Moreover, cell cycle progression and apoptosis are under the control of clock genes, and clock genes are expressed in cancer cells, and mutations/polymorphisms in clock genes are associated with cancer risk and development (Lahti, Merikanto, & Partonen, 2012).

7. CONCLUSION

Despite the pleiotropic and crucial roles described for melatonin during normal life, and despite the discovery of numerous developmental regulators of pineal gland formation, there is, to the best of our knowledge, no evidence for congenital malformations of the pineal gland. In contrast, dysregulation of pineal function is suspected to act as a cause for increased cancer risks. Abnormal pineal activity (in particular linked to impaired pineal afferences) is also responsible for a number of health problems, in particular in the cases of AD and PD patients. While a causal role for pineal malfunction in AD and PD has yet to be established, abnormal functioning of the pineal gland considerably worsens clinical prognosis and the patient's quality of life.

The significance of the remnant expression of opsins and phototransduction molecules in the mammalian pineal gland is a surprisingly unexplored field. Recent studies highlighting a more widespread role for opsins in sensing environmental changes such as temperature (Shen et al., 2011) and possibly sounds (Senthilan et al., 2012) have opened new possibilities to explore.

Finally, while both melatonin and its receptors are present in the embryo, the effects of this hormone on developmental processes remain poorly explored. No doubt that future studies will reveal new roles for the "the chemical expression of darkness" during embryogenesis.

ACKNOWLEDGMENTS

We are indebted to Dina Arvanitis, Nick Foulkes, and members of our lab for critical comments on the manuscript. This work was supported by the CNRS, Inserm, Université Paul Sabatier, Fondation Retina France, and ARC.

REFERENCES

Acuna-Castroviejo, D., Coto-Montes, A., Gaia Monti, M., Ortiz, G. G., & Reiter, R. J. (1997). Melatonin is protective against MPTP-induced striatal and hippocampal lesions. *Life Sciences*, *60*(2), L23–L29.

Adi, N., Mash, D. C., Ali, Y., Singer, C., Shehadeh, L., & Papapetropoulos, S. (2010). Melatonin MT1 and MT2 receptor expression in Parkinson's disease. *Medical Science Monitor*, *16*(2), BR61–BR67.

Ames, B. N., Shigenaga, M. K., & Hagen, T. M. (1993). Oxidants, antioxidants, and the degenerative diseases of aging. In: *Proceedings of the National Academy of Sciences of the United States of America*, *90*(17), 7915–7922.

Antolin, I., Mayo, J. C., Sainz, R. M., del Brio Mde, L., Herrera, F., Martin, V., et al. (2002). Protective effect of melatonin in a chronic experimental model of Parkinson's disease. *Brain Research*, *943*(2), 163–173.

Appelbaum, L., Anzulovich, A., Baler, R., & Gothilf, Y. (2005). Homeobox–clock protein interaction in zebrafish. A shared mechanism for pineal-specific and circadian gene expression. *The Journal of Biological Chemistry*, *280*(12), 11544–11551.

Appelbaum, L., & Gothilf, Y. (2006). Mechanism of pineal-specific gene expression: The role of E-box and photoreceptor conserved elements. *Molecular and Cellular Endocrinology*, *252*(1–2), 27–33.

Appelbaum, L., Toyama, R., Dawid, I. B., Klein, D. C., Baler, R., & Gothilf, Y. (2004). Zebrafish serotonin-N-acetyltransferase-2 gene regulation: Pineal-restrictive downstream module contains a functional E-box and three photoreceptor conserved elements. *Molecular Endocrinology*, *18*(5), 1210–1221.

Araki, M. (1992). Cellular mechanism for norepinephrine suppression of pineal photoreceptor-like cell differentiation in rat pineal cultures. *Developmental Biology*, *149*(2), 440–447.

Asaoka, Y., Mano, H., Kojima, D., & Fukada, Y. (2002). Pineal expression-promoting element (PIPE), a cis-acting element, directs pineal-specific gene expression in zebrafish. In:

Proceedings of the National Academy of Sciences of the United States of America, 99(24), 15456–15461.
Bailey, M. J., & Cassone, V. M. (2005). Melanopsin expression in the chick retina and pineal gland. *Brain Research Molecular Brain Research, 134*(2), 345–348.
Barth, K. A., Kishimoto, Y., Rohr, K. B., Seydler, C., Schulte-Merker, S., & Wilson, S. W. (1999). Bmp activity establishes a gradient of positional information throughout the entire neural plate. *Development, 126*(22), 4977–4987.
Becker-Andre, M., Wiesenberg, I., Schaeren-Wiemers, N., Andre, E., Missbach, M., Saurat, J. H., et al. (1994). Pineal gland hormone melatonin binds and activates an orphan of the nuclear receptor superfamily. *The Journal of Biological Chemistry, 269*(46), 28531–28534.
Beskonakli, E., Palaoglu, S., Aksaray, S., Alanoglu, G., Turhan, T., & Taskin, Y. (2001). Effect of pinealectomy on immune parameters in rats with Staphylococcus aureus infection. *Neurosurgical Review, 24*(1), 26–30.
Blackshaw, S., & Snyder, S. H. (1997). Parapinopsin, a novel catfish opsin localized to the parapineal organ, defines a new gene family. *The Journal of Neuroscience, 17*(21), 8083–8092.
Blask, D. E., Dauchy, R. T., Sauer, L. A., Krause, J. A., & Brainard, G. C. (2003). Growth and fatty acid metabolism of human breast cancer (MCF-7) xenografts in nude rats: Impact of constant light-induced nocturnal melatonin suppression. *Breast Cancer Research and Treatment, 79*(3), 313–320.
Blickenstaff, R. T., Reddy, S., & Witt, R. (1994). Potential radioprotective agents—V. Melatonin analogs. Oral activity of p-aminopropiophenone and its ethylene ketal. *Bioorganic & Medicinal Chemistry, 2*(10), 1057–1060.
Borbely, A. A. (1982). A two process model of sleep regulation. *Human Neurobiology, 1*(3), 195–204.
Bordet, R., Devos, D., Brique, S., Touitou, Y., Guieu, J. D., Libersa, C., et al. (2003). Study of circadian melatonin secretion pattern at different stages of Parkinson's disease. *Clinical Neuropharmacology, 26*(2), 65–72.
Brainard, G. C., Hanifin, J. P., Greeson, J. M., Byrne, B., Glickman, G., Gerner, E., et al. (2001). Action spectrum for melatonin regulation in humans: Evidence for a novel circadian photoreceptor. *The Journal of Neuroscience, 21*(16), 6405–6412.
Breder, C. M., & Rasquin, P. (1950). A preliminary report on the role of the pineal organ in the control of pigment cells and light reactions in recent teleost fishes. *Science, 111*(2871), 10–12, illust.
Bubenik, G. A. (2002). Gastrointestinal melatonin: Localization, function, and clinical relevance. *Digestive Diseases and Sciences, 47*(10), 2336–2348.
Carpizo, R., Martinez, A., Mediavilla, D., Gonzalez, M., Abad, A., & Sanchez-Barcelo, E. J. (2006). Smith–Magenis syndrome: A case report of improved sleep after treatment with beta1-adrenergic antagonists and melatonin. *The Journal of Pediatrics, 149*(3), 409–411.
Carskadon, M. A., Brown, E. D., & Dement, W. C. (1982). Sleep fragmentation in the elderly: Relationship to daytime sleep tendency. *Neurobiology of Aging, 3*(4), 321–327.
Cau, E., & Blader, P. (2009). Notch activity in the nervous system: To switch or not switch? *Neural Development, 4*, 36.
Cau, E., Quillien, A., & Blader, P. (2008). Notch resolves mixed neural identities in the zebrafish epiphysis. *Development, 135*(14), 2391–2401.
Cau, E., & Wilson, S. W. (2003). Ash1a and Neurogenin1 function downstream of Floating head to regulate epiphysial neurogenesis. *Development, 130*(11), 2455–2466.
Cermakian, N., Whitmore, D., Foulkes, N. S., & Sassone-Corsi, P. (2000). Asynchronous oscillations of two zebrafish CLOCK partners reveal differential clock control and function. In: *Proceedings of the National Academy of Sciences of the United States of America, 97*(8), 4339–4344.

Chaudhury, D., & Colwell, C. S. (2002). Circadian modulation of learning and memory in fear-conditioned mice. *Behavioural Brain Research, 133*(1), 95–108.

Chaurasia, S. S., Rollag, M. D., Jiang, G., Hayes, W. P., Haque, R., Natesan, A., et al. (2005). Molecular cloning, localization and circadian expression of chicken melanopsin (Opn4): Differential regulation of expression in pineal and retinal cell types. *Journal of Neurochemistry, 92*(1), 158–170.

Clanton, J. A., Hope, K. D., & Gamse, J. T. (2013). Fgf signaling governs cell fate in the zebrafish pineal complex. *Development, 140*(2), 323–332.

Concha, M. L., Russell, C., Regan, J. C., Tawk, M., Sidi, S., Gilmour, D. T., et al. (2003). Local tissue interactions across the dorsal midline of the forebrain establish CNS laterality. *Neuron, 39*(3), 423–438.

Concha, M. L., & Wilson, S. W. (2001). Asymmetry in the epithalamus of vertebrates. *Journal of Anatomy, 199*(Pt 1–2), 63–84.

Costa, E. J., Lopes, R. H., & Lamy-Freund, M. T. (1995). Permeability of pure lipid bilayers to melatonin. *Journal of Pineal Research, 19*(3), 123–126.

Csaba, G., & Barath, P. (1975). Morphological changes of thymus and the thyroid gland after postnatal extirpation of pineal body. *Endocrinologia Experimentalis, 9*(1), 59–67.

Danilova, N., Krupnik, V. E., Sugden, D., & Zhdanova, I. V. (2004). Melatonin stimulates cell proliferation in zebrafish embryo and accelerates its development. *FASEB Journal, 18*(6), 751–753.

Dauchy, R. T., Blask, D. E., Sauer, L. A., Brainard, G. C., & Krause, J. A. (1999). Dim light during darkness stimulates tumor progression by enhancing tumor fatty acid uptake and metabolism. *Cancer Letters, 144*(2), 131–136.

Dauchy, R. T., Sauer, L. A., Blask, D. E., & Vaughan, G. M. (1997). Light contamination during the dark phase in "photoperiodically controlled" animal rooms: Effect on tumor growth and metabolism in rats. *Laboratory Animal Science, 47*(5), 511–518.

D'Autilia, S., Broccoli, V., Barsacchi, G., & Andreazzoli, M. (2010). Xenopus Bsx links daily cell cycle rhythms and pineal photoreceptor fate. In: *Proceedings of the National Academy of Sciences of the United States of America, 107*(14), 6352–6357.

de Borsetti, N. H., Dean, B. J., Bain, E. J., Clanton, J. A., Taylor, R. W., & Gamse, J. T. (2011). Light and melatonin schedule neuronal differentiation in the habenular nuclei. *Developmental Biology, 358*(1), 251–261.

De Leersnyder, H., Claustrat, B., Munnich, A., & Verloes, A. (2006). Circadian rhythm disorder in a rare disease: Smith–Magenis syndrome. *Molecular and Cellular Endocrinology, 252*(1–2), 88–91.

Doi, M., Nakajima, Y., Okano, T., & Fukada, Y. (2001). Light-induced phase-delay of the chicken pineal circadian clock is associated with the induction of cE4bp4, a potential transcriptional repressor of cPer2 gene. In: *Proceedings of the National Academy of Sciences of the United States of America, 98*(14), 8089–8094.

Dubocovich, M. L., Masana, M. I., & Benloucif, S. (1999). Molecular pharmacology and function of melatonin receptor subtypes. *Advances in Experimental Medicine and Biology, 460*, 181–190.

Dufourcq, P., Rastegar, S., Strahle, U., & Blader, P. (2004). Parapineal specific expression of gfi1 in the zebrafish epithalamus. *Gene Expression Patterns, 4*(1), 53–57.

Dykens, J. A., Stern, A., & Trenkner, E. (1987). Mechanism of kainate toxicity to cerebellar neurons in vitro is analogous to reperfusion tissue injury. *Journal of Neurochemistry, 49*(4), 1222–1228.

Edgar, D. M., Dement, W. C., & Fuller, C. A. (1993). Effect of SCN lesions on sleep in squirrel monkeys: Evidence for opponent processes in sleep–wake regulation. *The Journal of Neuroscience, 13*(3), 1065–1079.

Ekstrom, P., & Meissl, H. (1988). Intracellular staining of physiologically identified photoreceptor cells and hyperpolarizing interneurons in the teleost pineal organ. *Neuroscience*, *25*(3), 1061–1070.

Ekstrom, P., & Meissl, H. (2003). Evolution of photosensory pineal organs in new light: The fate of neuroendocrine photoreceptors. *Philosophical Transactions of the Royal Society of London Series B: Biological Sciences*, *358*(1438), 1679–1700.

Engbretson, G. A., & Anderson, K. J. (1990). Neuronal structure of the lacertilian parietal eye, I: A retrograde label and electron-microscopic study of the ganglion cells in the photoreceptor layer. *Visual Neuroscience*, *5*(4), 395–404.

Engbretson, G. A., & Hutchison, V. H. (1976). Parietalectomy and thermal selection in the lizard *Sceloporus magister*. *The Journal of Experimental Zoology*, *198*(1), 29–38.

Engbretson, G. A., & Linser, P. J. (1991). Glial cells of the parietal eye: Structural and biochemical similarities to retinal Muller cells. *The Journal of Comparative Neurology*, *314*(4), 799–806.

Estivill-Torrus, G., Vitalis, T., Fernandez-Llebrez, P., & Price, D. J. (2001). The transcription factor Pax6 is required for development of the diencephalic dorsal midline secretory radial glia that form the subcommissural organ. *Mechanisms of Development*, *109*(2), 215–224.

Falcon, J., & Meissl, H. (1981). The photosensory function of the pineal organ of the pike (esox lucius l.). Correlation between structure and function. *Journal of Comparative Physiology*, *144*, 127–137.

Fernandes, A. M., Fero, K., Arrenberg, A. B., Bergeron, S. A., Driever, W., & Burgess, H. A. (2012). Deep brain photoreceptors control light-seeking behavior in zebrafish larvae. *Current Biology*, *22*(21), 2042–2047.

Firth, B. T., Ralph, C. L., & Boardman, T. J. (1980). Independent effects of the pineal and a bacterial pyrogen in behavioural thermoregulation in lizards. *Nature*, *285*(5764), 399–400.

Flynn-Evans, E. E., Stevens, R. G., Tabandeh, H., Schernhammer, E. S., & Lockley, S. W. (2009). Total visual blindness is protective against breast cancer. *Cancer Causes & Control*, *20*(9), 1753–1756.

Forsell, J., Ekstrom, P., Flamarique, I. N., & Holmqvist, B. (2001). Expression of pineal ultraviolet- and green-like opsins in the pineal organ and retina of teleosts. *The Journal of Experimental Biology*, *204*(Pt 14), 2517–2525.

Foster, R. G., & Roberts, A. (1982). The pineal eye in Xenopus laevis embryos and larvae: A photoreceptor with a direct excitatory effect on behaviour. *Journal of Comparative Physiology*, *145*, 413–419.

Frigato, E., Vallone, D., Bertolucci, C., & Foulkes, N. S. (2006). Isolation and characterization of melanopsin and pinopsin expression within photoreceptive sites of reptiles. *Naturwissenschaften*, *93*(8), 379–385.

Furukawa, T., Morrow, E. M., Li, T., Davis, F. C., & Cepko, C. L. (1999). Retinopathy and attenuated circadian entrainment in Crx-deficient mice. *Nature Genetics*, *23*(4), 466–470.

Gamse, J. T., Shen, Y. C., Thisse, C., Thisse, B., Raymond, P. A., Halpern, M. E., et al. (2002). Otx5 regulates genes that show circadian expression in the zebrafish pineal complex. *Nature Genetics*, *30*(1), 117–121.

Gamse, J. T., Thisse, C., Thisse, B., & Halpern, M. E. (2003). The parapineal mediates left–right asymmetry in the zebrafish diencephalon. *Development*, *130*(6), 1059–1068.

Ganguly, S., Gastel, J. A., Weller, J. L., Schwartz, C., Jaffe, H., Namboodiri, M. A., et al. (2001). Role of a pineal cAMP-operated arylalkylamine N-acetyltransferase/14-3-3-binding switch in melatonin synthesis. In: *Proceedings of the National Academy of Sciences of the United States of America*, *98*(14), 8083–8088.

Ganguly, S., Weller, J. L., Ho, A., Chemineau, P., Malpaux, B., & Klein, D. C. (2005). Melatonin synthesis: 14-3-3-dependent activation and inhibition of arylalkylamine N-acetyltransferase mediated by phosphoserine-205. In: *Proceedings of the National Academy of Sciences of the United States of America, 102*(4), 1222–1227.

Gundy, G. C., Ralph, C. L., & Wurst, G. Z. (1975). Parietal eyes in lizards: Zoogeographical correlates. *Science, 190*(4215), 671–673.

Hansen, J. (2001). Light at night, shiftwork, and breast cancer risk. *Journal of the National Cancer Institute, 93*(20), 1513–1515.

Hardeland, R., Madrid, J. A., Tan, D. X., & Reiter, R. J. (2012). Melatonin, the circadian multioscillator system and health: The need for detailed analyses of peripheral melatonin signaling. *Journal of Pineal Research, 52*(2), 139–166.

Hardeland, R., & Poeggeler, B. (2003). Non-vertebrate melatonin. *Journal of Pineal Research, 34*(4), 233–241.

Harman, D. (1993). Free radical involvement in aging. Pathophysiology and therapeutic implications. *Drugs & Aging, 3*(1), 60–80.

Harman, D. (1994). Free-radical theory of aging. Increasing the functional life span. *Annals of the New York Academy of Sciences, 717*, 1–15.

Haupt, Y., Rowan, S., Shaulian, E., Kazaz, A., Vousden, K., & Oren, M. (1997). p53 mediated apoptosis in HeLa cells: Transcription dependent and independent mechanisms. *Leukemia, 11*(Suppl. 3), 337–339.

Hofman, M. A. (2000). The human circadian clock and aging. *Chronobiology International, 17*(3), 245–59.

Huang, S. K., Klein, D. C., & Korf, H. W. (1992). Immunocytochemical demonstration of rod-opsin, S-antigen, and neuron-specific proteins in the human pineal gland. *Cell and Tissue Research, 267*(3), 493–498.

Hughes, R. J., & Badia, P. (1997). Sleep-promoting and hypothermic effects of daytime melatonin administration in humans. *Sleep, 20*(2), 124–131.

Jamieson, D., & Roberts, A. (2000). Responses of young Xenopus laevis tadpoles to light dimming: Possible roles for the pineal eye. *The Journal of Experimental Biology, 203*(Pt 12), 1857–1867.

Jengeleski, C. A., Powers, R. E., O'Connor, D. T., & Price, D. L. (1989). Noradrenergic innervation of human pineal gland: Abnormalities in aging and Alzheimer's disease. *Brain Research, 481*(2), 378–382.

Karolczak, M., Korf, H. W., & Stehle, J. H. (2005). The rhythm and blues of gene expression in the rodent pineal gland. *Endocrine, 27*(2), 89–100.

Kliukiene, J., Tynes, T., & Andersen, A. (2001). Risk of breast cancer among Norwegian women with visual impairment. *British Journal of Cancer, 84*(3), 397–399.

Korf, H. W., Korf, B., Schachenmayr, W., Chader, G. J., & Wiggert, B. (1992). Immunocytochemical demonstration of interphotoreceptor retinoid-binding protein in cerebellar medulloblastoma. *Acta Neuropathologica, 83*(5), 482–487.

Koyanagi, M., Kawano, E., Kinugawa, Y., Oishi, T., Shichida, Y., Tamotsu, S., et al. (2004). Bistable UV pigment in the lamprey pineal. In: *Proceedings of the National Academy of Sciences of the United States of America, 101*(17), 6687–6691.

Krauss, A., & Neumeyer, C. (2003). Wavelength dependence of the optomotor response in zebrafish (Danio rerio). *Vision Research, 43*(11), 1273–1282.

Kulman, G., Lissoni, P., Rovelli, F., Roselli, M. G., Brivio, F., & Sequeri, P. (2000). Evidence of pineal endocrine hypofunction in autistic children. *Neuro Endocrinology Letters, 21*(1), 31–34.

Kunz, D., Schmitz, S., Mahlberg, R., Mohr, A., Stoter, C., Wolf, K. J., et al. (1999). A new concept for melatonin deficit: On pineal calcification and melatonin excretion. *Neuropsychopharmacology, 21*(6), 765–772.

Kusmic, C. B. L., Passarelli, V., & Gualtieri, P. (1993). Photoreceptor in the pineal morphology and visual pigment content organ in the retina of juvenile and adult trout, Salmo irideus. *Micron, 24*, 279–286.

Lahti, T., Merikanto, I., & Partonen, T. (2012). Circadian clock disruptions and the risk of cancer. *Annals of Medicine, 44*(8), 847–853.

Lavie, P. (1997). Melatonin: Role in gating nocturnal rise in sleep propensity. *Journal of Biological Rhythms, 12*(6), 657–665.

Lewy, A. J., Tetsuo, M., Markey, S. P., Goodwin, F. K., & Kopin, I. J. (1980). Pinealectomy abolishes plasma melatonin in the rat. *The Journal of Clinical Endocrinology and Metabolism, 50*(1), 204–205.

Lincoln, G., Messager, S., Andersson, H., & Hazlerigg, D. (2002). Temporal expression of seven clock genes in the suprachiasmatic nucleus and the pars tuberalis of the sheep: Evidence for an internal coincidence timer. In: *Proceedings of the National Academy of Sciences of the United States of America, 99*(21), 13890–13895.

Liu, C., Weaver, D. R., Jin, X., Shearman, L. P., Pieschl, R. L., Gribkoff, V. K., et al. (1997). Molecular dissection of two distinct actions of melatonin on the suprachiasmatic circadian clock. *Neuron, 19*(1), 91–102.

Lyssenko, V., Nagorny, C. L., Erdos, M. R., Wierup, N., Jonsson, A., Spegel, P., et al. (2009). Common variant in MTNR1B associated with increased risk of type 2 diabetes and impaired early insulin secretion. *Nature Genetics, 41*(1), 82–88.

Machado, C. R., Wragg, L. E., & Machado, A. B. (1969). Circadian rhythm of serotonin in the pineal body of immunosympathectomized immature rats. *Science, 164*(3878), 442–443.

Mano, H., & Fukada, Y. (2007). A median third eye: Pineal gland retraces evolution of vertebrate photoreceptive organs. *Photochemistry and Photobiology, 83*(1), 11–18.

Mano, H., Kojima, D., & Fukada, Y. (1999). Exo-rhodopsin: A novel rhodopsin expressed in the zebrafish pineal gland. *Brain Research Molecular Brain Research, 73*(1–2), 110–118.

Marchiafava, P. L., & Kusmic, C. (1993). The electrical responses of the trout pineal photoreceptors to brief and prolonged illumination. *Progress in Brain Research, 95*, 3–13.

Maronde, E., & Stehle, J. H. (2007). The mammalian pineal gland: Known facts, unknown facets. *Trends in Endocrinology and Metabolism, 18*(4), 142–149.

Masai, I., Heisenberg, C. P., Barth, K. A., Macdonald, R., Adamek, S., & Wilson, S. W. (1997). Floating head and masterblind regulate neuronal patterning in the roof of the forebrain. *Neuron, 18*(1), 43–57.

McCurry, S. M., Pike, K. C., Vitiello, M. V., Logsdon, R. G., Larson, E. B., & Teri, L. (2011). Increasing walking and bright light exposure to improve sleep in community-dwelling persons with Alzheimer's disease: Results of a randomized, controlled trial. *Journal of the American Geriatrics Society, 59*(8), 1393–1402.

Mediavilla, M. D., Cos, S., & Sanchez-Barcelo, E. J. (1999). Melatonin increases p53 and p21WAF1 expression in MCF-7 human breast cancer cells in vitro. *Life Sciences, 65*(4), 415–420.

Meiniel, A. (1976). The embryonic pineal organ of *Lacerta vivipara* J. I. Differentiation of the cells of the sensory line (CSL). The gradient of morphological differentiation. *Journal of Neural Transmission, 39*(1–2), 139–174.

Meiniel, A. (1980). Ultrastructure of serotonin-containing cells in the pineal organ of Lampetra planeri (Petromyzontidae). *Cell and Tissue Research, 207*(3), 407–427.

Meissl, H., Nakamura, T., & Thiele, G. (1986). Neural response mechanisms in the photoreceptive pineal organ of goldfish. *Comparative Biochemistry and Physiology A, Comparative Physiology, 84*(3), 467–473.

Melke, J., Goubran Botros, H., Chaste, P., Betancur, C., Nygren, G., Anckarsater, H., et al. (2008). Abnormal melatonin synthesis in autism spectrum disorders. *Molecular Psychiatry, 13*(1), 90–98.

Mendelson, W. B., Gillin, J. C., Dawson, S. D., Lewy, A. J., & Wyatt, R. J. (1980). Effects of melatonin and propranolol on sleep of the rat. *Brain Research, 201*(1), 240–244.

Menendez-Pelaez, A., & Reiter, R. J. (1993). Distribution of melatonin in mammalian tissues: The relative importance of nuclear versus cytosolic localization. *Journal of Pineal Research, 15*(2), 59–69.

Mishima, K., Okawa, M., Shimizu, T., & Hishikawa, Y. (2001). Diminished melatonin secretion in the elderly caused by insufficient environmental illumination. *The Journal of Clinical Endocrinology and Metabolism, 86*(1), 129–134.

Moller, M. (1979). Presence of a pineal nerve (*nervus pinealis*) in fetal mammals. *Progress in Brain Research, 52*, 103–106.

Moller, M. (1997). Peptidergic cells in the mammalian pineal gland. Morphological indications for a paracrine regulation of the pinealocyte. *Biology of the Cell, 89*(9), 561–567.

Moller, M., & Baeres, F. M. (2002). The anatomy and innervation of the mammalian pineal gland. *Cell and Tissue Research, 309*(1), 139–150.

Moore, R. Y., & Sibony, P. (1988). Enkephalin-like immunoreactivity in neurons in the human pineal gland. *Brain Research, 457*(2), 395–398.

Morris, M., Lack, L., & Barrett, J. (1990). The effect of sleep/wake state on nocturnal melatonin excretion. *Journal of Pineal Research, 9*(2), 133–138.

Muller, H. L., Handwerker, G., Wollny, B., Faldum, A., & Sorensen, N. (2002). Melatonin secretion and increased daytime sleepiness in childhood craniopharyngioma patients. *The Journal of Clinical Endocrinology and Metabolism, 87*(8), 3993–3996.

Murphy, P. J., & Campbell, S. S. (1996). Physiology of the circadian system in animals and humans. *Journal of Clinical Neurophysiology, 13*(1), 2–16.

Nelson, R. J., & Demas, G. E. (1996). Seasonal changes in immune function. *The Quarterly Review of Biology, 71*(4), 511–548.

Neuhauss, S. C., Biehlmaier, O., Seeliger, M. W., Das, T., Kohler, K., Harris, W. A., et al. (1999). Genetic disorders of vision revealed by a behavioral screen of 400 essential loci in zebrafish. *The Journal of Neuroscience, 19*(19), 8603–8615.

Nir, I., Meir, D., Zilber, N., Knobler, H., Hadjez, J., & Lerner, Y. (1995). Brief report: Circadian melatonin, thyroid-stimulating hormone, prolactin, and cortisol levels in serum of young adults with autism. *Journal of Autism and Developmental Disorders, 25*(6), 641–654.

Nishida, A., Furukawa, A., Koike, C., Tano, Y., Aizawa, S., Matsuo, I., et al. (2003). Otx2 homeobox gene controls retinal photoreceptor cell fate and pineal gland development. *Nature Neuroscience, 6*(12), 1255–1263.

Nishio, S., Kakizawa, T., Chatelain, G., Triqueneaux, G., Brunet, F., Rambaud, J., et al. (2008). OTX5 regulates pineal expression of the zebrafish REV-ERB alpha through a new DNA binding site. *Molecular Endocrinology, 22*(1), 23–32.

Nordlund, J. J., & Lerner, A. B. (1977). The effects of oral melatonin on skin color and on the release of pituitary hormones. *The Journal of Clinical Endocrinology and Metabolism, 45*(4), 768–774.

Nosjean, O., Ferro, M., Coge, F., Beauverger, P., Henlin, J. M., Lefoulon, F., et al. (2000). Identification of the melatonin-binding site MT3 as the quinone reductase 2. *The Journal of Biological Chemistry, 275*(40), 31311–31317.

Oksche, A. (1965). Survey of the development and comparative morphology of the pineal organ. *Progress in Brain Research, 10*, 3–29.

Oren, M. (2003). Decision making by p53: Life, death and cancer. *Cell Death and Differentiation, 10*(4), 431–442.

Pandi-Perumal, S. R., Trakht, I., Srinivasan, V., Spence, D. W., Maestroni, G. J., Zisapel, N., et al. (2008). Physiological effects of melatonin: Role of melatonin receptors and signal transduction pathways. *Progress in Neurobiology, 85*(3), 335–353.

Pappolla, M. A., Chyan, Y. J., Poeggeler, B., Frangione, B., Wilson, G., Ghiso, J., et al. (2000). An assessment of the antioxidant and the antiamyloidogenic properties of melatonin: Implications for Alzheimer's disease. *Journal of Neural Transmission, 107*(2), 203–231.

Perentes, E., Rubinstein, L. J., Herman, M. M., & Donoso, L. A. (1986). S-antigen immunoreactivity in human pineal glands and pineal parenchymal tumors. A monoclonal antibody study. *Acta Neuropathologica, 71*(3–4), 224–227.

Peschke, E., Muhlbauer, E., Musshoff, U., Csernus, V. J., Chankiewitz, E., & Peschke, D. (2002). Receptor (MT(1)) mediated influence of melatonin on cAMP concentration and insulin secretion of rat insulinoma cells INS-1. *Journal of Pineal Research, 33*(2), 63–71.

Pevet, P. (1980). The pineal gland of the mole (Talpa europaea L.). *Cell and Tissue Research, 206*(3), 417–430.

Pevet, P., & Collin, J. P. (1976). Mammalian pinealocytes. Differences, homologies, origin. Study in the adult mole (talpa europaea L.). *Journal of Ultrastructure Research, 57*(1), 22–31.

Pevet, P., Kappers, J. A., & Voute, A. M. (1977). The pineal gland of nocturnal mammals. I. The pinealocytes of the bat (Nyctalus noctula, Schreber). *Journal of Neural Transmission, 40*(1), 47–68.

Phillips, J. A., & Howes, K. A. (1988). The pineal complex, aggressive behavior and thermoregulation in curly-tailed lizards, Leiocephalus carinatus. *Physiology & Behavior, 42*(1), 103–108.

Pierce, L. X., Noche, R. R., Ponomareva, O., Chang, C., & Liang, J. O. (2008). Novel functions for Period 3 and Exo-rhodopsin in rhythmic transcription and melatonin biosynthesis within the zebrafish pineal organ. *Brain Research, 1223*, 11–24.

Plouhinec, J. L., Granier, C., Le Mentec, C., Lawson, K. A., Saberan-Djoneidi, D., Aghion, J., et al. (2004). Identification of the mammalian Not gene via a phylogenomic approach. *Gene Expression Patterns, 5*(1), 11–22.

Plouhinec, J. L., Sauka-Spengler, T., Germot, A., Le Mentec, C., Cabana, T., Harrison, G., et al. (2003). The mammalian Crx genes are highly divergent representatives of the Otx5 gene family, a gnathostome orthology class of orthodenticle-related homeogenes involved in the differentiation of retinal photoreceptors and circadian entrainment. *Molecular Biology and Evolution, 20*(4), 513–521.

Poeggeler, B. (1993). Introduction. Melatonin and the light–dark zeitgeber in vertebrates, invertebrates and unicellular organisms. *Experientia, 49*(8), 611–613.

Poirel, V. J., Boggio, V., Dardente, H., Pevet, P., Masson-Pevet, M., & Gauer, F. (2003). Contrary to other non-photic cues, acute melatonin injection does not induce immediate changes of clock gene mRNA expression in the rat suprachiasmatic nuclei. *Neuroscience, 120*(3), 745–755.

Quay, W. B. (1965). Histological structure and cytology of the pineal organ in birds and mammals. *Progress in Brain Research, 10*, 49–86.

Quillien, A., Blanco-Sanchez, B., Halluin, C., Moore, J. C., Lawson, N. D., Blader, P., et al. (2011). BMP signaling orchestrates photoreceptor specification in the zebrafish pineal gland in collaboration with Notch. *Development, 138*(11), 2293–2302.

Rawashdeh, O., de Borsetti, N. H., Roman, G., & Cahill, G. M. (2007). Melatonin suppresses nighttime memory formation in zebrafish. *Science, 318*(5853), 1144–1146.

Rawashdeh, O., & Maronde, E. (2012). The hormonal Zeitgeber melatonin: Role as a circadian modulator in memory processing. *Frontiers in Molecular Neuroscience, 5*, 27.

Regan, J. C., Concha, M. L., Roussigne, M., Russell, C., & Wilson, S. W. (2009). An Fgf8-dependent bistable cell migratory event establishes CNS asymmetry. *Neuron, 61*(1), 27–34.

Reid, K., Van den Heuvel, C., & Dawson, D. (1996). Day-time melatonin administration: Effects on core temperature and sleep onset latency. *Journal of Sleep Research, 5*(3), 150–154.

Reiter, R. J. (1980). The pineal and its hormones in the control of reproduction in mammals. *Endocrine Reviews*, *1*(2), 109–131.
Reiter, R. J. (1991). Melatonin: The chemical expression of darkness. *Molecular and Cellular Endocrinology*, *79*(1–3), C153–C158.
Reiter, R. J. (2002). Potential biological consequences of excessive light exposure: Melatonin suppression, DNA damage, cancer and neurodegenerative diseases. *Neuro Endocrinology Letters*, *23*(Suppl. 2), 9–13.
Reiter, R. J., & Richardson, B. A. (1992). Some perturbations that disturb the circadian melatonin rhythm. *Chronobiology International*, *9*(4), 314–321.
Reppert, S. M., Godson, C., Mahle, C. D., Weaver, D. R., Slaugenhaupt, S. A., & Gusella, J. F. (1995). Molecular characterization of a second melatonin receptor expressed in human retina and brain: The Mel1b melatonin receptor. In: *Proceedings of the National Academy of Sciences of the United States of America*, *92*(19), 8734–8738.
Reppert, S. M., Weaver, D. R., & Ebisawa, T. (1994). Cloning and characterization of a mammalian melatonin receptor that mediates reproductive and circadian responses. *Neuron*, *13*(5), 1177–1185.
Reppert, S. M., Weaver, D. R., & Godson, C. (1996). Melatonin receptors step into the light: Cloning and classification of subtypes. *Trends in Pharmacological Sciences*, *17*(3), 100–102.
Roussigne, M., Bianco, I. H., Wilson, S. W., & Blader, P. (2009). Nodal signalling imposes left–right asymmetry upon neurogenesis in the habenular nuclei. *Development*, *136*(9), 1549–1557.
Sakai, K., Imamoto, Y., Su, C. Y., Tsukamoto, H., Yamashita, T., Terakita, A., et al. (2012). Photochemical nature of parietopsin. *Biochemistry*, *51*(9), 1933–1941.
Santoro, R., Marani, M., Blandino, G., Muti, P., & Strano, S. (2012). Melatonin triggers p53Ser phosphorylation and prevents DNA damage accumulation. *Oncogene*, *31*(24), 2931–2942.
Sauer, L. A., Dauchy, R. T., & Blask, D. E. (1997). Dietary linoleic acid intake controls the arterial blood plasma concentration and the rates of growth and linoleic acid uptake and metabolism in hepatoma 7288CTC in Buffalo rats. *The Journal of Nutrition*, *127*(7), 1412–1421.
Schernhammer, E. S., Berrino, F., Krogh, V., Secreto, G., Micheli, A., Venturelli, E., et al. (2008). Urinary 6-sulfatoxymelatonin levels and risk of breast cancer in postmenopausal women. *Journal of the National Cancer Institute*, *100*(12), 898–905.
Schernhammer, E. S., Kroenke, C. H., Laden, F., & Hankinson, S. E. (2006). Night work and risk of breast cancer. *Epidemiology*, *17*(1), 108–111.
Schernhammer, E. S., Laden, F., Speizer, F. E., Willett, W. C., Hunter, D. J., Kawachi, I., et al. (2001). Rotating night shifts and risk of breast cancer in women participating in the nurses' health study. *Journal of the National Cancer Institute*, *93*(20), 1563–1568.
Schmid, H. A. (1993). Decreased melatonin biosynthesis, calcium flux, pineal gland calcification and aging: A hypothetical framework. *Gerontology*, *39*(4), 189–199.
Senthilan, P. R., Piepenbrock, D., Ovezmyradov, G., Nadrowski, B., Bechstedt, S., Pauls, S., et al. (2012). Drosophila auditory organ genes and genetic hearing defects. *Cell*, *150*(5), 1042–1054.
Shen, W. L., Kwon, Y., Adegbola, A. A., Luo, J., Chess, A., & Montell, C. (2011). Function of rhodopsin in temperature discrimination in Drosophila. *Science*, *331*(6022), 1333–1336.
Shiraki, T., Kojima, D., & Fukada, Y. (2010). Light-induced body color change in developing zebrafish. *Photochemical & Photobiological Sciences*, *9*(11), 1498–1504.
Sigurdardottir, L. G., Valdimarsdottir, U. A., Fall, K., Rider, J. R., Lockley, S. W., Schernhammer, E., et al. (2012). Circadian disruption, sleep loss, and prostate cancer risk: A systematic review of epidemiologic studies. *Cancer Epidemiology, Biomarkers & Prevention*, *21*(7), 1002–1011.

Skene, D. J., Vivien-Roels, B., Sparks, D. L., Hunsaker, J. C., Pevet, P., Ravid, D., et al. (1990). Daily variation in the concentration of melatonin and 5-methoxytryptophol in the human pineal gland: Effect of age and Alzheimer's disease. *Brain Research, 528*(1), 170–174.

Slominski, A., Tobin, D. J., Zmijewski, M. A., Wortsman, J., & Paus, R. (2008). Melatonin in the skin: Synthesis, metabolism and functions. *Trends in Endocrinology and Metabolism, 19*(1), 17–24.

Snelson, C. D., Santhakumar, K., Halpern, M. E., & Gamse, J. T. (2008). Tbx2b is required for the development of the parapineal organ. *Development, 135*(9), 1693–1702.

Srinivasan, V., Maestroni, G. J., Cardinali, D. P., Esquifino, A. I., Perumal, S. R., & Miller, S. C. (2005). Melatonin, immune function and aging. *Immunity & Ageing, 2*, 17.

Staiger, H., Machicao, F., Schafer, S. A., Kirchhoff, K., Kantartzis, K., Guthoff, M., et al. (2008). Polymorphisms within the novel type 2 diabetes risk locus MTNR1B determine beta-cell function. *PLoS One, 3*(12), e3962.

Stehle, J. H., Saade, A., Rawashdeh, O., Ackermann, K., Jilg, A., Sebesteny, T., et al. (2011). A survey of molecular details in the human pineal gland in the light of phylogeny, structure, function and chronobiological diseases. *Journal of Pineal Research, 51*(1), 17–43.

Stoykova, A., Fritsch, R., Walther, C., & Gruss, P. (1996). Forebrain patterning defects in Small eye mutant mice. *Development, 122*(11), 3453–3465.

Su, C. Y., Luo, D. G., Terakita, A., Shichida, Y., Liao, H. W., Kazmi, M. A., et al. (2006). Parietal-eye phototransduction components and their potential evolutionary implications. *Science, 311*(5767), 1617–1621.

Takanaka, Y., Okano, T., Iigo, M., & Fukada, Y. (1998). Light-dependent expression of pinopsin gene in chicken pineal gland. *Journal of Neurochemistry, 70*(3), 908–913.

Takanaka, Y., Okano, T., Yamamoto, K., & Fukada, Y. (2002). A negative regulatory element required for light-dependent pinopsin gene expression. *The Journal of Neuroscience, 22*(11), 4357–4363.

Tan, D. X., Manchester, L. C., Hardeland, R., Lopez-Burillo, S., Mayo, J. C., Sainz, R. M., et al. (2003). Melatonin: A hormone, a tissue factor, an autocoid, a paracoid, and an antioxidant vitamin. *Journal of Pineal Research, 34*(1), 75–78.

Tan, D. X., Poeggeler, B., Reiter, R. J., Chen, L. D., Chen, S., Manchester, L. C., et al. (1993). The pineal hormone melatonin inhibits DNA-adduct formation induced by the chemical carcinogen safrole in vivo. *Cancer Letters, 70*(1–2), 65–71.

Tan, D., Reiter, R. J., Chen, L. D., Poeggeler, B., Manchester, L. C., & Barlow-Walden, L. R. (1994). Both physiological and pharmacological levels of melatonin reduce DNA adduct formation induced by the carcinogen safrole. *Carcinogenesis, 15*(2), 215–218.

Tansey, M. G., McCoy, M. K., & Frank-Cannon, T. C. (2007). Neuroinflammatory mechanisms in Parkinson's disease: Potential environmental triggers, pathways, and targets for early therapeutic intervention. *Experimental Neurology, 208*(1), 1–25.

Thapan, K., Arendt, J., & Skene, D. J. (2001). An action spectrum for melatonin suppression: Evidence for a novel non-rod, non-cone photoreceptor system in humans. *The Journal of Physiology, 535*(Pt 1), 261–267.

Tordjman, S., Anderson, G. M., Pichard, N., Charbuy, H., & Touitou, Y. (2005). Nocturnal excretion of 6-sulphatoxymelatonin in children and adolescents with autistic disorder. *Biological Psychiatry, 57*(2), 134–138.

Tynes, T., Hannevik, M., Andersen, A., Vistnes, A. I., & Haldorsen, T. (1996). Incidence of breast cancer in Norwegian female radio and telegraph operators. *Cancer Causes & Control, 7*(2), 197–204.

Tzischinsky, O., & Lavie, P. (1994). Melatonin possesses time-dependent hypnotic effects. *Sleep, 17*(7), 638–645.

Underwood, H., & Calaban, M. (1987). Pineal melatonin rhythms in the lizard *Anolis carolinensis*: II. Photoreceptive inputs. *Journal of Biological Rhythms*, *2*(3), 195–206.

Unger, T., Juven-Gershon, T., Moallem, E., Berger, M., Vogt Sionov, R., Lozano, G., et al. (1999). Critical role for Ser20 of human p53 in the negative regulation of p53 by Mdm2. *The EMBO Journal*, *18*(7), 1805–1814.

Vatine, G., Vallone, D., Appelbaum, L., Mracek, P., Ben-Moshe, Z., Lahiri, K., et al. (2009). Light directs zebrafish period2 expression via conserved D and E boxes. *PLoS Biology*, 7(10), e1000223.

Vatine, G., Vallone, D., Gothilf, Y., & Foulkes, N. S. (2011). It's time to swim! Zebrafish and the circadian clock. *FEBS Letters*, *585*(10), 1485–1494.

Vigh, B., Manzano, M. J., Zadori, A., Frank, C. L., Lukats, A., Rohlich, P., et al. (2002). Nonvisual photoreceptors of the deep brain, pineal organs and retina. *Histology and Histopathology*, *17*(2), 555–590.

Vigh, B., & Vigh-Teichmann, I. (1993). Development of the photoreceptor outer segment-like cilia of the CSF-contacting pinealocytes of the ferret (Putorius furo). *Archives of Histology and Cytology*, *56*(5), 485–493.

Vigh-Teichmann, I., Vigh, B., & Wirtz, G. H. (1989). Immunoelectron microscopy of rhodopsin and vitamin A in the pineal organ and lateral eye of the lamprey. *Experimental Biology*, *48*(4), 203–213.

Wada, S., Kawano-Yamashita, E., Koyanagi, M., & Terakita, A. (2012). Expression of UV-sensitive parapinopsin in the iguana parietal eyes and its implication in UV-sensitivity in vertebrate pineal-related organs. *PLoS One*, 7(6), e39003.

Weaver, D. R., Stehle, J. H., Stopa, E. G., & Reppert, S. M. (1993). Melatonin receptors in human hypothalamus and pituitary: Implications for circadian and reproductive responses to melatonin. *The Journal of Clinical Endocrinology and Metabolism*, *76*(2), 295–301.

Whitmore, D., Foulkes, N. S., Strahle, U., & Sassone-Corsi, P. (1998). Zebrafish Clock rhythmic expression reveals independent peripheral circadian oscillators. *Nature Neuroscience*, *1*(8), 701–707.

Wiesenberg, I., Missbach, M., Kahlen, J. P., Schrader, M., & Carlberg, C. (1995). Transcriptional activation of the nuclear receptor RZR alpha by the pineal gland hormone melatonin and identification of CGP 52608 as a synthetic ligand. *Nucleic Acids Research*, *23*(3), 327–333.

Williams, S. R., Zies, D., Mullegama, S. V., Grotewiel, M. S., & Elsea, S. H. (2012). Smith–Magenis syndrome results in disruption of CLOCK gene transcription and reveals an integral role for RAI1 in the maintenance of circadian rhythmicity. *American Journal of Human Genetics*, *90*(6), 941–949.

Wilson, S. W., & Easter, S. S., Jr. (1991). Stereotyped pathway selection by growth cones of early epiphysial neurons in the embryonic zebrafish. *Development*, *112*(3), 723–746.

Wilson, B. W., Lynch, H. J., & Ozaki, Y. (1978). 5-Methoxytryptophol in rat serum and pineal: Detection, quantitation, and evidence for daily rhythmicity. *Life Sciences*, *23*(10), 1019–1023.

Wilson, S. W., Ross, L. S., Parrett, T., & Easter, S. S., Jr. (1990). The development of a simple scaffold of axon tracts in the brain of the embryonic zebrafish, Brachydanio rerio. *Development*, *108*(1), 121–145.

Wyatt, J. K., Dijk, D. J., Ritz-de Cecco, A., Ronda, J. M., & Czeisler, C. A. (2006). Sleep-facilitating effect of exogenous melatonin in healthy young men and women is circadian-phase dependent. *Sleep*, *29*(5), 609–618.

Yoshizawa, M., & Jeffery, W. R. (2008). Shadow response in the blind cavefish Astyanax reveals conservation of a functional pineal eye. *The Journal of Experimental Biology*, *211*(Pt 3), 292–299.

Zawilska, J. B., Vivien-Roels, B., Skene, D. J., Pevet, P., & Nowak, J. Z. (2000). Phase-shifting effects of light on the circadian rhythms of 5-methoxytryptophol and melatonin in the chick pineal gland. *Journal of Pineal Research*, *29*(1), 1–7.

Zeitzer, J. M., Ayas, N. T., Shea, S. A., Brown, R., & Czeisler, C. A. (2000). Absence of detectable melatonin and preservation of cortisol and thyrotropin rhythms in tetraplegia. *The Journal of Clinical Endocrinology and Metabolism*, *85*(6), 2189–2196.

Zhang, J., Perry, G., Smith, M. A., Robertson, D., Olson, S. J., Graham, D. G., et al. (1999). Parkinson's disease is associated with oxidative damage to cytoplasmic DNA and RNA in substantia nigra neurons. *The American Journal of Pathology*, *154*(5), 1423–1429.

Zhang, C., Song, Y., Thompson, D. A., Madonna, M. A., Millhauser, G. L., Toro, S., et al. (2010). Pineal-specific agouti protein regulates teleost background adaptation. In: *Proceedings of the National Academy of Sciences of the United States of America*, *107*(47), 20164–20171.

Zhdanova, I. V., Wurtman, R. J., Lynch, H. J., Ives, J. R., Dollins, A. B., Morabito, C., et al. (1995). Sleep-inducing effects of low doses of melatonin ingested in the evening. *Clinical Pharmacology and Therapeutics*, *57*(5), 552–558.

Ziv, L., Tovin, A., Strasser, D., & Gothilf, Y. (2007). Spectral sensitivity of melatonin suppression in the zebrafish pineal gland. *Experimental Eye Research*, *84*(1), 92–99.

CHAPTER SIX

From Pancreas Morphogenesis to β-Cell Regeneration

Fabio Avolio[*,†,‡], Anja Pfeifer[*,†,‡], Monica Courtney[*,†,‡], Elisabet Gjernes[*,†,‡], Nouha Ben-Othman[*,†,‡], Andhira Vieira[*,†,‡], Noémie Druelle[*,†,‡], Biljana Faurite[*,†,‡], Patrick Collombat[*,†,‡,1]

[*]Univ. Nice Sophia Antipolis, iBV, UMR 7277, Nice, France
[†]Inserm, iBV, U1091, Nice, France
[‡]CNRS, iBV, UMR 7277, Nice, France
[1]Corresponding author: e-mail address: collombat@unice.fr

Contents

Abstract

Type 1 diabetes is a metabolic disease resulting in the selective loss of pancreatic insulin-producing β-cells and affecting millions of people worldwide. The side effects of diabetes are varied and include cardiovascular, neuropathologic, and kidney diseases. Despite the most recent advances in diabetes care, patients suffering from type 1 diabetes still display a shortened life expectancy compared to their healthy counterparts. In an effort to improve β-cell-replacement therapies, numerous approaches are currently being pursued, most of these aiming at finding ways to differentiate stem/progenitor cells into β-like cells by mimicking embryonic development. Unfortunately, these efforts have hitherto not allowed the generation of fully functional β-cells. This chapter summarizes recent findings, allowing a better insight into the molecular mechanisms underlying the genesis of β-cells during the course of pancreatic morphogenesis. Furthermore, a focus is made on new research avenues concerning the conversion of pre-existing pancreatic cells into β-like cells, such approaches holding great promise for the development of type 1 diabetes therapies.

Current Topics in Developmental Biology, Volume 106
ISSN 0070-2153
http://dx.doi.org/10.1016/B978-0-12-416021-7.00006-7

1. ANATOMY OF THE PANCREAS

The mature pancreas is an abdominal gland connected to the duodenum, which consists of two functionally and morphologically distinct compartments deriving from the endoderm. The exocrine compartment (98% of the total organ mass) mostly contains enzyme-secreting acinar cells organized into clusters surrounding a network system of ducts. Mature duct cells actively secrete bicarbonate and mucins. The duct network also collects digestive enzymes from acinar cells that are emptied into the duodenum where they contribute to food processing (Slack, 1995). The endocrine compartment (<2% of the pancreatic tissue) is organized into highly vascularized and innervated cell clusters termed islets of Langerhans comprising five different hormone-secreting cell subtypes: α-, β-, δ-, ε-, and PP-cells secreting glucagon, insulin, somatostatin, ghrelin, and PP (pancreatic polypeptide), respectively. The intimate interaction between islets and vascular cells regulates hormone release, establishing a fine control of the glucose homeostasis in the body (Konstantinova et al., 2007; Prado, Pugh-Bernard, Elghazi, Sosa-Pineda, & Sussel, 2004; Slack, 1995). Insulin and glucagon coordinately act to maintain glycemic homeostasis by regulating the storage, metabolism, and neogenesis of glucose. Insulin is released in response to an increase in blood glucose levels and acts by reducing glycemia by promoting the storage and/or metabolism of glucose. Glucagon has the opposite function by promoting glucose release and/or neogenesis from glycogen, in the case of hypoglycemia. Somatostatin and PP have been implicated in the regulation of other hormones and exocrine enzyme secretion (Adrian, Bloom, Hermansen, & Iversen, 1978; Csaba & Dournaud, 2001; Roncoroni, Violi, Montanari, & Muri, 1983). The function of ghrelin, secreted by ε-cells, is still unclear regarding glycemic homeostasis (Heller et al., 2005; Vignjevic et al., 2012; Wierup, Svensson, Mulder, & Sundler, 2002).

2. DIABETES

Diabetes mellitus represents one of the most common endocrine diseases in all populations and all age groups. Diabetes can be subdivided into two conditions: type 1 diabetes provoked by the progressive loss of insulin-producing β-cells and type 2 diabetes that involves either a failure in insulin production or target-organ insulin resistance. For the latter, it is important to note that a loss of β-cells can also occur in later stages of the disease (Donath & Halban, 2004; Rhodes, 2005).

Without proper treatment, type 1 diabetes could result in dramatic complications, including nephropathy, loss of vision, foot ulcers, amputation, or even cardiovascular diseases (Nathan, 1993). Therefore, in order to counteract the loss of insulin-producing cells, the administration of exogenous insulin represents the principal therapy for type 1 diabetes treatment. However, environmental variations, such as exercise, pregnancy, diet, and age, may lead to strong variations of the glycemia and thereby result in the aforementioned complications. Alternatively, the transplantation of islets of Langerhans can alleviate insulin dependence in type 1 diabetic patients. However, this approach remains seldom used due to the need for immunosuppression and the scarcity of organ donors (two to four donors are required to treat a single patient). Despite the most recent advances in diabetes care, patients suffering from type 1 diabetes still display a shortened life expectancy (about 8–10 years reduced) in regard to healthy people (International Diabetes Federation, Diabetes Atlas) (Diabetes UK, 2010; Pan & Wright, 2011). Therefore, in the search for alternative treatments, most current efforts aim at finding alternative sources of β-cells. The generation of functional insulin-producing cells from stem/precursor cells represents one of the most promising research avenues together with the reprogramming/transdifferentiation of other pancreatic cells into β-cells.

The use of embryonic stem cells or induced pluripotent stem cells may lend themselves well as an unlimited source of replacement β-cells. However, to date, despite a number of established protocols, both the efficiency of *in vitro* differentiation/programming and the function of the resulting β-like cells remain limited (Van Hoof, D'Amour, & German, 2009).

The plasticity of the adult pancreatic organ is an area of great interest. Plasticity of a fully differentiated cell is described as the ability of a cell to change its epigenome and adopt a functionally distinct fate. Recent lineage tracing experiments performed in mice provides direct evidence that the "irreversibly terminally differentiated" pancreatic cells are surprisingly more plastic than previously assumed, given the appropriate context. A better understanding of the biological and molecular mechanisms that regulate each of the developmental steps could allow us to further improve cell-replacement therapies for diabetes.

3. PANCREATIC MORPHOGENESIS

The first obvious morphological sign of pancreatic development in mice appears at ~E8.5 as two protrusions at the dorsal and ventral portion of foregut/midgut junction (Ackermann & Gannon, 2007; Gittes, 2009; Murtaugh, 2007; Pictet, Clark, Williams, & Rutter, 1972).

Ventral and dorsal outgrowths receive signals from the surrounding mesenchyme (Fig. 6.1A and B) (Gittes, 2009; Murtaugh & Melton, 2003; Oliver-Krasinski & Stoffers, 2008; Slack, 1995). Rodent pancreatic formation has been classically divided into two waves of development: the "primary" and "secondary" transitions (Pictet et al., 1972). Several members of common signaling pathways, such as RA (retinoic acid), BMP (bone morphogenetic protein), FGF (fibroblast growth factor), and sonic hedgehog, are involved in the mechanisms governing early pancreatic genesis (Bayha, Jorgensen, Serup, & Grapin-Botton, 2009; Hebrok, 2003; Jensen, 2004; Kumar & Melton, 2003; Lau & Hebrok, 2010; Pan, Chen, Bayha, & Pieler, 2007; Yamaoka & Itakura, 1999; Zaret, 2000, 2001). The fate of pancreatic progenitors is then specified by a complex program of transcription factor gene expression. Lineage-tracing studies coupled to gain-/loss-of-function analyses allowed us to gain insight into the molecular mechanisms regulating endocrine pancreatic development (Ackermann & Gannon, 2007; Collombat, Hecksher-Sorensen, Serup, & Mansouri, 2006; Gittes, 2009; Murtaugh, 2007; Oliver-Krasinski & Stoffers, 2008; Rieck, Bankaitis, & Wright, 2012).

In mice, the "primary transition" occurs between E9.5 and E12.5 when the dorsal foregut endoderm thickens and bulges into the surrounding mesenchyme (Fig. 6.1B). This period involves active proliferation of pancreatic progenitors to generate a stratified epithelium, followed by the formation of multiple microlumens and their subsequent coalescence (Kesavan et al., 2009). At E11.5, the gut tube begins to undergo its first rotation movement, which brings the dorsal and ventral buds into close proximity for their fusion into a single organ. By ~E12.5, the epithelium starts to undergo active plexus remodeling, epithelial expansion, and production of more plexuses, sending protrusions into the mesenchyme and the entire organ (Fig. 6.1C). We have to introduce a recently proposed alteration to the scheme of organ outgrowth: rather than a peripheral growth and epithelial extension via classical branching morphogenesis, it is now important to consider that a complex process of plexus formation, proliferation, remodeling, and further

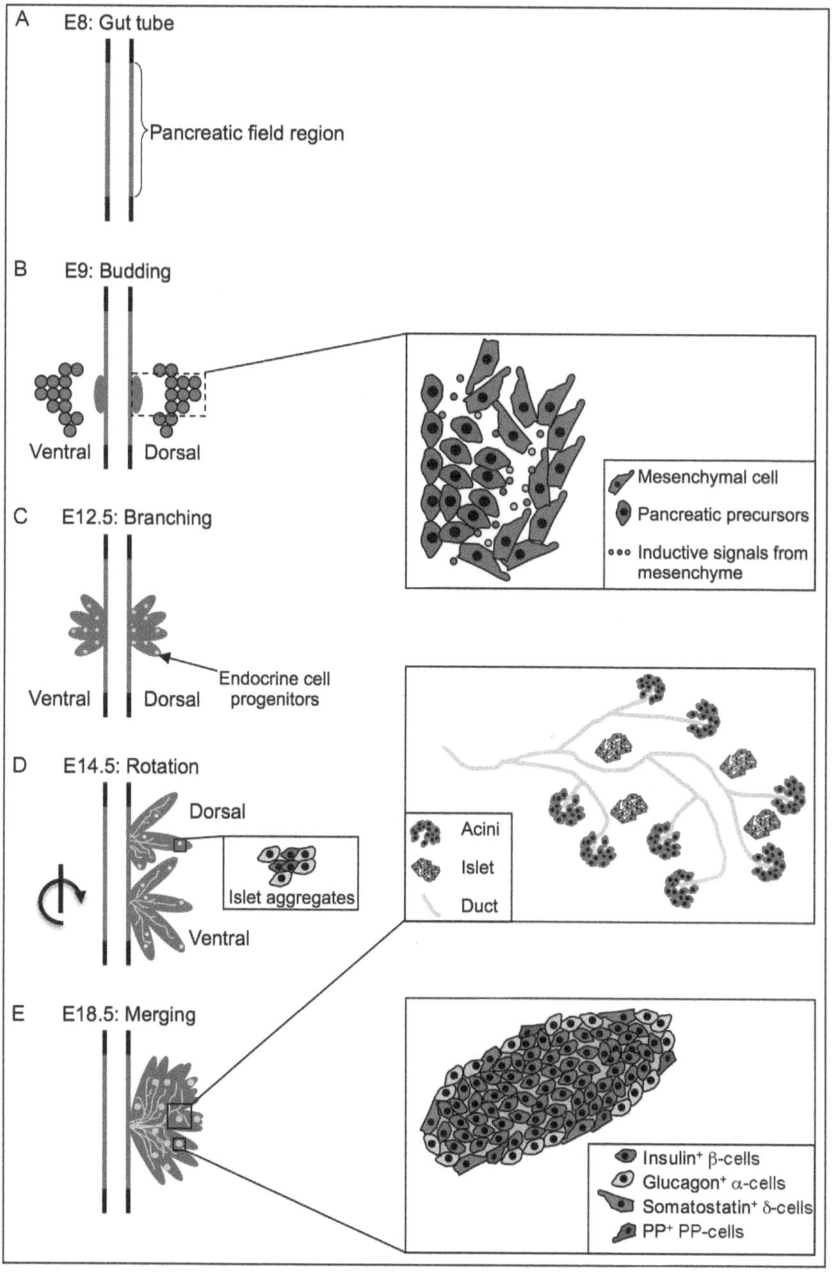

Figure 6.1 Schematic representation of the chronological progression of the embryonic mouse pancreatic morphogenesis. (A) Highlight of the pancreatic field of foregut tube. This region is composed of cells that will bulge from the foregut to give rise to the pancreas. (B) At ~E9, the dorsal and ventral foregut endoderm thickens and evaginates

(Continued)

expansion might afford growth and differentiation over the entirety of the organ. This process would include a massive amount of "outward pushing" that arises from internal growth and remodeling (Pan & Wright, 2011). Compartmentalization of the pancreatic epithelium commences, segregating the epithelium into distinct "tip" and "trunk" domains. This distinction is now better defined on the basis of molecular markers, including some of the transcription factors that are responsible for the gene expression programs of the cells in these regions. Considering these new molecular markers, the tip domain appears to contain multipotent pancreatic cells, which are believed to change into acinar-fated progenitors, whereas the adjacent trunk epithelial region consists of an endocrine/duct bipotent progenitor pool (Stanger, Tanaka, & Melton, 2007; Zhou et al., 2007). During the "secondary transition" starting at E13.5, the epithelium undergoes a massive differentiation into the three main pancreatic lineages (Fig. 6.1C and D).

The delaminating endocrine cells go through a currently poorly defined epithelial exit process probably involving an epithelial to mesenchymal transition (Rukstalis & Habener, 2007). Endocrine cells leaving the epithelium assemble into clustered endocrine islets but often stay in proximity to, or even associated with, one or the other of their parent ducts.

After the secondary transition, at ~E16.5, the epithelium expands further, probably largely driven by acinar proliferation (Fig. 6.1D). During late gestation and in the first few weeks of postnatal life, endocrine cells start to coalesce into mature islets (Fig. 6.1E). About 80% of the islet cell mass present at birth is generated by the proliferation and differentiation of endocrine progenitors, with the other 20% arising from islet cell proliferation (Bouwens & Rooman, 2005).

Around E18.5, ventral and dorsal parts of the pancreas fuse in a single organ structure (Fig. 6.1E). After birth, the predominant mechanism for islet cell mass maintenance is self-duplication. There is a short refractory period for β-cell proliferation observed in the first 4 weeks postnatally and a longer

Figure 6.1—Cont'd into the surrounding mesenchyme. (C) Active proliferation of pancreatic progenitors and formation of microlumens. The arrow outlines the emergence of endocrine precursor cells. (D) Rotation that starts at ~E14.5 and leads to the close contact of the ventral and dorsal counterparts prior to their fusion into a single organ. Subsequently, the first islet aggregates are detected. (E) Ventral and dorsal parts of the pancreas fuse in a single organ assuming the final organ structure. Concurrently, well-formed islets are detected within the exocrine population. (See color plate.)

refractory period after weaning (Bouwens & Rooman, 2005; Salpeter, Klein, Huangfu, Grimsby, & Dor, 2010; Teta, Rankin, Long, Stein, & Kushner, 2007). In order to compensate for body growth, acinar differentiation, maturation, and proliferation also continue *postpartum* before gradually decreasing until weaning (Desai et al., 2007; Salpeter et al., 2010).

4. THE ENDOCRINE DIFFERENTIATION PROGRAM

The first differentiated endocrine cells, mainly glucagon-producing cells, appear in the dorsal bud around E8.5 (Herrera, 2000; Kesavan et al., 2009; Pictet et al., 1972; Villasenor, Chong, Henkemeyer, & Cleaver, 2010). Subsequently, at ~E10, a small number of insulin-expressing cells arise, most of them coexpressing glucagon (Pictet et al., 1972; Teitelman, Alpert, Polak, Martinez, & Hanahan, 1993). However, it seems that such cells, expressing both glucagon and insulin, do not contribute to the mature endocrine compartment (Herrera, 2000). At approximately E13–15, during the "secondary transition," a major increase in endocrine cell numbers occurs and results in the generation of fully differentiated insulin-expressing β-cells and glucagon-expressing α-cells. Within the next 24 h, the first somatostatin-expressing δ-cells emerge and, over the next few days (E14–18), the endocrine cells coalesce into small aggregates (Pictet et al., 1972). Before birth, at E18, PP-expressing cells differentiate and the endocrine cells begin to form the islets of Langerhans.

Several endodermal or organ-specific transcription factors are involved in general organogenesis, including Sox2 (stomach) (Que et al., 2007), Pdx1 (pancreas, duodenum, posterior stomach, and common bile duct) (Offield et al., 1996), Nkx2.1 (lung) (Minoo, Su, Drum, Bringas, & Kimura, 1999), Six1 and Six4 (pharyngeal endoderm) (Zou et al., 2006), Hhex (liver) (Keng et al., 2000; Martinez Barbera et al., 2000), and the caudal homeobox genes Cdx1 and Cdx2 (intestine) (Beck et al., 2003; Beck, Erler, Russell, & James, 1995). Interestingly, a number of these also exhibit additional pancreas-specific roles.

Pdx1, a homeodomain-containing transcription factor, is expressed as early as E8.5 in the dorsal and ventral endoderm distal to the stomach and duodenal epithelium. The expression of Pdx1 progressively becomes restricted to the β-cell lineage in which it controls glucose-induced insulin secretion (Guz et al., 1995; Ohlsson, Karlsson, & Edlund, 1993; Stoffers, Heller, Miller, & Habener, 1999) and the expression of genes linked to β-cell integrity (Leonard et al., 1993; Miller, McGehee, & Habener, 1994; Ohlsson et al., 1993), such as the expression of Glut2, islet amyloid polypeptide, and glucokinase (Chakrabarti, James, & Mirmira, 2002; Macfarlane et al., 2000; Waeber, Thompson, Nicod, & Bonny, 1996; Watada, Kajimoto, Kaneto, et al., 1996; Watada, Kajimoto, Umayahara, et al., 1996). Lineage-tracing experiments have demonstrated that mature pancreatic cells derive from *Pdx1*-expressing progenitor cells (Gu, Brown, & Melton, 2003), and this was further confirmed by the pancreatic agenesis detected in *Pdx1*-deficient mice (Offield et al., 1996). A conditional inactivation of this factor during development, as well as in mature β-cells, outlined its requirement to not only specify but also maintain the β-cell phenotype (Ahlgren, Jonsson, Jonsson, Simu, & Edlund, 1998; Holland, Gonez, Naselli, Macdonald, & Harrison, 2005). The inactivation of *Pdx1* in late-gestational β-cells leads to a decrease in the proliferation of insulin-producing cells with an increase in the proliferation of glucagon-producing cells (Gannon et al., 2008). These results suggest that Pdx1 function is necessary not only for the genesis and maintenance of β-cells but also for late-gestational regulation of the proper endocrine cell numbers (Gannon et al., 2008). Using an inducible tet-off approach, the conditional downregulation of *Pdx1* was found to provoke not only an arrest of β-cell development but also alterations in acinar cell genesis. One explanation may be the absence of Ptf1a, a downstream factor of Pdx1, which plays a crucial role in acinar cell development (Hale et al., 2005). Ptf1a, during early pancreatic development, is expressed throughout the pancreatic epithelium, whereas, at later stages, it becomes exclusively restricted to acinar cells (Krapp et al., 1996). Transgenic mice lacking *Ptf1a* do not develop any acinar cells, whereas the four endocrine cell types are present but misallocated in the spleen (Krapp et al., 1998). Based on these results, it was assumed that Ptf1a was exclusively required for the formation of acinar cells, whereas endocrine and duct cells would derive from different progenitors. However, it has been demonstrated that a majority of duct and endocrine cells derive from *Ptf1a*-expressing precursor cells (Kawaguchi et al., 2002). Both Pdx1 and Ptf1a are required to drive pancreatic multipotent precursors toward a pancreatic cell fate. Indeed, the phenotype of animals with the combined loss of both factors was analyzed to gain further insight into the hierarchy of Pdx1 and Ptf1a (Burlison, Long, Fujitani, Wright, & Magnuson, 2008). These double-mutant mice displayed a similar phenotype of that seen in *Pdx1* single mutant mice, suggesting that Ptf1a acts downstream of Pdx1. However, recent findings revealed that Ptf1a interacts with the *Pdx1* promoter, thereby potentiating *Pdx1* expression in early pancreatic progenitors and later in acinar cells (Wiebe et al., 2007).

Unlike Pdx1, the bHLH transcription factor, Ngn3 exclusively acts in the differentiation of the endocrine tissue, all endocrine cells deriving from Ngn3-expressing cells (Gu, Dubauskaite, & Melton, 2002). Ngn3 expression is detected between E8.5 and E15.5, then it is subsequently reduced, and, under normal conditions, it is barely detectable in adults. Ngn3, through loss-of-function experiments, has been found required for the specification of all entero- and endocrine lineages (Gradwohl, Dierich, LeMeur, & Guillemot, 2000; Gu et al., 2003; Jenny et al., 2002). Gain-of-function analyses further established its potential in promoting endocrine cell differentiation (Heremans et al., 2002; Schwitzgebel et al., 2000). The ectopic expression of *Ngn3* in *Pdx1*-expressing cells led to a premature onset of endocrine differentiation resulting in the nearly exclusive genesis of glucagon-producing cells (Apelqvist et al., 1999; Grapin-Botton, Majithia, & Melton, 2001). Recent data have brought additional insight into the function of Ngn3 demonstrating that its expression occurs in two distinct temporal waves, correlating with the first and secondary transitions described by Pictet et al. (Pictet et al., 1972; Villasenor, Chong, & Cleaver, 2008). This distinct temporal expression generates "early"- and "late"-born endocrine cells with different developmental potentials: early $Ngn3^+$ cells exclusively give rise to α-cells, whereas the activation of Ngn3 at later developmental stages promotes the genesis of β- and PP- (after E11.5) or δ- (after E14.5) cells (Johansson et al., 2007).

The zinc-finger-containing transcription factor insulinoma-associated antigen 1 (IA-1) was found to be a direct downstream target of Ngn3 and to act in the initial steps underlying endocrine cell differentiation. Endocrine cells were still detected in *IA-1* mutant pancreata, but the majority of these did not express any hormones (Gierl, Karoulias, Wende, Strehle, & Birchmeier, 2006). Interestingly, the ectopic expression of *IA-1* in duct cells is not sufficient to induce the endocrine specification program, although the coexpression of Ngn3 and IA-1 was found to significantly enhance the efficiency of duct-to-endocrine transdifferentiation compared to ectopic Ngn3-expression only (Mellitzer et al., 2006).

Arx and *Pax4* represent two additional targets of Ngn3 that were found to play key roles in the specification of the endocrine cell subtypes: *Arx* promotes the α-/PP-cell fates, whereas *Pax4* induces the β-/δ-cell lineages. Indeed, in *Pax4*-deficient pancreata, β-/δ-cells do not develop, while a proportional increase in α-cell numbers is observed (Sosa-Pineda, Chowdhury, Torres, Oliver, & Gruss, 1997). Conversely, upon the loss of *Arx*, an increase in both β- and δ-cell populations is observed at the expense of α-cells (Collombat et al., 2003). It is worth noting that, in both instances, the total islet cell number is found unchanged as compared to wild-type pancreata. Arx and Pax4 factors exhibit antagonistic roles in the specification of the endocrine subtype destinies through an inhibitory, cross-regulatory circuit that controls the transcriptional state of these two genes. The combined loss of *Pax4* and *Arx* does not result in an arrest in the differentiation of all four endocrine cell types, but rather in a depletion of β- and α-cells, while the δ-cell population is increased and the PP-cell content unaltered (Collombat et al., 2005). It was therefore concluded that Pax4 is not required to directly specify the β-/δ-cell destinies, but rather acts by inhibiting the α-cell fate through the inhibition of *Arx* expression. In addition to *Pax4*, another member of the *Pax* gene family, *Pax6*, plays an important role in islet cell differentiation. *Pax6* is expressed in all endocrine hormone-producing cells throughout pancreatic morphogenesis. Conventional knockout of *Pax6* in mice revealed its requirement for the development of all four endocrine cell subtypes, as well as its involvement in forming well-structured islets of Langerhans (St-Onge, Sosa-Pineda, Chowdhury, Mansouri, & Gruss, 1997). By means of conditional ablation of *Pax6* in mice, an additional role for this gene in maintaining the postnatal β-cell phenotype was observed with the development of a diabetic condition associated with hypoinsulinemia (Ashery-Padan et al., 2004).

In an effort to further define the transcriptional network driving the development of the pancreas, the role of Nkx2.2, a homeodomain-containing factor belonging to the NK class and acting early during development, was characterized. Nkx2.2 was found to play a preponderant role in the specification of the β-cell lineage. In mature islets, Nkx2.2 is detected in α-, β-, and PP-cells. Mice depleted of *Nkx2.2* display a loss of β-cells and reduced α- and PP-cell numbers, whereas the number of δ-cells remains unaffected. It is worth noting that the ghrelin-expressing cell population appears considerably augmented in such animals (Prado et al., 2004; Sussel et al., 1998). The analysis of *Pax4/Nkx2.2* double knockout mice revealed alterations similar to those observed in pancreata of *Nkx2.2*$^{-/-}$ mutant mice, suggesting that Pax4 may act downstream of Nkx2.2. This notion was further supported by the lack of alteration in the expression levels of Nkx2.2 in *Pax4*-deficient mice (Prado et al., 2004; Sussel et al., 1998).

Nkx6.1, another early marker of the pancreatic epithelium, also belongs to the Nkx subfamily and is first detected at E9.5. Its expression is initially detected in $Ngn3^+$ endocrine precursor cells but becomes restricted to mature β-cells where it was found to be required for proper insulin secretion (Oster, Jensen, Edlund, & Larsson, 1998; Sander et al., 2000). Mice lacking *Nkx6.1* exhibit a dramatic reduction in mature β-cells, while other islet cell subtypes develop normally (Henseleit et al., 2005). Early $insulin^+$ cells emerging during the first transition are not affected, suggesting a decisive role for Nkx6.1 in the generation of "late-born" mature β-cells originating from the second transition (Pedersen et al., 2005). Nkx6.2, a paralog to Nkx6.1, displays a similar expression pattern compared to Nkx6.1, but in contrast to the latter, it is not detected in adult β-cells (Nelson, Schaffer, & Sander, 2007; Pedersen et al., 2005). *Nkx6.2* mutant mice do not display any obvious phenotype, whereas mice deficient for both *Nkx6.1* and *Nkx6.2* exhibit not only similar alterations as seen in *Nkx6.1* mutant mice (loss of mature β-cells), but also a drastic reduction in $glucagon^+$ cells, indicating an additional involvement of the Nkx factors in α-cell formation (Henseleit et al., 2005). Notably, the misexpression of *Nkx6.2* in *Pdx1*-producing cells of $Nkx6.1^{-/-}$ mice was found to rescue β-cell development, indicating that both factors possess redundant activities for β-cell differentiation (Nelson et al., 2007).

Additional factors were characterized for their involvement in the maintenance of the β-cell identity, including representatives of the Maf gene family (MafA, MafB, and c.Maf) that were shown to act in the terminal differentiation of β-cells as well as α-cells. For instance, MafA was found to directly interact with the promoter of the insulin gene and transactivate *insulin* expression (Aramata, Han, Yasuda, & Kataoka, 2005; Matsuoka et al., 2003; Zhao et al., 2005). Its expression is initiated at E13.5 and is restricted to $insulin^+$ cells during embryogenesis as well as throughout adulthood (Matsuoka et al., 2004). Mice deficient for *MafA* develop diabetes mellitus accompanied by a significant decrease in blood insulin levels and, in addition, display an abnormal islet architecture. *MafA*-deficient $insulin^+$ cells are unable to secrete insulin upon glucose challenge, indicating a β-cell defect in glucose-induced insulin secretion (Zhang et al., 2005). The ectopic expression of *MafA* in embryonic chicken endoderm, as well as in non-pancreatic cell cultures, was found to be sufficient to promote insulin secretion (Artner, Hang, Guo, Gu, & Stein, 2008; Matsuoka et al., 2007). Altogether, this nonextensive description of the major players acting to specify the different endocrine cell subtypes outlines the complexity of the mechanisms involved, some of these factors displaying different roles depending on the age or the cell subtypes.

5. REGENERATIVE MEDICINE

The field of regenerative medicine has seen great progresses concerning the conversion of various differentiated cells into β-like cells *in vivo*. Indeed, pancreatic cells seemingly can dedifferentiate into a progenitor-like state and be redifferentiated into β-like cells or directly transdifferentiate in β-like cells. Specifically, it has been shown that newly formed β-cells could originate from the duct epithelium, endocrine cells, or acinar cells (Fig. 6.2).

Mature duct cells, or a subset of this population, are able to reacquire the ability to generate endocrine progenitors after injury, such as pancreatic duct ligation (PDL). PDL consists of a ligature of the main pancreatic duct resulting in distal damage, acinar involution, and associated inflammatory response. PDL is considered to be an injury model that generates tissue damage and cellular responses similar to those observed in acute pancreatitis. Using this approach, Xu et al. demonstrated that $Ngn3^+$ cells could be mobilized in the ductal epithelium after PDL, thereby identifying an *in vivo* potential source of new endocrine cells. These $Ngn3^+$ duct cells, upon isolation, could produce all endocrine cell types when cultured in *Ngn3* null pancreatic bud explants *in vitro* (Xu et al., 2008). From other studies, however, using independent PDL and β-cell destruction strategies

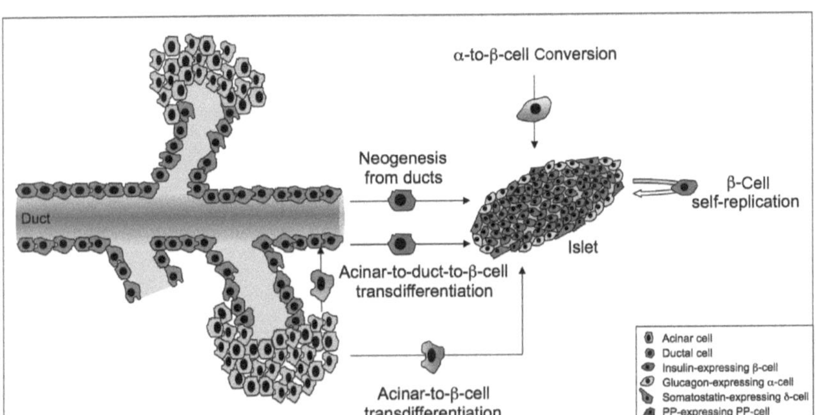

Figure 6.2 Different mechanisms leading to β-cell neogenesis. Several documented examples of reprogramming/regeneration induced by genetic- or injury-based approaches. Newly formed β-like cells could originate from the duct epithelium, endocrine cells, or acinar cells or be generated by self-replication. (See color plate.)

(combining alloxan and EGF/gastrin treatment), it has been shown that adult HNF1β$^+$ mature duct cells could not produce β-cells upon injury (Solar et al., 2009). One limitation with this study, which is common to all experiments dealing with inducible CreER systems, is that only 40–60% of the HNF1β-expressing duct cells were labeled; thus, it remains a possibility that small subpopulations within the ducts still maintain the ability to transdifferentiate into β-cells.

Endocrine α-to-β-cell conversion represents another potential strategy for a regenerative medicine approach in the treatment of diabetes. By expressing the diphtheria toxin (DT) receptor in insulin-expressing cells and using DT administration, it has been demonstrated that it is possible to achieve near total β-cell destruction in mice (Thorel et al., 2010). Lineage tracing after DT ablation was used to conclude that under ~99% β-cell destruction, α-cells move through a bihormonal glucagon$^+$/insulin$^+$ state on their way to generate insulin$^+$ β-like cells. It is important to note that a deep lesion is required; in fact, α-to-β-cell conversion was not seen under ~95% destruction (Thorel et al., 2010). Understanding the molecular mechanisms of α-to-β-cell transdifferentiation in this model could elucidate the endogenous signaling pathways that could be activated/deactivated to convert α-cells into β-cells under normal physiological or *in vitro* culture conditions. An α-to-β-cell transdifferentiation was also found in response to PDL coupled with β-cell destruction by alloxan treatment. However, in this work, genetic lineage tracing was not used to validate the authenticity of the observed α-to-β-cell conversion (Chung, Hao, Piran, Keinan, & Levine, 2010).

Under normal physiological conditions, the β-cell turnover is slow and decreases progressively with age (Teta, Long, Wartschow, Rankin, & Kushner, 2005; Teta et al., 2007). However, upon injury, it appears that at least some β-cells can re-enter the cell cycle. Indeed, under partial pancreatectomy (Ppx), it has been demonstrated that duplication of pre-existing β-cells represents the predominant mechanism for generating new β-cells (Dor, Brown, Martinez, & Melton, 2004; Teta et al., 2007). The ability of β-cells to proliferate under such acute surgical injury is impaired in *FoxM1* null mutants, leading to a dramatic decrease in β-cell mass (Ackermann Misfeldt, Costa, & Gannon, 2008). However, these studies did not rule out the possible contribution of other sources to the β-cell compartment.

Acinar cells could represent an interesting source for reprogramming strategies toward β-cells because of their vast number and developmental relationship to the pancreatic endocrine lineage. Acinar cells, *in vitro*, demonstrated a potential for endocrine/duct conversion when treated with nicotinamide, EGF, or TGF-α (Means et al., 2005; Minami et al., 2005). Acinar cells, *in vivo*, can transiently dedifferentiate to a progenitor-like state upon cerulein treatment (Jensen et al., 2005). This transient dedifferentiation of acinar cells is then followed by acinar regeneration with no contribution of acinar lineage to ductal and endocrine lineages. The same conclusion was obtained from studies using Ppx in combination with acinar-specific lineage tracing (Desai et al., 2007). Elastase-CreERT2-labeled acini were only capable of producing acinar cells, but not ductal or endocrine cells. Furthermore, via lineage tracing of acinar cells, with the same elastase-CreERT2 in a model of hyperplastic duct formation, it was concluded that acini-to-duct transdifferentiation was a minor contributor, but the endocrine lineage was not analyzed (Strobel et al., 2007). Once again, the low percentage (~30% of the total acinar pool) of positive cells when CreERT2-induced labeling is used could be the reason behind these results. When TGF-α overexpression was used to generate duct hyperplasia, lineage tracing of acinar cells using elastase-Cre or villin-Cre suggested that most hyperplastic ducts were transdifferentiated from mature acini and that any endocrine cells detected in/near these duct structures did not come from acini or hyperplastic duct cells (Blaine et al., 2010). All these studies have led to the same conclusion that, upon injury-induced regeneration, acinar cells can only give rise to acinar and ductal cells, but not to endocrine cells.

Reprogramming can also be obtained via the forced expression of key developmental transcription factors in adult cells. Although such a method does not hold at the moment, any therapeutic feasibility in humans, these manipulations can provide information on the mechanisms of plasticity required to impact cell fate. Acinar cells were shown to be capable of conversion into β-cells *in vivo* when given an adenovirus cocktail of three transcription factors (Ngn3, Pdx1, and MafA) (Zhou, Brown, Kanarek, Rajagopal, & Melton, 2008): Ngn3 promoting the acquisition of general endocrine competence; Pdx1 establishing a stable β-cell fate rather than other endocrine fates; and MafA inducting Pdx1 function and driving the newly generated β-cells toward achieving a final, mature, functional state (Ahlgren et al., 1998; Gannon et al., 2008; Gradwohl et al., 2000; Vanhoose et al., 2008). The induction of β-cells was sporadic, apparently occurring from acinar cells without replication or acquisition of a dedifferentiated state. The acini-derived β-cells did not cluster, perhaps because only a few cells received all factors at appropriate relative doses or because only certain acinar cells are competent to respond. These acini-derived β-cells could improve glucose homeostasis in diabetic mice, but the full physiological function was not analyzed. The remaining question is whether there is a contribution of inflammation that arises as a consequence of the virus-induced pancreatitis-like condition (Chen, Chong, Darwiche, Thomas, & Kay, 2004).

Reprogramming has also been observed when enforcing the expression of endocrine lineage transcription factors. This approach gives rise to the interconversion between closely linked cell types, such as α-to-β or the opposite β-to-α identity switch. Persistent expression of *Arx*, one of the key determinants of α-cell specification, in the $Pax6^{+}$ endocrine precursor population during the embryogenesis causes a dramatic reduction in β- and δ-cells, concurrent with increased α- and PP-cell populations in the adult pancreas. Importantly, it was also found that the selective induction of *Arx* production in adult β-cells forces their conversion into α-cells (Collombat et al., 2007). The complementary conversion of α-to-β-cells was reported in response to persistent *Pax4* production in endocrine precursors or glucagon-producing cells. In this latter situation, it was also shown that, upon α-to-β-cell conversion, duct-lining cells could be mobilized, re-express the developmental gene *Ngn3*, and successively adopt an α- and a β-like cell fate (Collombat et al., 2009).

As aforementioned, adult pancreatic β-cells can replicate during body growth and after injury to maintain glucose homeostasis. It has recently been reported that β-cells deficient in *Dnmt1*, an enzyme that propagates DNA methylation patterns during cell division, were converted to α-cells. *Arx* has been reported as methylated and repressed in β-cells and hypomethylated and expressed in α-cells and *Dnmt1*-deficient β-cells. The methylated region of the *Arx* locus in β-cells was bound by methyl-binding protein MeCP2, which recruited PRMT6, an enzyme that methylates histone H3R2 resulting in the repression of *Arx*. This suggests that propagation of DNA methylation during cell division also ensures recruitment of enzymatic machinery capable of modifying and transmitting histone marks. These results revealed that propagation of DNA methylation during cell division is essential for the repression of α-cell lineage-determining genes in order to maintain a pancreatic β-cell identity (Dhawan, Georgia, Tschen, Fan, & Bhushan, 2011).

Overall, these lineage-switching and induced neogenesis reports are generating significant excitement over the possibility of these processes being therapeutically manageable, and that we might, one day, be able to turn the process on and off as desired. In this context, the genetic reprogramming of endogenous and regenerating α-to-β-like cells holds much more promise in that, if it can be optimally controlled, it could represent a new avenue toward a putative treatment of type 1 diabetes.

ACKNOWLEDGMENTS

Due to space constraints, we apologize to many colleagues in the field whose work we could not cite. The authors are supported by the Juvenile Diabetes Research Foundation (17-2011-16, 2-2010-567, 26-2008-639, and 17-2013-426), the INSERM Avenir Program, the INSERM, the European Research Council (StG-2011-281265), the FMR (DRC20091217179), the ANR/BMBF (2009 GENO 105 01/01KU0906), Club Isatis, Mr. and Mrs. Dorato, the Fondation Générale de Santé, and the Foundation Schlumberger pour l'Education et la Recherche.

REFERENCES

Ackermann, A. M., & Gannon, M. (2007). Molecular regulation of pancreatic beta-cell mass development, maintenance, and expansion. *Journal of Molecular Endocrinology*, *38*, 193–206.

Ackermann Misfeldt, A., Costa, R. H., & Gannon, M. (2008). Beta-cell proliferation, but not neogenesis, following 60% partial pancreatectomy is impaired in the absence of FoxM1. *Diabetes*, *57*, 3069–3077.

Adrian, T. E., Bloom, S. R., Hermansen, K., & Iversen, J. (1978). Pancreatic polypeptide, glucagon and insulin secretion from the isolated perfused canine pancreas. *Diabetologia*, *14*, 413–417.

Ahlgren, U., Jonsson, J., Jonsson, L., Simu, K., & Edlund, H. (1998). Beta-cell-specific inactivation of the mouse Ipf1/Pdx1 gene results in loss of the beta-cell phenotype and maturity onset diabetes. *Genes & Development*, *12*, 1763–1768.

Apelqvist, A., Li, H., Sommer, L., Beatus, P., Anderson, D. J., Honjo, T., et al. (1999). Notch signalling controls pancreatic cell differentiation. *Nature*, *400*, 877–881.

Aramata, S., Han, S. I., Yasuda, K., & Kataoka, K. (2005). Synergistic activation of the insulin gene promoter by the beta-cell enriched transcription factors MafA, Beta2, and Pdx1. *Biochimica et Biophysica Acta*, *1730*, 41–46.

Artner, I., Hang, Y., Guo, M., Gu, G., & Stein, R. (2008). MafA is a dedicated activator of the insulin gene in vivo. *The Journal of Endocrinology*, *198*, 271–279.

Ashery-Padan, R., Zhou, X., Marquardt, T., Herrera, P., Toube, L., Berry, A., et al. (2004). Conditional inactivation of Pax6 in the pancreas causes early onset of diabetes. *Developmental Biology*, *269*, 479–488.

Bayha, E., Jorgensen, M. C., Serup, P., & Grapin-Botton, A. (2009). Retinoic acid signaling organizes endodermal organ specification along the entire antero-posterior axis. *PLoS One*, *4*, e5845.

Beck, F., Chawengsaksophak, K., Luckett, J., Giblett, S., Tucci, J., Brown, J., et al. (2003). A study of regional gut endoderm potency by analysis of Cdx2 null mutant chimaeric mice. *Developmental Biology*, *255*, 399–406.

Beck, F., Erler, T., Russell, A., & James, R. (1995). Expression of Cdx-2 in the mouse embryo and placenta: Possible role in patterning of the extra-embryonic membranes. *Developmental Dynamics*, *204*, 219–227.

Blaine, S. A., Ray, K. C., Anunobi, R., Gannon, M. A., Washington, M. K., & Means, A. L. (2010). Adult pancreatic acinar cells give rise to ducts but not endocrine cells in response to growth factor signaling. *Development, 137*, 2289–2296.

Bouwens, L., & Rooman, I. (2005). Regulation of pancreatic beta-cell mass. *Physiological Reviews, 85*, 1255–1270.

Burlison, J. S., Long, Q., Fujitani, Y., Wright, C. V., & Magnuson, M. A. (2008). Pdx-1 and Ptf1a concurrently determine fate specification of pancreatic multipotent progenitor cells. *Developmental Biology, 316*, 74–86.

Chakrabarti, S. K., James, J. C., & Mirmira, R. G. (2002). Quantitative assessment of gene targeting in vitro and in vivo by the pancreatic transcription factor, Pdx1. Importance of chromatin structure in directing promoter binding. *The Journal of Biological Chemistry, 277*, 13286–13293.

Chen, Y., Chong, M. M. W., Darwiche, R., Thomas, H. E., & Kay, T. W. H. (2004). Severe pancreatitis with exocrine destruction and increased islet neogenesis in mice with suppressor of cytokine signaling-1 deficiency. *The American Journal of Pathology, 165*, 913–921.

Chung, C.-H., Hao, E., Piran, R., Keinan, E., & Levine, F. (2010). Pancreatic β-cell neogenesis by direct conversion from mature α-cells. *Stem Cells, 28*, 1630–1638.

Collombat, P., Hecksher-Sorensen, J., Broccoli, V., Krull, J., Ponte, I., Mundiger, T., et al. (2005). The simultaneous loss of Arx and Pax4 genes promotes a somatostatin-producing cell fate specification at the expense of the alpha- and beta-cell lineages in the mouse endocrine pancreas. *Development, 132*, 2969–2980.

Collombat, P., Hecksher-Sorensen, J., Krull, J., Berger, J., Riedel, D., Herrera, P. L., et al. (2007). Embryonic endocrine pancreas and mature beta cells acquire alpha and PP cell phenotypes upon Arx misexpression. *The Journal of Clinical Investigation, 117*, 961–970.

Collombat, P., Hecksher-Sorensen, J., Serup, P., & Mansouri, A. (2006). Specifying pancreatic endocrine cell fates. *Mechanisms of Development, 123*, 501–512.

Collombat, P., Mansouri, A., Hecksher-Sorensen, J., Serup, P., Krull, J., Gradwohl, G., et al. (2003). Opposing actions of Arx and Pax4 in endocrine pancreas development. *Genes & Development, 17*, 2591–2603.

Collombat, P., Xu, X., Ravassard, P., Sosa-Pineda, B., Dussaud, S., Billestrup, N., et al. (2009). The ectopic expression of Pax4 in the mouse pancreas converts progenitor cells into alpha and subsequently beta cells. *Cell, 138*, 449–462.

Csaba, Z., & Dournaud, P. (2001). Cellular biology of somatostatin receptors. *Neuropeptides, 35*, 1–23.

Desai, B. M., Oliver-Krasinski, J., De Leon, D. D., Farzad, C., Hong, N., Leach, S. D., et al. (2007). Preexisting pancreatic acinar cells contribute to acinar cell, but not islet beta cell, regeneration. *The Journal of Clinical Investigation, 117*, 971–977.

Dhawan, S., Georgia, S., Tschen, S.-I., Fan, G., & Bhushan, A. (2011). Pancreatic β cell identity is maintained by DNA methylation-mediated repression of Arx. *Developmental Cell, 20*, 419–429.

Diabetes UK (2010). *Diabetes in the UK 2010: Key statistics on diabetes* (pp. 1–21) (online).

Donath, M. Y., & Halban, P. A. (2004). Decreased beta-cell mass in diabetes: Significance, mechanisms and therapeutic implications. *Diabetologia, 47*, 581–589.

Dor, Y., Brown, J., Martinez, O. I., & Melton, D. A. (2004). Adult pancreatic beta-cells are formed by self-duplication rather than stem-cell differentiation. *Nature, 429*, 41–46.

Gannon, M., Ables, E. T., Crawford, L., Lowe, D., Offield, M. F., Magnuson, M. A., et al. (2008). pdx-1 function is specifically required in embryonic beta cells to generate appropriate numbers of endocrine cell types and maintain glucose homeostasis. *Developmental Biology, 314*, 406–417.

Gierl, M. S., Karoulias, N., Wende, H., Strehle, M., & Birchmeier, C. (2006). The zinc-finger factor Insm1 (IA-1) is essential for the development of pancreatic beta cells and intestinal endocrine cells. *Genes & Development, 20*, 2465–2478.

Gittes, G. K. (2009). Developmental biology of the pancreas: A comprehensive review. *Developmental Biology, 326*, 4–35.

Gradwohl, G., Dierich, A., LeMeur, M., & Guillemot, F. (2000). Neurogenin3 is required for the development of the four endocrine cell lineages of the pancreas. *In: Proceedings of the National Academy of Sciences of the United States of America, 97*, 1607–1611.

Grapin-Botton, A., Majithia, A. R., & Melton, D. A. (2001). Key events of pancreas formation are triggered in gut endoderm by ectopic expression of pancreatic regulatory genes. *Genes & Development, 15*, 444–454.

Gu, G., Brown, J. R., & Melton, D. A. (2003). Direct lineage tracing reveals the ontogeny of pancreatic cell fates during mouse embryogenesis. *Mechanisms of Development, 120*, 35–43.

Gu, G., Dubauskaite, J., & Melton, D. A. (2002). Direct evidence for the pancreatic lineage: NGN3+ cells are islet progenitors and are distinct from duct progenitors. *Development, 129*, 2447–2457.

Guz, Y., Montminy, M. R., Stein, R., Leonard, J., Gamer, L. W., Wright, C. V., et al. (1995). Expression of murine STF-1, a putative insulin gene transcription factor, in beta cells of pancreas, duodenal epithelium and pancreatic exocrine and endocrine progenitors during ontogeny. *Development, 121*, 11–18.

Hale, M. A., Kagami, H., Shi, L., Holland, A. M., Elsasser, H. P., Hammer, R. E., et al. (2005). The homeodomain protein PDX1 is required at mid-pancreatic development for the formation of the exocrine pancreas. *Developmental Biology, 286*, 225–237.

Hebrok, M. (2003). Hedgehog signaling in pancreas development. *Mechanisms of Development, 120*, 45–57.

Heller, R. S., Jenny, M., Collombat, P., Mansouri, A., Tomasetto, C., Madsen, O. D., et al. (2005). Genetic determinants of pancreatic epsilon-cell development. *Developmental Biology, 286*, 217–224.

Henseleit, K. D., Nelson, S. B., Kuhlbrodt, K., Hennings, J. C., Ericson, J., & Sander, M. (2005). NKX6 transcription factor activity is required for alpha- and beta-cell development in the pancreas. *Development, 132*, 3139–3149.

Heremans, Y., Van De Casteele, M., in't Veld, P., Gradwohl, G., Serup, P., Madsen, O., et al. (2002). Recapitulation of embryonic neuroendocrine differentiation in adult human pancreatic duct cells expressing neurogenin 3. *The Journal of Cell Biology, 159*, 303–312.

Herrera, P. L. (2000). Adult insulin- and glucagon-producing cells differentiate from two independent cell lineages. *Development, 127*, 2317–2322.

Holland, A. M., Gonez, L. J., Naselli, G., Macdonald, R. J., & Harrison, L. C. (2005). Conditional expression demonstrates the role of the homeodomain transcription factor Pdx1 in maintenance and regeneration of beta-cells in the adult pancreas. *Diabetes, 54*, 2586–2595.

Jenny, M., Uhl, C., Roche, C., Duluc, I., Guillermin, V., Guillemot, F., et al. (2002). Neurogenin3 is differentially required for endocrine cell fate specification in the intestinal and gastric epithelium. *The EMBO Journal, 21*, 6338–6347.

Jensen, J. (2004). Gene regulatory factors in pancreatic development. *Developmental Dynamics, 229*, 176–200.

Jensen, J. N., Cameron, E., Garay, M. V. R., Starkey, T. W., Gianani, R., & Jensen, J. (2005). Recapitulation of elements of embryonic development in adult mouse pancreatic regeneration. *Gastroenterology, 128*, 728–741.

Johansson, K. A., Dursun, U., Jordan, N., Gu, G., Beermann, F., Gradwohl, G., et al. (2007). Temporal control of neurogenin3 activity in pancreas progenitors reveals competence windows for the generation of different endocrine cell types. *Developmental Cell, 12*, 457–465.

Kawaguchi, Y., Cooper, B., Gannon, M., Ray, M., MacDonald, R. J., & Wright, C. V. (2002). The role of the transcriptional regulator Ptf1a in converting intestinal to pancreatic progenitors. *Nature Genetics*, *32*, 128–134.

Keng, V. W., Yagi, H., Ikawa, M., Nagano, T., Myint, Z., Yamada, K., et al. (2000). Homeobox gene Hex is essential for onset of mouse embryonic liver development and differentiation of the monocyte lineage. *Biochemical and Biophysical Research Communications*, *276*, 1155–1161.

Kesavan, G., Sand, F. W., Greiner, T. U., Johansson, J. K., Kobberup, S., Wu, X., et al. (2009). Cdc42-mediated tubulogenesis controls cell specification. *Cell*, *139*, 791–801.

Konstantinova, I., Nikolova, G., Ohara-Imaizumi, M., Meda, P., Kucera, T., Zarbalis, K., et al. (2007). EphA-Ephrin-A-mediated beta cell communication regulates insulin secretion from pancreatic islets. *Cell*, *129*, 359–370.

Krapp, A., Knofler, M., Frutiger, S., Hughes, G. J., Hagenbuchle, O., & Wellauer, P. K. (1996). The p48 DNA-binding subunit of transcription factor PTF1 is a new exocrine pancreas-specific basic helix-loop-helix protein. *The EMBO Journal*, *15*, 4317–4329.

Krapp, A., Knofler, M., Ledermann, B., Burki, K., Berney, C., Zoerkler, N., et al. (1998). The bHLH protein PTF1-p48 is essential for the formation of the exocrine and the correct spatial organization of the endocrine pancreas. *Genes & Development*, *12*, 3752–3763.

Kumar, M., & Melton, D. (2003). Pancreas specification: A budding question. *Current Opinion in Genetics & Development*, *13*, 401–407.

Lau, J., & Hebrok, M. (2010). Hedgehog signaling in pancreas epithelium regulates embryonic organ formation and adult beta-cell function. *Diabetes*, *59*, 1211–1221.

Leonard, J., Peers, B., Johnson, T., Ferreri, K., Lee, S., & Montminy, M. R. (1993). Characterization of somatostatin transactivating factor-1, a novel homeobox factor that stimulates somatostatin expression in pancreatic islet cells. *Molecular Endocrinology*, *7*, 1275–1283.

Macfarlane, W. M., Campbell, S. C., Elrick, L. J., Oates, V., Bermano, G., Lindley, K. J., et al. (2000). Glucose regulates islet amyloid polypeptide gene transcription in a PDX1- and calcium-dependent manner. *The Journal of Biological Chemistry*, *275*, 15330–15335.

Martinez Barbera, J. P., Clements, M., Thomas, P., Rodriguez, T., Meloy, D., Kioussis, D., et al. (2000). The homeobox gene Hex is required in definitive endodermal tissues for normal forebrain, liver and thyroid formation. *Development*, *127*, 2433–2445.

Matsuoka, T. A., Artner, I., Henderson, E., Means, A., Sander, M., & Stein, R. (2004). The MafA transcription factor appears to be responsible for tissue-specific expression of insulin. *In: Proceedings of the National Academy of Sciences of the United States of America*, *101*, 2930–2933.

Matsuoka, T. A., Kaneto, H., Stein, R., Miyatsuka, T., Kawamori, D., Henderson, E., et al. (2007). MafA regulates expression of genes important to islet beta-cell function. *Molecular Endocrinology*, *21*, 2764–2774.

Matsuoka, T. A., Zhao, L., Artner, I., Jarrett, H. W., Friedman, D., Means, A., et al. (2003). Members of the large Maf transcription family regulate insulin gene transcription in islet beta cells. *Molecular and Cellular Biology*, *23*, 6049–6062.

Means, A. L., Meszoely, I. M., Suzuki, K., Miyamoto, Y., Rustgi, A. K., Coffey, R. J., et al. (2005). Pancreatic epithelial plasticity mediated by acinar cell transdifferentiation and generation of nestin-positive intermediates. *Development*, *132*, 3767–3776.

Mellitzer, G., Bonne, S., Luco, R. F., Van De Casteele, M., Lenne-Samuel, N., Collombat, P., et al. (2006). IA1 is NGN3-dependent and essential for differentiation of the endocrine pancreas. *The EMBO Journal*, *25*, 1344–1352.

Miller, C. P., McGehee, R. E., & Habener, J. F. (1994). IDX-1: A new homeodomain transcription factor expressed in rat pancreatic islets and duodenum that transactivates the somatostatin gene. *The EMBO Journal*, *13*, 1145–1156.

Minami, K., Okuno, M., Miyawaki, K., Okumachi, A., Ishizaki, K., Oyama, K., et al. (2005). Lineage tracing and characterization of insulin-secreting cells generated from adult pancreatic acinar cells. *In: Proceedings of the National Academy of Sciences of the United States of America, 102*, 15116–15121.

Minoo, P., Su, G., Drum, H., Bringas, P., & Kimura, S. (1999). Defects in tracheoesophageal and lung morphogenesis in Nkx2.1(−/−) mouse embryos. *Developmental Biology, 209*, 60–71.

Murtaugh, L. C. (2007). Pancreas and beta-cell development: From the actual to the possible. *Development, 134*, 427–438.

Murtaugh, L. C., & Melton, D. A. (2003). Genes, signals, and lineages in pancreas development. *Annual Review of Cell and Developmental Biology, 19*, 71–89.

Nathan, D. M. (1993). Long-term complications of diabetes mellitus. *The New England Journal of Medicine, 328*, 1676–1685.

Nelson, S. B., Schaffer, A. E., & Sander, M. (2007). The transcription factors Nkx6.1 and Nkx6.2 possess equivalent activities in promoting beta-cell fate specification in Pdx1+ pancreatic progenitor cells. *Development, 134*, 2491–2500.

Offield, M. F., Jetton, T. L., Labosky, P. A., Ray, M., Stein, R. W., Magnuson, M. A., et al. (1996). PDX-1 is required for pancreatic outgrowth and differentiation of the rostral duodenum. *Development, 122*, 983–995.

Ohlsson, H., Karlsson, K., & Edlund, T. (1993). IPF1, a homeodomain-containing transactivator of the insulin gene. *The EMBO Journal, 12*, 4251–4259.

Oliver-Krasinski, J. M., & Stoffers, D. A. (2008). On the origin of the beta cell. *Genes & Development, 22*, 1998–2021.

Oster, A., Jensen, J., Edlund, H., & Larsson, L. I. (1998). Homeobox gene product Nkx 6.1 immunoreactivity in nuclei of endocrine cells of rat and mouse stomach. *The Journal of Histochemistry and Cytochemistry, 46*, 717–721.

Pan, F. C., Chen, Y., Bayha, E., & Pieler, T. (2007). Retinoic acid-mediated patterning of the pre-pancreatic endoderm in Xenopus operates via direct and indirect mechanisms. *Mechanisms of Development, 124*, 518–531.

Pan, F. C., & Wright, C. (2011). Pancreas organogenesis: From bud to plexus to gland. *Developmental Dynamics, 240*, 530–565.

Pedersen, J. K., Nelson, S. B., Jorgensen, M. C., Henseleit, K. D., Fujitani, Y., Wright, C. V., et al. (2005). Endodermal expression of Nkx6 genes depends differentially on Pdx1. *Developmental Biology, 288*, 487–501.

Pictet, R. L., Clark, W. R., Williams, R. H., & Rutter, W. J. (1972). An ultrastructural analysis of the developing embryonic pancreas. *Developmental Biology, 29*, 436–467.

Prado, C. L., Pugh-Bernard, A. E., Elghazi, L., Sosa-Pineda, B., & Sussel, L. (2004). Ghrelin cells replace insulin-producing beta cells in two mouse models of pancreas development. *In: Proceedings of the National Academy of Sciences of the United States of America, 101*, 2924–2929.

Que, J., Okubo, T., Goldenring, J. R., Nam, K.-T., Kurotani, R., Morrisey, E. E., et al. (2007). Multiple dose-dependent roles for Sox2 in the patterning and differentiation of anterior foregut endoderm. *Development, 134*, 2521–2531.

Rhodes, C. J. (2005). Type 2 diabetes-a matter of beta-cell life and death? *Science, 307*, 380–384.

Rieck, S., Bankaitis, E. D., & Wright, C. V. E. (2012). Lineage determinants in early endocrine development. *Seminars in Cell & Developmental Biology, 23*, 673–684.

Roncoroni, L., Violi, V., Montanari, M., & Muri, M. (1983). Effect of somatostatin on exocrine pancreas evaluated on a total external pancreatic fistula of neoplastic origin. *The American Journal of Gastroenterology, 78*, 425–428.

Rukstalis, J. M., & Habener, J. F. (2007). Snail2, a mediator of epithelial-mesenchymal transitions, expressed in progenitor cells of the developing endocrine pancreas. *Gene Expression Patterns, 7*, 471–479.

Salpeter, S. J., Klein, A. M., Huangfu, D., Grimsby, J., & Dor, Y. (2010). Glucose and aging control the quiescence period that follows pancreatic beta cell replication. *Development, 137*, 3205–3213.

Sander, M., Sussel, L., Conners, J., Scheel, D., Kalamaras, J., Dela Cruz, F., et al. (2000). Homeobox gene Nkx6.1 lies downstream of Nkx2.2 in the major pathway of beta-cell formation in the pancreas. *Development, 127*, 5533–5540.

Schwitzgebel, V. M., Scheel, D. W., Conners, J. R., Kalamaras, J., Lee, J. E., Anderson, D. J., et al. (2000). Expression of neurogenin3 reveals an islet cell precursor population in the pancreas. *Development, 127*, 3533–3542.

Slack, J. M. (1995). Developmental biology of the pancreas. *Development, 121*, 1569–1580.

Solar, M., Cardalda, C., Houbracken, I., Martin, M., Maestro, M. A., De Medts, N., et al. (2009). Pancreatic exocrine duct cells give rise to insulin-producing beta cells during embryogenesis but not after birth. *Developmental Cell, 17*, 849–860.

Sosa-Pineda, B., Chowdhury, K., Torres, M., Oliver, G., & Gruss, P. (1997). The Pax4 gene is essential for differentiation of insulin-producing beta cells in the mammalian pancreas. *Nature, 386*, 399–402.

Stanger, B. Z., Tanaka, A. J., & Melton, D. A. (2007). Organ size is limited by the number of embryonic progenitor cells in the pancreas but not the liver. *Nature, 445*, 886–891.

Stoffers, D. A., Heller, R. S., Miller, C. P., & Habener, J. F. (1999). Developmental expression of the homeodomain protein IDX-1 in mice transgenic for an IDX-1 promoter/lacZ transcriptional reporter. *Endocrinology, 140*, 5374–5381.

St-Onge, L., Sosa-Pineda, B., Chowdhury, K., Mansouri, A., & Gruss, P. (1997). Pax6 is required for differentiation of glucagon-producing alpha-cells in mouse pancreas. *Nature, 387*, 406–409.

Strobel, O., Dor, Y., Alsina, J., Stirman, A., Lauwers, G., Trainor, A., et al. (2007). In vivo lineage tracing defines the role of acinar-to-ductal transdifferentiation in inflammatory ductal metaplasia. *Gastroenterology, 133*, 1999–2009.

Sussel, L., Kalamaras, J., Hartigan-O'Connor, D. J., Meneses, J. J., Pedersen, R. A., Rubenstein, J. L., et al. (1998). Mice lacking the homeodomain transcription factor Nkx2.2 have diabetes due to arrested differentiation of pancreatic beta cells. *Development, 125*, 2213–2221.

Teitelman, G., Alpert, S., Polak, J. M., Martinez, A., & Hanahan, D. (1993). Precursor cells of mouse endocrine pancreas coexpress insulin, glucagon and the neuronal proteins tyrosine hydroxylase and neuropeptide Y, but not pancreatic polypeptide. *Development, 118*, 1031–1039.

Teta, M., Long, S. Y., Wartschow, L. M., Rankin, M. M., & Kushner, J. A. (2005). Very slow turnover of beta-cells in aged adult mice. *Diabetes, 54*, 2557–2567.

Teta, M., Rankin, M. M., Long, S. Y., Stein, G. M., & Kushner, J. A. (2007). Growth and regeneration of adult beta cells does not involve specialized progenitors. *Developmental Cell, 12*, 817–826.

Thorel, F., Nepote, V., Avril, I., Kohno, K., Desgraz, R., Chera, S., et al. (2010). Conversion of adult pancreatic alpha-cells to beta-cells after extreme beta-cell loss. *Nature, 464*, 1149–1154.

Van Hoof, D., D'Amour, K. A., & German, M. S. (2009). Derivation of insulin-producing cells from human embryonic stem cells. *Stem Cell Research, 3*, 73–87.

Vanhoose, A. M., Samaras, S., Artner, I., Henderson, E., Hang, Y., & Stein, R. (2008). MafA and MafB regulate Pdx1 transcription through the Area II control region in pancreatic beta cells. *The Journal of Biological Chemistry, 283*, 22612–22619.

Vignjevic, S., Todorovic, V., Damjanovic, S., Budec, M., Mitrovic, O., Djikic, D., et al. (2012). Similar developmental patterns of ghrelin- and glucagon-expressing cells in the human pancreas. *Cells, Tissues, Organs, 196*, 362–373.

Villasenor, A., Chong, D. C., & Cleaver, O. (2008). Biphasic Ngn3 expression in the developing pancreas. *Developmental Dynamics*, *237*, 3270–3279.

Villasenor, A., Chong, D. C., Henkemeyer, M., & Cleaver, O. (2010). Epithelial dynamics of pancreatic branching morphogenesis. *Development*, *137*, 4295–4305.

Waeber, G., Thompson, N., Nicod, P., & Bonny, C. (1996). Transcriptional activation of the GLUT2 gene by the IPF-1/STF-1/IDX-1 homeobox factor. *Molecular Endocrinology*, *10*, 1327–1334.

Watada, H., Kajimoto, Y., Kaneto, H., Matsuoka, T., Fujitani, Y., Miyazaki, J. i., et al. (1996). Involvement of the homeodomain-containing transcription factor PDX-1 in islet amyloid polypeptide gene transcription. *Biochemical and Biophysical Research Communications*, *229*, 746–751.

Watada, H., Kajimoto, Y., Umayahara, Y., Matsuoka, T., Kaneto, H., Fujitani, Y., et al. (1996). The human glucokinase gene beta-cell-type promoter: An essential role of insulin promoter factor 1/PDX-1 in its activation in HIT-T15 cells. *Diabetes*, *45*, 1478–1488.

Wiebe, P. O., Kormish, J. D., Roper, V. T., Fujitani, Y., Alston, N. I., Zaret, K. S., et al. (2007). Ptf1a binds to and activates area III, a highly conserved region of the Pdx1 promoter that mediates early pancreas-wide Pdx1 expression. *Molecular and Cellular Biology*, *27*, 4093–4104.

Wierup, N., Svensson, H., Mulder, H., & Sundler, F. (2002). The ghrelin cell: A novel developmentally regulated islet cell in the human pancreas. *Regulatory Peptides*, *107*, 63–69.

Xu, X., D'Hoker, J., Stange, G., Bonne, S., De Leu, N., Xiao, X., et al. (2008). Beta cells can be generated from endogenous progenitors in injured adult mouse pancreas. *Cell*, *132*, 197–207.

Yamaoka, T., & Itakura, M. (1999). Development of pancreatic islets (review). *International Journal of Molecular Medicine*, *3*, 247–261.

Zaret, K. S. (2000). Liver specification and early morphogenesis. *Mechanisms of Development*, *92*, 83–88.

Zaret, K. S. (2001). Hepatocyte differentiation: From the endoderm and beyond. *Current Opinion in Genetics & Development*, *11*, 568–574.

Zhang, C., Moriguchi, T., Kajihara, M., Esaki, R., Harada, A., Shimohata, H., et al. (2005). MafA is a key regulator of glucose-stimulated insulin secretion. *Molecular and Cellular Biology*, *25*, 4969–4976.

Zhao, L., Guo, M., Matsuoka, T.-A., Hagman, D. K., Parazzoli, S. D., Poitout, V., et al. (2005). The islet beta cell-enriched MafA activator is a key regulator of insulin gene transcription. *The Journal of Biological Chemistry*, *280*, 11887–11894.

Zhou, Q., Brown, J., Kanarek, A., Rajagopal, J., & Melton, D. A. (2008). In vivo reprogramming of adult pancreatic exocrine cells to beta-cells. *Nature*, *455*, 627–632.

Zhou, Q., Law, A. C., Rajagopal, J., Anderson, W. J., Gray, P. A., & Melton, D. A. (2007). A multipotent progenitor domain guides pancreatic organogenesis. *Developmental Cell*, *13*, 103–114.

Zou, D., Silvius, D., Davenport, J., Grifone, R., Maire, P., & Xu, P.-X. (2006). Patterning of the third pharyngeal pouch into thymus/parathyroid by Six and Eya1. *Developmental Biology*, *293*, 499–512.

CHAPTER SEVEN

Adrenocortical Development, Maintenance, and Disease

Rebecca Yates[*], Harshini Katugampola[†], Dominic Cavlan[†], Katy Cogger[†], Eirini Meimaridou[†], Claire Hughes[†], Louise Metherell[†], Leonardo Guasti[†], Peter King[†,1]

[*]Barts and the London School of Medicine and Dentistry, Queen Mary, University of London, London, United Kingdom

[†]Centre for Endocrinology, William Harvey Research Institute, Barts and the London School of Medicine and Dentistry, Queen Mary, University of London, London, United Kingdom

[1]Corresponding author: e-mail address: p.j.king@qmul.ac.uk

Contents

Current Topics in Developmental Biology, Volume 106
ISSN 0070-2153
http://dx.doi.org/10.1016/B978-0-12-416021-7.00007-9

Abstract

The adrenal gland controls a plethora of crucial physiological functions, and dysfunction is associated with severe morbidity. Because of the vital importance of appropriate adrenal function, the development and function of the gland have been intensively studied, and these investigations have revealed fascinating developmental origins and a remarkable remodeling and regenerative capacity in the adult. This chapter, focusing on the adrenal cortex, will describe our current understanding of the development and maintenance of the adrenal gland, which has been advanced over recent years by the use of sophisticated genetic models in the study of both normal function and disease. This work has shed light on the transcriptional networks and signaling pathways involved in development and maintenance of the gland and in its pathology; these are discussed in the light of the wealth of physiological information gathered in studies of human and rodent adrenal development and function.

1. INTRODUCTION

The earliest descriptions of the adrenal gland were made by Galen in the second century, but the gland was definitively described and illustrated by in 1563 by Eustachio (1563) who recognized that the structures he observed were discrete organs, albeit of unknown function. It was not until nineteenth-century advances in lens manufacture and microscopy that Kölliker (1852) could distinguish the cortex from the medulla and conclude on the basis of their separate early development and different appearance and that these two structures had different functions. Thomas Addison was the first to recognize the importance of the adrenal gland for well-being

(Addison, 1855) and Charles-Edouard Brown-Sequard demonstrated in 1856 that experimental animals could not survive bilateral adrenalectomy (Brown-Sequard, 1856). The cortical zones were later identified and named by Arnold (1886). The adrenal gland has been the subject of continuous investigation since these discoveries, and the chapter will discuss the physiology of the gland, in particular the cortex, and more recent mechanistic insights into development, growth, maintenance, and disease of this organ.

2. THE ADRENAL GLAND: STRUCTURE AND FUNCTION

2.1. Anatomy

The adrenal glands are a pair of composite endocrine organs, with the outer cortex and inner medulla having different functions and developmental origins. The glands are located anterior to and superior to the upper pole of the kidney (Fig. 7.1). They are yellowish in color and asymmetrical, with the right adrenal gland being pyramidal in shape and the larger left adrenal gland being crescent-shaped. Adult human adrenal glands weigh 8–10 g, up to 90% of which is cortex. The adrenal cortex is the principal steroid-producing organ in the body and the medulla secretes catecholamines that function as "fight or flight" hormones.

The adrenal gland is enclosed within a capsule of fibroblasts and myofibroblasts, which contains an arterial plexus and sends trabeculae containing blood vessels and nerves into the underlying cortex. Capsular cells are flattened with elongated nuclei arranged in the plane of the capsule. Lying underneath the capsule is the outermost zone of the cortex, the zona glomerulosa (ZG). This narrow zone contains relatively small columnar epithelial cells arranged in round clusters. The ZG cells have intensely staining nuclei and basophilic cytoplasms and have low lipid content. Their mitochondria are typified by the lamelliform ultrastructural appearance of their cristae. This zone constitutes up to 15% of the cortex and produces the mineralocorticoid aldosterone, responsible for controlling blood volume and salt/water balance. Beneath the ZG is the zona fasciculata (ZF), which contains lipid droplet-rich polygonal epithelial cells arranged in radial columns separated by fenestrated capillaries. The mitochondria of these cells have tubulovesicular cristae. The ZF predominantly secretes glucocorticoids—cortisol in humans and corticosterone in mice and rats—which have influential effects on metabolism, the cardiovascular system, and the immune system. The innermost cortical zone is the zona reticularis (ZR), made up of polyhedral cells with prominent lysosomes arranged in a mesh-like

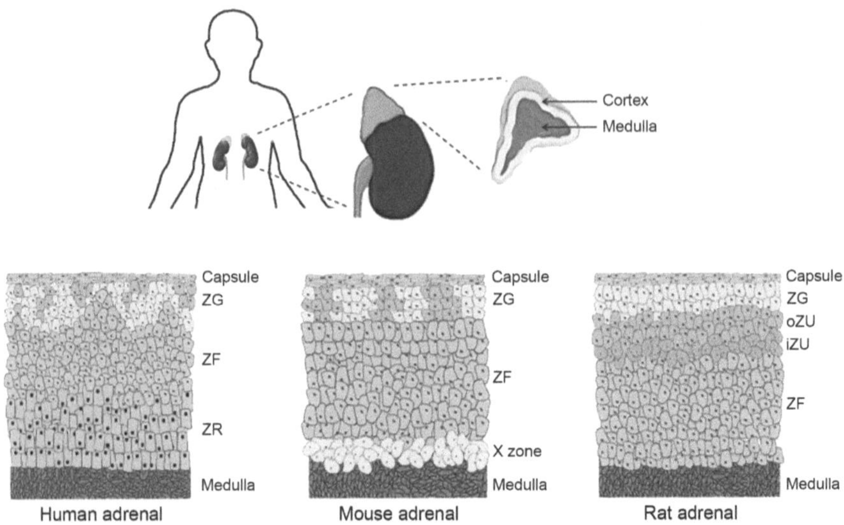

Figure 7.1 Anatomy and structure of the adrenal gland. (A) The adrenal glands are located at the upper poles of the kidneys. (B) Zonation differs between species. The adult human adrenal gland has three recognizable cortical zones, the ZG, ZF, and ZR, which surrounds the medulla. Clusters of CYP11B2-positive cells are observed in the ZG, separated from the CYP11B1-positive ZF cells, which can extend up to the capsule. SHH-expressing cells are also observed in clusters (pink), and by analogy with the mouse, these are depicted as separate from the CYP11B2 clusters, although this has not been directly proved. There is no sharp boundary between the ZF and the ZR. In the mouse, the CYP11B2-positive ZG and CYP11B1-positive ZF cells are in direct contact, with the CYP11B2-positive ZG cells arranged in clusters separated by clusters of SHH-positive, CYP11B2-negative cells (pink). The X zone (green) forms around the medulla following birth and persists until puberty in males or first pregnancy in females. In the rat, the zones are arranged as concentric shells with the undifferentiated zone (ZU) separating CYP11B2-positive cells from CYP11B1-positive cells. The ZU can be further divided into the Shh-positive, CYP11B2-negative outer ZU (oZU) and the inner ZU (iZU), which does not express Shh or CYP11B2. The DZ and TZ develop into the ZG and ZF, respectively, and the FZ regresses at birth to be replaced by the ZR in infancy. The islands of chromaffin cells coalesce to form the medulla after birth. (See color plate.)

structure of anastomosing cords. The ZR makes glucocorticoids and, in humans and higher primates, adrenal androgens such as androstenedione, dehydroepiandrosterone (DHEA), and its sulfate, DHEAS. The cortical zones were identified and named by Arnold (1886).

The medulla makes up the inner mass of the gland and is composed of large, polygonal, epithelial cells surrounded by venous sinusoids, and these cells produce catecholamines, not only principally adrenaline but also some noradrenaline depending upon the species. There is minimal, if any, intervening connective tissue between the cortex and the medulla, leaving cortical and medullary cells in direct contact.

Although the general arrangement of the gland is the same across mammalian species, some differences in zonation can be observed, particularly between those most studied, humans, mice, and rats. In the rat, the zones are organized in concentric shells, with the ZG a continuous zone beneath the capsule. However, the ZG does not immediately abut the ZF. Instead, there is a sudanophobic zone three to four cells thick referred to variously as the zona intermedia, white zone, or undifferentiated zone (ZU) (Cater & Lever, 1954; Guasti, Paul, Laufer, & King, 2011; Mitani et al., 1994). These cells have mitochondria similar to those of the ZF with tubulovesicular cristae (Domoto, Boyd, Mulrow, & Kashgarian, 1973). In the mouse, however, there is no UZ and the innermost ZG and outermost ZF cells directly contact each other (King, Paul, & Laufer, 2009; Paul & Laufer, 2011). The human ZG is arranged in discrete clusters such that some cords of ZF extend between them as far as the capsule (Boulkroun et al., 2010, 2011; Nishimoto et al., 2010) (Fig. 7.1).

2.2. Blood supply

Superior, middle, and inferior adrenal arteries arise from the aorta, renal, and inferior phrenic arteries and ramify over the capsule before forming a subcapsular arteriolar plexus. From here, capsular capillaries feed thin-walled venous sinusoids that traverse the cortex and form another plexus in the ZR in the adult, while venules pass between the medulla chromaffin cells to enter the adrenal vein. The fully formed medulla is also supplied by direct branches from the subcapsular plexus, which bypass the ZR plexus. This serves to ensure that most cells in the adrenal gland are only one to two cells away from a vascular endothelial cell, resulting in efficient delivery of adrenal hormones to the blood stream. Venous drainage from the adrenal gland is usually via a single vein, directly to the inferior vena cava on the right and the renal vein on the left. There are subcapsular and medullary lymphatic plexus, draining to lumbar and para-aortic lymph nodes (Dobbie & Symington, 1966).

2.3. Nerve supply

The autonomic nerve supply of the adrenal gland is the largest of any organ, relative to its size. Myelinated sympathetic preganglionic fibers from the intermediolateral cell column or lateral horn of T10–11 synapse on the medullary chromaffin cells. The cortex receives afferent nerve supply from the medulla, with sympathetic fibers traveling along the blood vessels (Parker, Kesse, Mohamed, & Afework, 1993).

2.4. Steroid synthesis

The adrenal steroids are synthesized from cholesterol by a series of reactions that are catalyzed by two classes of enzymes, cytochrome P450 (CYP) mixed function oxidases and short-chain dehydrogenases, and take place in the inner membrane of mitochondria or in the smooth endoplasmic reticulum (SER) with the intermediate compounds shuttling back and forth (for review, see Payne & Hales, 2004). Cholesterol, principally obtained in esterified form from LDL (humans) and HDL (rodents) via LDL receptor endocytic or SR-B1 selective transport pathways, respectively (for review, see Hu, Zhang, Shen, & Azhar, 2010), is initially cleaved from its esterified form by cholesteryl ester hydrolase and translocated from the outer to the inner mitochondrial membrane by the transporter enzyme steroidogenic acute regulatory protein (StAR). This occurs preferentially at sites of close contact between the membranes that are more numerous in tubulovesicular than lamelliform mitochondria (Vinson, 2003). The preponderance of tubulovesicular mitochondria in the ZF compared to the ZG reflects the higher levels of steroidogenesis in the ZF than the ZG. Once translocated, cholesterol is converted by side-chain cleavage (CYP11A1) into pregnenolone. This is then transported to the SER where, in humans, it is converted to 17OH pregnenolone by CYP17 (17α-hydroxylase) in the ZF and ZR but not in the ZG, where CYP17 is not expressed. 3β-hydroxysteroid dehydrogenase (3βHSD) converts 17OH pregnenolone to 17OH progesterone in the ZF and ZR and pregnenolone to progesterone in the ZG. Still in the SER, CYP21 (21-hydroxylase) converts progesterone and 17OH progesterone to 11-deoxycorticosterone and 11-deoxycortisol, respectively. These compounds are transported back to the inner mitochondrial membrane, and CYP11B1 (11β-hydroxylase) converts 11-deoxycortisol to cortisol, the major human glucocorticoid, in the ZF. In the ZG, 11-deoxycorticosterone is converted to aldosterone, the major mineralocorticoid, by CYP11B2 (aldosterone synthase). CYP17 is not expressed in the adult cortex in rats and mice, and 11-deoxycorticosterone is the substrate for both CYP11B2 and B1, resulting in corticosterone rather than cortisol being the glucocorticoid produced in these species. CYP17 has two activities, acting as both a 17α-hydroxylase and a 17,20-lyase. The hydroxylase activity is required for cortisol production but cytochrome b5, expressed in the ZR, promotes the lyase activity that cleaves two carbon atoms to produce DHEA from 17OH pregnenolone, such that cortisol production is suppressed in the ZR and adrenal androgens are formed instead (Endoh, Kristiansen, Casson, Buster, & Hornsby, 1996). DHEA can then be sulfated by SULT2A1 to DHEAS or converted to androstenedione by 3βHSD. These are the major androgens produced in the human adrenal gland but small amounts of testosterone can also be produced from androstenedione by 17-ketosteroid reductase (17βHSD) (Fig. 7.2). Because adult rat and mouse adrenal glands do not express CYP17, they do not synthesize adrenal androgens and the ZF and ZR are considered to be functionally equivalent.

2.5. Glucocorticoid synthesis and function

Glucocorticoid production is under the control of the hypothalamic–pituitary–adrenal (HPA) axis. Upon activation of the HPA axis, parvocellular neurons of the hypothalamic paraventricular nucleus synthesize and secrete corticotrophin-releasing hormone (CRH) and arginine vasopressin (AVP) into the hypophyseal portal circulation (Carrasco & Van de Kar, 2003; Chrousos, 1995; Sapolsky, Romero, & Munck, 2000; Van de Kar & Blair, 1999). CRH is the main regulator of the release of adrenocorticotropic hormone (ACTH) from the anterior lobe of the pituitary gland into the systemic circulation (Carrasco & Van de Kar, 2003). AVP has a synergistic role in CRH-induced ACTH release but is ineffective in the absence of CRH (Chrousos, 1995). CRH neurons projecting from the paraventricular nucleus of the hypothalamus onto proopiomelanocortin (POMC)-containing neurons in the hypothalamic arcuate nucleus can also stimulate ACTH production. They stimulate POMC release, which is cleaved to form ACTH (Pritchard & White, 2007).

ACTH binds to the ACTH receptor MC2R (melanocortin receptor 2), a 7-transmembrane G protein-coupled receptor (GPCR), causing the α-subunit of the stimulatory heterotrimeric G protein ($G_{\alpha s}$) to associate with adenylate cyclase. This catalyzes the conversion of ATP to cAMP, leading to the activation of downstream signaling pathways, including the activation of the cAMP-dependent protein kinase (PKA). PKA phosphorylates and activates cholesteryl ester hydrolases and StAR, increasing the amount of cholesterol delivered to the inner mitochondrial membrane. cAMP also induces the transcription of StAR and CYP11A1, and CYP17 to promote cortisol

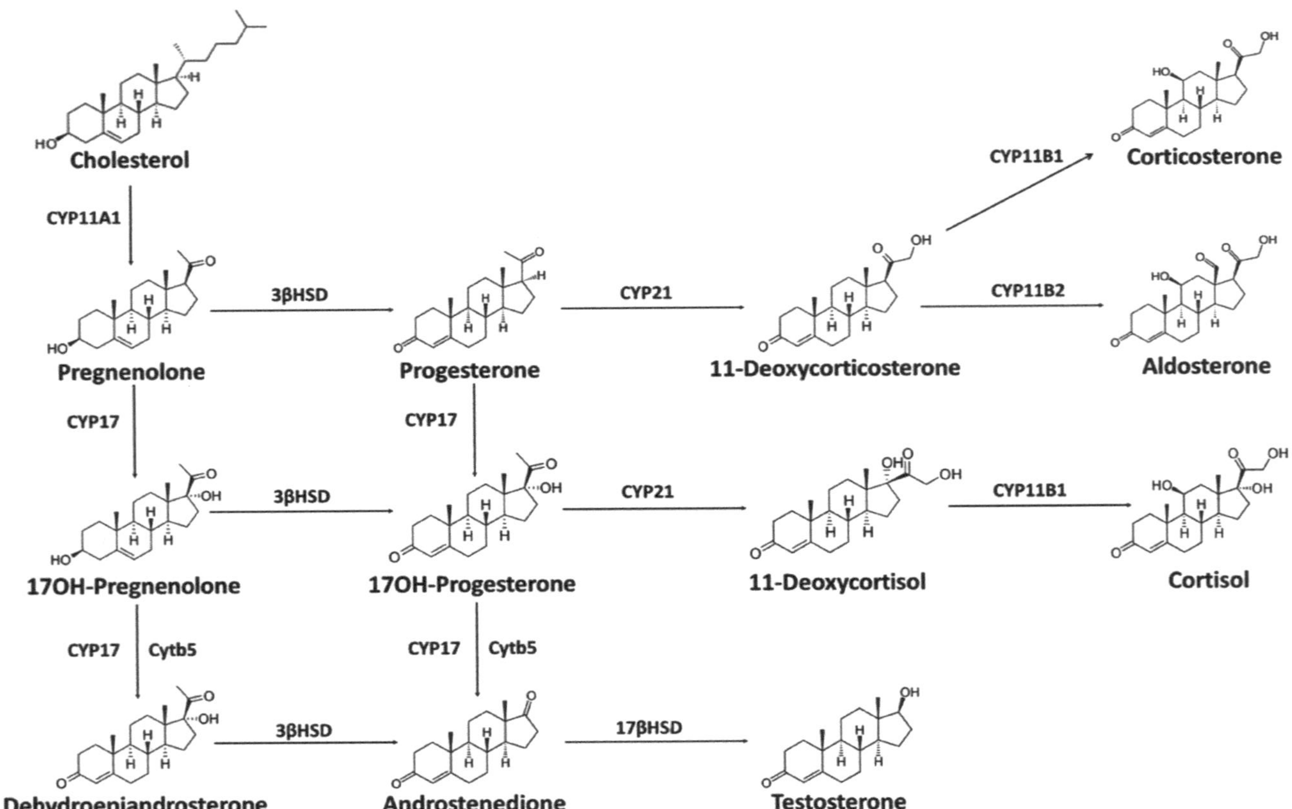

Figure 7.2 Steroidogenesis pathway. The major steroid products secreted by the adrenal cortex, mineralocorticoids, glucocorticoids, and adrenal androgens are synthesized in sequential steps by a cohort of steroidogenic enzymes. Adult rodents do not express CYP17 and so do not make adrenal androgens (dehydroepiandrosterone and androstenedione). They are also unable to make cortisol for this reason and synthesize corticosterone as the major glucocorticoid. All species make aldosterone as the major mineralocorticoid. See text for details.

production (Aumo, Rusten, Mellgren, Bakke, & Lewis, 2010; Sewer, Dammer, & Jagarlapudi, 2007; Waterman & Bischof, 1996). Glucocorticoids inhibit the secretion of CRH and AVP by the hypothalamus and POMC cleavage in the pituitary gland, thus reducing the production of ACTH and therefore cortisol itself in a negative feedback loop (de Kloet, 1995).

Cortisol is the principal glucocorticoid produced by the human ZF, with influential effects on metabolism, the cardiovascular system, and the immune system. Although at basal levels its role is a predominantly permissive one, allowing the effects of other agents, it is released in response to stress. Glucocorticoids increase blood glucose levels by stimulating gluconeogenesis, inhibiting glucose storage, and stimulating lipolysis in adipose tissue to release free fatty acids. They can also cause proteolysis in some muscle tissues. In the cardiovascular system, cortisol increases the transcription of receptors for angiotensin II (AngII), epinephrine, and norepinephrine, to regulate blood pressure, contractility, and tone of the heart (Sakaue & Hoffman, 1991).

Glucocorticoids also have important anti-inflammatory properties, inhibiting the actions of immune cells (T and B lymphocytes) and the synthesis and release of cytokines (including interleukins (IL-)1–6, IL-12, IFN-γ, TNF-α, and chemokines) and other inflammatory mediators, such as histamine and bradykinin. In this way, cortisol prevents excessive inflammation and tissue and organ damage (Franchimont, 2004).

2.6. Aldosterone synthesis and function

Aldosterone production is induced by increases in extracellular potassium levels and by ACTH, although AngII is the primary regulator, via the renin–angiotensin–aldosterone system (RAAS). The RAAS has little control over arterial blood pressure under normal circumstances but plays a pivotal role when sodium levels fall or hemorrhage occurs, as it has major vasoconstrictor capabilities (Collier, Robinson, & Vane, 1973; Scornik & Paladini, 1964). Renin is a proteolytic enzyme that cleaves angiotensinogen, synthesized by the liver, into angiotensin I (Miller, 1981). It is released from the juxtaglomerular cells of the kidney into the blood, stimulated by sympathetic nerve inputs, catecholamines, decreased renal perfusion pressure, and decreased sodium delivery to the distal tubule (Johnson & Davis, 1973; Tobian, Tomboulian, & Janecek, 1959). Angiotensin I is hydrolyzed by angiotensin-converting enzyme (ACE), found at the surface of pulmonary and renal endothelium, forming AngII (Oparil, Sanders, & Haber, 1970). This provokes vasoconstriction of arteriolar smooth muscle, increases the contraction of the heart (positive inotropic effect), elicits the release of catecholamines, and stimulates aldosterone production by the ZG of the adrenal cortex.

AngII binds the AT1 cell surface receptor expressed on ZG cells, which is a GPCR coupled to phosphoinositidase C (PI3C) (Bird et al., 1993). PI3C causes hydrolysis of phosphatidylinositol 4,5-bisphosphate, generating inositol 1,4,5-trisphosphate (Ins(1,4,5)P3) and 1,2-diacylglycerol (DAG). Ins(1,4,5)P3 is an intracellular second messenger that opens calcium channels on intracellular stores, resulting in the release of calcium into the cytoplasm (Wojcikiewicz & Nahorski, 1993). Elevated intracellular calcium levels activate calmodulin and CaMK (calmodulin-dependent protein kinases), and DAG activates protein kinase C (PKC), leading to the phosphorylation and activation of other second messenger cascades, which modulate aldosterone production, possibly by phosphorylation of transcription factors regulating CYP11B2 transcription (Nogueira, Bollag, & Rainey, 2009; Pezzi, Clyne, Ando, Mathis, & Rainey, 1997).

Aldosterone is the primary mineralocorticoid produced by the ZG of the adrenal cortex, regulating sodium retention, water balance, and blood pressure. Sodium is actively reabsorbed in the distal nephron of the kidney via ENaC sodium channels, whose activity is upregulated by aldosterone. This is accompanied by passive water reabsorption to increase the extracellular and blood fluid volumes and hence blood pressure (Miller, 1981). Aldosterone also has effects on the cardiovascular system and can cause cardiac fibrosis due to the activation of an inflammatory cascade (Fuller & Young, 2005; Wehling et al., 1998).

2.7. Catecholamine synthesis and function

Tyrosine is converted to L-DOPA and then dopamine in medullary chromaffin cells by the successive actions of tyrosine hydroxylase (a mixed function oxidase, which can also hydroxylate phenylalanine to produce tyrosine) and DOPA decarboxylase (Fukami, Haavik, & Flatmark, 1990). Dopamine β-hydroxylase, another mixed function oxidase, then converts dopamine to noradrenaline, and this is converted to adrenaline by phenylethanolamine N-methyltransferase (PNMT) in adrenergic chromaffin cells. Dopamine, noradrenaline, and adrenaline are stored in vesicles along with chromogranin A. In humans, most chromaffin cells express PNMT, and therefore, almost all the catecholamine output is adrenaline, but in rodents, some chromaffin cells do not express PNMT and approximately 20% is in the form of noradrenaline (Schinner & Bornstein, 2005; Wong, 2003).

The catecholamines bind to adrenergic receptors, with noradrenaline binding preferentially to α-receptors, while adrenaline binds to both α- and β-receptors. Activation of these receptors causes rapid and diverse effects including vasoconstriction, piloerection increased heart rate, and increased blood glucose, all part of the "fight or flight" response.

3. ADRENAL DEVELOPMENT

3.1. Early adrenal development

The adrenal cortex develops from a thickening of the coelomic epithelium located at the base of the groove between the urogenital ridge and the dorsal mesentery that can be identified in humans from about 4 weeks postconception (wpc) (Else & Hammer, 2005; Parker et al., 2002) and 9.5–10 days postconception (dpc) in the mouse (Hatano, Takakusu, Nomura, & Morohashi, 1996; Ikeda, Shen, Ingraham, & Parker, 1994), referred to as the adrenogonadal primordium (AGP), a group of cells identified by their expression of the transcription factor steroidogenic factor-1 (SF-1, Ad4BP, and NR5A1) (Hatano et al., 1996; Luo, Ikeda, & Parker, 1994) (for reviews, see Else & Hammer, 2005; Kim et al., 2009; Laufer, Kesper, Vortkamp, & King, 2012) (Fig. 7.3). As this primordium grows, the cells delaminate from the coelomic epithelium and invade the underlying mesenchyme, with the cells adjacent to the mesonephros migrating dorsolaterally to form the gonadal primordium, while the more medial cells that express the highest levels of SF-1 migrate and condense to form the adrenal primordium at the cranial end of the mesonephros, ventrolateral to the dorsal aorta. This occurs at 33 dpc (Goto et al., 2006; Hanley, Rainey, Wilson, Ball, & Parker, 2001) (10.5 dpc in the mouse). The cells of the adrenal primordium continue to proliferate and CYP17 expression is detectable from 41 to 44 dpc (Goto et al., 2006; Hanley et al., 2001) with vascularization of the gland apparent at this time. By 50–52 dpc, two discrete zones are detectable in the adrenal cortex, the inner zone, referred to as the fetal zone (FZ), composed of large polyhedral eosinophilic cells with tubulovesicular mitochondrial cristae with high levels of expression of steroidogenic enzymes, and the smaller outer zone, or definitive zone (DZ), which is composed of small, tightly packed basophilic cells (Goto et al., 2006) with lamelliform mitochondrial cristae with much lower levels of steroidogenic enzyme

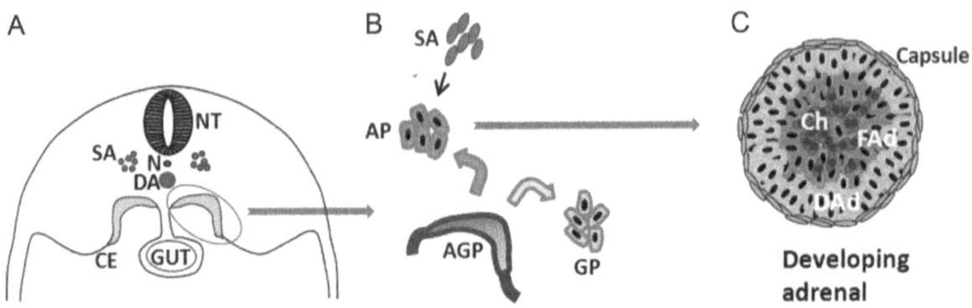

Figure 7.3 Development of the adrenal gland. (A) The adrenogonadal primordium (AGP) develops as a thickening of the coelomic epithelium (CE) adjacent to the dorsal mesentery (DM). (B) These cells express SF-1 and the more medial population, which expresses higher levels of SF-1, delaminates into the mesenchyme to form the adrenal primordium (AP), while the cells adjacent to the mesonephros (M), expressing lower levels of SF-1, delaminate to form the gonadal primordium (GP). Migrating sympathoadrenal (SA) cells enter the AP. (C) The adrenal gland grows and surrounding mesenchymal cells form a capsule and the chromaffin (Ch) cells form the medulla. The AP cells form the fetal adrenal (FAd) gland and this is surrounded by the definitive adrenal gland (DAd). NT, neural tube; N, notochord; DA, dorsal aorta. (See color plate.)

expression (Goto et al., 2006; Hanley et al., 2001; Nussdorfer, 1986). The periphery of the gland is now heavily vascularized and the mesenchymal capsule is forming at this time in human adrenal development and is complete by the 9th week.

3.2. Early medulla development

The adrenal medulla develops from the trunk neural crest. Some of these cells migrate ventrally through the anterior sclerotome to reach the dorsal aorta and form the sympathoadrenal (SA) lineage (Anderson, Carnahan, Michelsohn, & Patterson, 1991). This sympathetic primordium gives rise to tyrosine hydroxylase-expressing catecholaminergic neuronal progenitor cells in response to bone morphogenetic protein cues from cells of the wall of the dorsal aorta and the surrounding mesenchyme (Reissmann et al., 1996; Shah, Groves, & Anderson, 1996). Some cells then migrate from the dorsal aorta along nerves and blood vessels to enter the adrenal primordium at the cranial end from 6 wpc (Ehrhart-Bornstein et al., 1997; Yamamoto, Yanai, & Arishima, 2004) and acquire the phenotype of chromaffin cells. It was thought that these were SA cells that lost expression of neuronal genes, but more recently, it has been shown that at least some of the presumptive chromaffin cell population enters the adrenal primordium still expressing neural crest markers and then gains TH expression but not neuronal markers (Ernsberger et al., 2005). Initially, the chromaffin cells are scattered within the cortex and are mostly noradrenergic, until 10 wpc when small islets of chromaffin cells express PNMT, the enzyme required to convert noradrenaline into adrenaline, to become adrenergic (Jozan et al., 2007).

3.3. Later development

By 9 wpc, the human adrenal gland is encapsulated with a cortex containing an outer DZ and inner FZ and islands of chromaffin cells in the center and has become heavily vascularized with adrenal arteries supplying blood to a subcapsular arteriolar plexus, and these give rise to capsular capillaries that feed thin-walled venous sinusoids and traverse the cortex. The gland continues to grow rapidly, increasing approximately 8- to 10-fold in size from this stage until birth, largely due to an increase in the size of the FZ, which accounts for 80–90% of the weight of the gland by midgestation (Keene, 1927; Spencer, Mesiano, Lee, & Jaffe, 1999), and a third cortical zone becomes evident by 14 weeks (Goto et al., 2006; Sucheston & Cannon, 1968). This transitional zone (TZ) (Mesiano, Coulter, & Jaffe, 1993; Mesiano, Mellon, & Jaffe, 1993) is located between the DZ and the FZ and contains cells whose histological appearance shares features of both DZ and FZ. By late gestation, the DZ has begun to resemble the ZG and the TZ, the ZF (Coulter & Jaffe, 1998; Sucheston & Cannon, 1968).

Steroid production from the fetal adrenal gland has important roles in maintaining intrauterine homeostasis and in the maturation of fetal organ systems (including the adrenal gland itself) in preparation for extrauterine life. The human placenta cannot make estrogens *de novo*, but the large amounts of DHEA and DHEAS produced by the FZ (100–200 mg per day; Carr & Simpson, 1981) are converted by placental aromatase to estrogens, which maintain pregnancy. The TZ produces cortisol, with an early peak from 8 to 9 wpc coinciding with transient 3βHSD expression (Goto et al., 2006). This is thought to protect female sexual development by negatively regulating the fetal HPA axis and thus inhibiting adrenal androgen production during this developmental window. Immediately before birth, there is a second peak in cortisol production that is required for fetal organ maturation, and this is at least partly regulated by ACTH from the fetal pituitary gland. Cortisol from the TZ also induces PNMT expression in chromaffin cells, thereby directing them to become adrenergic (Finotto et al., 1999; Wurtman & Axelrod, 1965).

3.4. Neonatal period and adult growth

Because of the rapid increase in the size of the FZ during gestation, the size of the neonatal human adrenal gland is relatively very large compared to the adult gland, weighing 3–5 g at term. After birth, the FZ rapidly involutes and remodels by a process involving apoptosis of cells in the inner region of the FZ (Jirasek, 1980), and there is a concomitant decrease in adrenal androgen secretion. The adult cortex develops from the DZ and TZ, giving rise to the ZG and ZF, respectively, and the weight of the adrenal gland drops by 50% within the first 2 weeks after birth. Cells with ZR morphology are detectable in humans from around 3 years of age until a continuous ZR forms at around 6 years of age and adrenal androgen synthesis recommences, a stage referred to as adrenarche. The ZR continues to increase in thickness until puberty (Hui et al., 2009; Nakamura, Gang, Suzuki, Sasano, & Rainey, 2009). After birth, the islands of chromaffin cells within the cortex coalesce to form a contiguous medulla (Wilburn, Goldsmith, Chang, & Jaffe, 1986).

In mice, there is a zone thought to be analogous to the FZ, namely, the X zone, next to the medulla. Unlike the FZ, however, the X zone starts to develop after birth with clusters of small lipid-poor cells with acidophilic cytoplasms appearing adjacent to the medulla. By weaning, the X zone has increased in size to occupy about one-third of the cortex, with a gradation of acidophilia from deep staining by the medulla to light staining adjacent to the ZF. The X zone degenerates in males at puberty (~45 days), whereas in females, it undergoes vacuolar degeneration at first pregnancy or gradually degenerates in unbred females (Jones, 1948; Sucheston & Cannon, 1972).

4. THEORIES OF GROWTH AND ZONATION

The adrenal cortex, as well as having a constant turnover of cells to replace dying cells, is a highly dynamic organ that undergoes rapid changes in response to hormonal demand and surgical injury, following which it shows remarkable regenerative capacity. Observations on adrenocortical cell growth from the late nineteenth century led to a number of theories of the origin of the cells of the different adrenocortical zones, but the weight of evidence favors the centripetal migration hypothesis, in which there is a continual production of new cortical cells at the periphery of the gland that migrate under this mitotic pressure to the inner cortical medulla boundary where they die. This theory was initially advanced by Gottschau (1883), and evidence was subsequently provided by observing that the majority of mitotic cells in mice and rats were present at the boundary between the ZG and the ZF and that most cell death occurred at the ZR/medulla boundary (for a review, see Vinson, 2003). Early lineage-tracing studies performed by subcutaneous injection of trypan blue in rats revealed that, after initial uptake by the capsule and ZG cells, dye could be observed moving into the ZF and eventually be detected in cells in the ZR (Salmon & Zwemer, 1941). More recently, mosaic transgenic studies have provided supporting evidence for this theory, with columns of cells expressing β-galactosidase under the control of the *Cyp21* promoter traversing the cortex from the capsule to the medulla boundary (Iannaccone, Morley, Skimina, Mullins, & Landini, 2003; Morley et al., 1996), and genetic lineage-tracing experiments have provided direct evidence for the centripetal migration (Huang, Miyagawa, Matsumaru, Parker, & Yao, 2010; King et al., 2009) (see later).

The precise origin of adrenocortical cells has been debated over the years. Zwemer, Wotton, and Norkus (1938) concluded from their intravital dye-labeling experiments that the newly synthesized cells were derived from the adrenal capsule, and this was supported by observations from enucleation studies in rodents (Baker & Baillif, 1939; Ingle & Higgins, 1939) (see later) and other remodeling protocols (Lombardo & Cortesini, 1988). However, others concluded from proliferation studies that the progenitor cells lie within the ZG (Mitani, Mukai, Miyamoto, Suematsu, & Ishimura, 2003; Taki & Nickerson, 1985). For example, the DZ cells possess characteristics of proliferating cells rather than steroidogenic cells; they express proliferating cell nuclear antigen (PCNA) and are relatively lipid-free and have mitochondria with lamelliform cristae (McNutt & Jones, 1970; Sucheston & Cannon, 1968), and their arrangement as caps of the of FZ cells with some cells entering the outer FZ as columns has led to the suggestion that these are the progenitor cells of the ZF (Mesiano, Coulter, & Jaffe, 1993; Mesiano, Mellon, & Jaffe, 1993; Spencer et al., 1999). Many studies have identified proliferating cells in the rat adrenal cortex at the periphery of the gland between the ZG and the ZF (Bertholet, 1980; Ford & Young, 1963; Kataoka, Ikehara, & Hattori, 1996; Mitani, Mukai, Miyamoto, Suematsu, & Ishimura, 1999). Mitani and coworkers have demonstrated that the majority of proliferating cells are located around the UZ, which they have suggested is a stem cell zone in this species (Mitani et al., 1999, 2003). Nevertheless, it is generally agreed that cells arise at the periphery and migrate through the cortex, with the implication that they change identity as they move between zones.

5. FACTORS INVOLVED IN ADRENAL DEVELOPMENT

Insights into the mechanisms controlling adrenal development have come from the studies of abnormal adrenal development in both humans and animal models. Genetic analysis of human adrenal insufficiency and observations from knockout mouse models have identified genes that are involved in broadly three different areas of development: first, those that are involved in the development of the intermediate mesoderm and urogenital ridge and affect kidney, adrenal, and gonadal development; second, those that affect the development of the AGP and affect both gonadal and adrenal development; and third, those that more specifically affect the development of the adrenal gland.

5.1. Urogenital ridge development

5.1.1 Odd-skipped related 1

Odd-skipped related 1 (Odd1) is a transcription factor crucial for embryonic patterning and early development. It is expressed throughout the intermediate mesoderm at 8.5 dpc and by 10.5 dpc in branchial arches and the limb buds (So & Danielian, 1999). *Odd1*-null mice die *in utero*, mostly between 11.5 and 12.5 dpc with no condensation of metanephric mesenchyme and impaired urogenital ridge formation, but those that survive to 15.5 dpc show complete absence of adrenal glands, gonads, and kidneys (Wang, Lan, Cho, Maltby, & Jiang, 2005).

5.1.2 WT1

The Wilms' tumor gene WT1 is a Zn finger transcription factor involved in urogenital development and tumorigenesis that is expressed in the coelomic epithelium and the adjacent mesenchyme. Germline mutations in humans have been shown to cause defects in gonad and kidney formation (Baird, Groves, Haber, Housman, & Cowell, 1992; Haber & Housman, 1992). *Wt1*-null mice die at 13.5 dpc and gonad and kidney development does not extend beyond thickening of the coelomic epithelium (Kreidberg et al., 1993), but mice partially rescued by a human *WT1* transgene survive to birth with urogenital abnormalities and much impaired adrenal development, indicating a role for WT1 in adrenal development (Moore, McInnes, Kreidberg, Hastie, & Schedl, 1999).

WT1 is expressed in the AGP and remains expressed in the bipotential gonad after separation from the adrenal primordium, but its expression is not observed in the adrenal primordium (Nachtigal et al., 1998; Val, Martinez-Barbera, & Swain, 2007).

5.1.3 *Sall1*

Sall1 is a member of the sal family of zinc finger-containing transcriptional regulators, and heterozygous mutations in this gene cause Townes–Brocks syndrome, which includes renal and genital abnormalities (Kohlhase, Wischermann, Reichenbach, Froster, & Engel, 1998). Sall1 is expressed in the intermediate mesoderm at 10.5 dpc and in the metanephric mesenchyme surrounding the ureteric bud and mesonephros at 11.5 dpc (Nishinakamura et al., 2001) in mice and is expressed in the FZ of the human fetal adrenal gland (Ma et al., 2002). *Sall1*-null mice have severe kidney dysgenesis or agenesis and hypoplastic adrenal glands at birth (Nishinakamura et al., 2001), whereas a mouse model that phenocopies the deletion seen in TBS has absent adrenal glands as well as kidneys and hypoplastic gonads at 16 dpc (Kiefer et al., 2003).

5.1.4 *Pbx1*

Pbx1 is a three-amino acid loop extension class homeodomain transcription factor that is essential for fetal development, with null mice dying at 15–16 dpc with multiple organ hypoplasia and aplasia (Selleri et al., 2001), including the kidney and adrenal gland (Schnabel, Selleri, & Cleary, 2003). At 10 dpc, Pbx1 is expressed in the nephrogenic cord and the coelomic epithelium, including the AGP (Zubair, Ishihara, Oka, Okumura, & Morohashi, 2006). By 14.5 dpc, Pbx1 is expressed throughout the urogenital structures, including the adrenal primordium, the interstitium of the gonads of both sexes, and the metanephric kidney (Schnabel et al., 2003). Pbx1 is detectable in human fetal adrenal glands by 5 weeks' gestation and is expressed throughout the cortex by 10 weeks' gestation (Ferraz-de-Souza et al., 2009). In *Pbx1*-null mice, by 12.0 dpc, the urogenital ridge is much reduced in size, and by 14.5 dpc, both the kidney and adrenal gland are absent, and gonadal development and sexual differentiation are poorly advanced (Schnabel et al., 2003). $Pbx1^{+/-}$ mice are viable and have smaller adrenal glands with a ZF containing fewer, but hypertrophied, cells and with fewer proliferative cells in the subcapsular region (Lichtenauer et al., 2007).

5.1.5 *Wnt4*

Wnt4 is a member of the Wnt family of secreted glycoproteins that have crucial roles in development and cell growth and differentiation. It is expressed in the metanephric mesenchyme on the medial side of the ureteric bud (Nishinakamura et al., 2001; Stark, Vainio, Vassileva, & McMahon, 1994) and the coelomic epithelium and mesenchyme of the genital ridge at 11.5 dpc in mice and is downregulated in the XY gonad from this time (Vainio, Heikkila, Kispert, Chin, & McMahon, 1999). By 14.5 dpc, its expression is localized in the subcapsular region of the adrenal primordium (Heikkila et al., 2002). A homozygous missense mutation in *WNT4* in humans that results in greatly reduced *WNT4* mRNA expression has been shown to be associated with renal agenesis, gonadal defects, and adrenal hypoplasia from 19 weeks' gestation (Mandel et al., 2008), and *Wnt4*-null mice demonstrate impaired kidney development from 15 dpc (Stark et al., 1994) along with defects of the female reproductive system (Vainio et al., 1999), Sertoli cell differentiation in the testes (Jeays-Ward, Dandonneau, & Swain, 2004), and adrenal glands with reduced CYP11B2 expression and Pref-1/Dlk1 expression (Heikkila et al., 2002).

5.1.6 *FoxD1/FoxD2*

FoxD1 and FoxD2 are members of the forkhead/winged helix family of transcription factors that play important roles in controlling cell proliferation and differentiation during development. FoxD1 is expressed in the stromal mesenchyme surrounding the nephrogenic mesenchyme at 12.5 dpc in early kidney development (Hatini, Huh, Herzlinger, Soares, & Lai, 1996) and in the adrenal capsule at 14.5 dpc (Itaranta, Viiri, Kaartinen, & Vainio, 2009). FoxD2 is expressed in the intermediate mesoderm in mice and in the metanephric and mesonephric mesenchyme at 11.5 dpc, in the stromal cells at 12.5 dpc, and in the capsule of the developing adrenal gland at 15.5 dpc (Kume, Deng, & Hogan, 2000). Homozygous null *FoxD1* and *FoxD2* mice have hypoplastic adrenal glands at birth as well as developmental defects of the kidneys and ureter.

5.2. AGP development

5.2.1 *SF-1*

SF-1 is an orphan nuclear receptor transcription factor that is a key regulator of both steroidogenic organ development, namely, the gonads and the adrenal glands, and steroidogenesis. It is expressed in all cells of the AGP, in both the DZ and FZ in humans, being the earliest detectable marker, with the highest expression levels in the cells at the rostral end of the AGP, which will form the adrenal primordium (Hanley et al., 1999; Hatano et al., 1996; Ikeda et al., 1994). Following separation of the primordia, SF-1 expression is maintained in all steroidogenic cells of the adrenal cortex, but is not expressed in the medulla or the capsule. During sexual differentiation of the gonads, SF-1 expression remains high in the developing testes in both the testicular cords and the Leydig cell interstitial compartment but declines in midgestation in the developing ovaries until it is reestablished in steroidogenic cells in late gestation (Ikeda et al., 2001).

Absence of SF-1 function causes severe defects in both adrenal and gonadal development in both humans and mice (for review, see El-Khairi & Achermann, 2012). Homozygous deletion of the *Sf-1* gene in mice causes loss of the adrenal glands and gonads by 12.5 dpc (Luo et al., 1994; Sadovsky et al., 1995). In *Sf-1*-null mice, the thickening of the coelomic epithelium that forms the AGP is observed but the cells undergo apoptosis by 11.5 dpc and the adrenal and gonadal primordia are lost by 12.5 dpc. Offspring die shortly after birth due to hypoglycemia as a result of adrenal insufficiency and exhibit XY sex reversal. *Sf-1*$^{+/-}$ mice are viable with normal gonadal development but have adrenal primordia with 12-fold smaller cross-sectional area at 12 dpc, which recovers to be only twofold smaller at 18.5 dpc, as a consequence of increased proliferation of the cortical cells from 13.5 dpc (Bland, Fowkes, & Ingraham, 2004). Adrenal glands in adult *Sf-1*$^{+/-}$ mice remain small (Babu et al., 2002) and have a reduced corticosterone output in response to stress (Bland et al., 2000). Heterozygous loss-of-function mutations have been identified in humans, but no cases of homozygous mutations have been reported, presumably indicating the importance of SF-1 for human fetal development (El-Khairi, Martinez-Aguayo, Ferraz-de-Souza, Lin, & Achermann, 2011; Ferraz-de-Souza, Lin, & Achermann, 2011). In contrast to the situation in mouse heterozygous knockouts, only two loss-of-function mutations have been reported that affect adrenal function, although this is a common cause of disorders of sex development in humans (Ferraz-de-Souza et al., 2011). The first was a patient who presented with adrenal failure and gonadal dysgenesis in infancy and was found to have a heterozygous G35E change in the DNA-binding domain of the protein (Achermann, Ito, Hindmarsh, & Jameson, 1999). Another mutation in the DNA-binding domain, R92Q, caused a similar phenotype but only when inherited homozygously, presumably because the mutation causes a less severe reduction in DNA-binding activity (Achermann et al., 2002).

5.2.2 *Dax1*

Dax1 (dosage-sensitive sex reversal adrenal hypoplasia congenita (AHC) critical region on the X chromosome, NR0B1) is also a member of the nuclear receptor transcription family but possesses an atypical DNA-binding domain. It negatively regulates transcription and is thought to act as a corepressor of other nuclear receptors, including SF-1, rather than directly binding to DNA (Iyer & McCabe, 2004). Like SF-1, it is expressed in the hypothalamus, pituitary gland, adrenal glands, and gonads (Ikeda et al., 1996; Swain, Zanaria, Hacker, Lovell-Badge, & Camerino, 1996). In human adrenal development, Dax1 is expressed from the earliest stages of the adrenal primordium with an overlapping expression pattern with SF-1, although at a lower level (Hanley et al., 2001). By 52 dpc, both are expressed throughout the entire cortex, and this is maintained through gestation but with lower expression of both by 18 weeks. In mice, Dax1 expression drops during gestation in males and is restricted to the ZG by 10 weeks of age but remains expressed throughout the cortex in females (Mukai et al., 2002).

Mutations in *DAX1* in humans cause a form of AHC (Muscatelli et al., 1994; Zanaria et al., 1994) in which patients present with adrenal insufficiency of both glucocorticoids and mineralocorticoids and an adrenal cortex with poorly developed adult zones but retention of cytomegalic FZ cells. Mice in which exon 2 of *Dax1* is deleted surprisingly display enhanced adrenal steroid output (Yu, Ito, Saunders, Camper, & Jameson, 1998) until the cortex starts to regress and cytomegalic cells appear as the animals age, suggesting that the mutation is hypermorphic in mice (Scheys, Heaton, & Hammer, 2011). However, it has been shown that Dax1 and Oct3/4 form a regulatory circuit in embryonic stem (ES) cells, perturbation of which can cause differentiation, and Dax1 has therefore been proposed to be involved in the maintenance of self-renewal (Sun et al., 2009). Given the subcapsular expression of Dax1 in mice, at least in males, it has been suggested that loss of Dax1 causes uncontrolled differentiation of adrenal stem/progenitor cells and their exhaustion may take longer in mice than humans, resulting in a delayed regression of the cortex (Scheys et al., 2011) (see later).

5.2.3 *Cbx2*

The mouse polycomb group protein Cbx2 (chromobox2 or M33) is required for adrenal and gonadal development. A mouse model in which the C-terminus of Cbx2 is lost displays gonadal developmental defects with indistinct genital ridges observed at 11.5 dpc and marked gonadal hypoplasia at 13.5 dpc, along with male-to-female sex reversal (Katoh-Fukui et al., 1998). Analysis of these mutants also detected a degree of adrenal hypoplasia at 18.5 dpc (Katoh-Fukui et al., 2005). Interestingly, adrenal levels of Sf-1 were reduced by approximately 50%, indicating a possible role for Cbx2 in the regulation of *Sf-1*.

5.2.4 Insr/IGFR1

The IGF1 receptor (IGFR1) is expressed in the human fetal adrenal gland, in both the DZ and the FZ (Shigematsu et al., 1989). IGF2 is expressed at high levels throughout the fetal adrenal gland, but IGF1 is restricted to the capsule (Mesiano, Coulter, & Jaffe, 1993; Mesiano, Mellon, & Jaffe, 1993).

Recently, it has been reported that a mouse model in which both the insulin receptor and Igf1r are constitutively deleted displays reduced growth, gonadal dysplasia with male-to-female sex reversal, and adrenal developmental defects ranging from severe hypoplasia to agenesis (Pitetti et al., 2013). Signaling through these receptors appears to impact specifically on the AGP with an approximately 40% reduction in the number of SF-1-positive progenitor cells in the genital ridge at 11.5 dpc and a greater overall decrease in *Sf-1* transcript levels, whereas other gross markers of development appeared normal, apart from a reduction in fetal size. The more deleterious effect on the adrenal gland than the gonads may result from this reduction in SF-1 expression (see later), and it is hypothesized that signaling through these receptors may upregulate SF-1 expression or phosphorylate SF-1 and enhance its transcriptional activity.

5.3. Adrenal primordium development

5.3.1 Cited2

Cited2 (CBP/p300-interacting transactivator with ED-rich tail 2) is a transcriptional cofactor that binds with high affinity to p300 and CBP transcriptional coactivators. It can both positively and negatively regulate transcription: it not only inhibits the binding of CBP/p300 to HIF-1α (Bhattacharya et al., 1999), thus inhibiting transcriptional responses to hypoxia, but also mediates CBP/p300 binding to AP-2 (TFAP2) to activate transcription of AP-2-responsive genes (Bamforth et al., 2001; Braganca et al., 2003). It positively regulates transcription by promoting interaction of CBP/p300 with transcription factors such as AP-2 (Braganca et al., 2003), Lhx2 (Glenn & Maurer, 1999), and PPARα (Tien, Davis, & Vanden Heuvel, 2004). Cited2 is expressed in the coelomic epithelium and adjacent mesenchyme of the urogenital ridge at 10 dpc and remains expressed at high levels in the adrenal cortex but at much lower levels in the gonads from 13.5 dpc (Val et al., 2007). Deletion of *Cited2* in mice is embryonically lethal (Bamforth et al., 2001; Val et al., 2007) with embryos exhibiting markedly reduced adrenal development by 12 dpc (Val et al., 2007) and adrenal agenesis by 17.5 dpc (Bamforth et al., 2001), as well as cardiac, neural tube (Bamforth et al., 2001), and left–right patterning (Bamforth et al., 2004) defects. Expression of some gonadal markers is reduced at 11.5 dpc, suggesting an early delay in development, but this recovers by 13.5 dpc and gonadal development proceeds grossly normally as judged by these transcriptional patterns. However, further inspection reveals disrupted testis morphology at all further stages and delayed ovarian differentiation (Combes et al., 2010).

5.3.2 Wnt/β-catenin signaling

β-Catenin is a bifunctional protein; the majority regulates the interaction of cadherin complexes with the cytoskeleton in adherens junctions, but a cytoplasmic pool responds to canonical Wnt signaling to act as a coactivator for transcription factors of the TCF/LEF. In the absence of Wnt signaling, cytoplasmic β-catenin is degraded by a complex containing GSK3β that phosphorylates β-catenin and targets it for degradation by the proteasome. Wnt signaling inhibits this degradation and allows nuclear accumulation of β-catenin (for reviews, see Berthon, Martinez, Bertherat, & Val, 2012; El Wakil & Lalli, 2011). This and/or the expression of β-catenin/TCF/LEF-responsive genes allows the detection of cells transducing canonical Wnt signals. Transcriptionally active β-catenin has been detected in the coelomic epithelium of the AGP and the mesonephros at 11.5 dpc (Usongo & Farookhi, 2012); by 12.5 dpc, it can be seen in some cells at the periphery of the adrenal primordium (Kim et al., 2008); and by 18.5 dpc, it localizes to subcapsular cell clusters where it remains expressed in adulthood. Subcapsular clusters of strongly β-catenin-expressing cells were identified immunohistochemically by 14.5 dpc (Kim et al., 2008). β-Catenin is crucial for embryogenesis and deletion causes early embryonic lethality. Deletion of β-catenin in mice in steroidogenic cells using an SF-1 cre driver demonstrated that at 12.5 dpc, the adrenal primordium is detectable and of approximately normal size. However, there are fewer SF-1-expressing cells and the levels of expression of the SF-1 target genes *CYP11a1* and *3βHSD* are reduced. By 13.5 dpc, these, along with CYP21, which is unaffected at 12.5 dpc, are virtually undetectable, and cell proliferation is greatly reduced.

The β-catenin-null adrenal glands are markedly smaller than wild-type adrenal glands at 14.5 dpc and also contain many fewer SF-1-positive cells. By 16.5 dpc, the vast majority of the cells in the adrenal gland are TH-positive with almost no detectable SF-1-positive cells, indicating that the majority of the gland consists of chromaffin cells. By 18.5 dpc, the gland is no longer detectable.

5.4. Transcriptional control of SF-1

As discussed earlier, SF-1 is the major driver of adrenal development and function, and it has been demonstrated that forced expression of SF-1 in mouse ES cells (Crawford, Sadovsky, & Milbrandt, 1997) or bone marrow cells (Gondo et al., 2004) causes their differentiation into steroidogenic cells that can produce deoxycorticosterone and 17OH progesterone in response to ACTH stimulation. Furthermore, the expression of SF-1 under the control of its fetal adrenal enhancer (FAdE) in transgenic mice (see later) results in the formation of larger adrenal glands and ectopic adrenal tissue, indicating that it has caused nonadrenal lineage cells to switch cell fate (Zubair, Oka, Parker, & Morohashi, 2009). Investigation of the transcriptional control of SF-1 during development has demonstrated that many of the defects described earlier can be viewed in the light of their role in activating *Sf-1* expression or modulating SF-1-dependent transcription. Both *Odd1* (Wang et al., 2005) and *Sall1* (Nishinakamura et al., 2001) deletion cause reduced expression of Wt1. Wt1 binds to sites within the *SF-1* promoter and together with the coactivator Cited2 drives the high levels of SF-1 expression required for adrenal primordium development (Val et al., 2007). SF-1 upregulates Dax1, which in turn represses SF-1 transcriptional activity and, hence, steroidogenesis (Clipsham & McCabe, 2003). It is not clear how disruption of SF-1 and its negative regulator can cause similar defects, but perhaps, it indicates a very fine control of SF-1 activity is required for adrenal development. However, Dax1 can function as a coactivator for SF-1 transcriptional activity in steroidogenic cells when expressed at high levels, and perhaps, the observations *in vivo* are complicated by the pleiotropic effects of these factors and not a simple consequence of their coactivity (Xu et al., 2009). SF-1 and β-catenin cooperate to transcriptionally activate *Dax1* expression (Mizusaki et al., 2003) and the expression of other genes (Gummow, Winnay, & Hammer, 2003). *Wnt4* expression is reduced in *Sall1*- (Nishinakamura et al., 2001) and *Wt1*-null mice, and it can upregulate Dax1 (Mizusaki et al., 2003), further demonstrating the complexity of the transcriptional network specifying adrenal development (Essafi et al., 2011).

6. SIGNALING PATHWAYS INVOLVED IN ADRENAL DEVELOPMENT

6.1. Shh signaling

Sonic hedgehog (Shh) is member of the vertebrate hedgehog family of secreted ligands, along with Desert hedgehog (Dhh) and Indian hedgehog (Ihh); these perform a multitude of crucial roles during embryonic development and are also required in the adult for tissue maintenance, differentiation, and the regulation of stem cell populations (for reviews, see King, Guasti, & Laufer, 2008; Laufer et al., 2012). Secreted Hh ligands bind to the 12-pass transmembrane protein Patched-1 (Ptch1) expressed on Hh signal-receiving cells (Stone et al., 1996). In the absence of ligand binding, Ptch1 inhibits the actions of the 7-transmembrane G protein-coupled receptor Smoothened (Smo) by a poorly characterized mechanism, and members of the zinc finger transcription factor family Gli3 and Gli2 are proteolytically processed, losing their C-terminal activation domain (Pan, Bai, Joyner, & Wang, 2006). In this form, these transcription factors act to repress transcription, although the majority of Gli2 is degraded by the proteasome. The binding of Hh to Ptch1 relieves the inhibition it exerts on Smo and allows it to prevent the processing of the Gli transcription factors. Full-length Gli3 and Gli2 act as transcriptional activators. Gli1, which only acts as a transcriptional activator, is not expressed in the absence of Hh, but is upregulated by the pathway, and so can be used as a marker for active Hh signaling (Vokes et al., 2007).

Shh, but not Dhh or Ihh, is expressed in the adrenal primordium following separation from the AGP (Bitgood & McMahon, 1995; Boulkroun et al., 2011; Ching & Vilain, 2009; Guasti et al., 2011; Huang et al., 2010; King et al., 2009; Paul & Laufer, 2011) in cells just beneath the capsule. Its expression remains restricted to this subcapsular cell population throughout gestation and adulthood. In rodents, Shh is expressed in relatively undifferentiated steroidogenic cells that express neither CYP11B1 nor CYP11B2 (Guasti et al., 2011; King et al., 2009). These cells are arranged in clusters interspersed between clusters of CYP11B2-expressing cells in mice (King et al., 2009), but in rats, the cells are in a continuous layer between the CYP11B2 and the CYP11B1 layers and separated from the CYP11B1 layer by the inner ZU (Guasti et al., 2011; Mitani et al., 1999) (see later). In humans, SHH-expressing cells are also observed in subcapsular clusters (Boulkroun et al., 2011). Although costaining with SHH has not been performed in humans, CYP11B2 is detected in clusters of cells (Boulkroun et al., 2011; Nishimoto et al., 2010), and by analogy with the mouse adrenal gland, it is likely that the SHH-expressing cells are between these clusters. There is also a layer of cells between the CYP11B2- and the CYP11B1-expressing populations that may be analogous to the ZU (Nishimoto et al., 2010), suggesting that SHH expression in the human adrenal gland bears features of both the rat and the mouse (Fig. 7.1).

Holoprosencephaly is a consequence of the inactivation of the Hh pathway, with defects observed in *SHH* (Nanni et al., 1999), *PTCH1* (Ming et al., 2002), and *GLI2* (Roessler et al., 2003), and is often associated with adrenal hypoplasia (Dubourg et al., 2007). Smith–Lemli–Opitz syndrome (SLOS), a condition in which defects in 7-dehydrocholesterol reductase, an enzyme required for the formation of cholesterol from 7-dehydrocholesterol, is associated with abnormalities that overlap those seen in cases of impaired Hh signaling (Kelley & Hennekam, 2000), including holoprosencephaly. SLOS is also associated with adrenal insufficiency (Chemaitilly et al., 2003). This may be due to the requirement of cholesterol as the substrate for steroidogenesis and consequently adrenal function or may be linked to two observations regarding Hh signaling. First, Hhs are covalently linked to cholesterol, the only such proteins known to undergo this modification, which is critical for their signaling (Riobo, 2012), and second, 7-dehydrocholesterol ((pro-) vitamin D3), which accumulates in the plasma of SLOS patients (Bijlsma et al., 2006), can be transported out of the cell by Ptch1 and act as a negative regulator of Smo. Hypoadrenalism was also reported in the first description of Pallister–Hall syndrome (Hall et al., 1980), which is caused by a protein-truncating mutation in *GLI3* (Hall et al., 1980). The introduction of a similar *Gli3* mutation, which creates a constitutive repressor form of Gli3, into mice was reported to result in adrenal agenesis (Bose, Grotewold, & Ruther, 2002), but this has recently been called into question (Laufer et al., 2012).

Homozygous deletion of *Shh* in mice is embryonically lethal, demonstrating the importance of this morphogen in development. Analysis of the adrenal gland at 14.5 and 16.5 dpc in *Shh*$^{-/-}$ embryos, however, indicates that the adrenal primordium forms but is much smaller than in the wild-type (Ching & Vilain, 2009; Huang et al., 2010; King et al., 2009). Conditional homozygous deletion of *Shh* from steroidogenic tissues by crossing floxed *Shh* mice with *Sf-1* cre mice is compatible with life, and these animals have small adrenal glands both during embryogenesis and in the adult but with grossly normal histological appearance and zonation with a normal medulla (Ching & Vilain, 2009; Huang et al., 2010; King et al., 2009).

6.2. FGF signaling

Fibroblast growth factor (FGF) signaling controls early developmental processes such as cell movement during gastrulation, mesodermal and neuroectoderm formation, anterior/posterior patterning, and organogenesis (Ornitz & Itoh, 2001; Turner & Grose, 2010). FGFs are a large family of secreted glycoproteins that bind to four signaling FGF receptors, FGFR1–4. The FGFRs are receptor tyrosine kinases with three extracellular immunoglobulin (Ig) domains, a single transmembrane domain, and an intracellular tyrosine kinase domain. FGFRs 1–3 are alternatively spliced in the third Ig domain such that an invariant exon IIIa is spliced to either exon IIIb or IIIc to produce IIIb or IIIc isoforms, which exhibit preferential binding to different FGFs (Turner & Grose, 2010). Binding of FGFs to their cognate receptors activates a variety of different pathways in different settings: most commonly, Ras/MAPK activation controls cell proliferation and differentiation, Akt activation is associated with cell survival, and PKC activation is associated with cell migration (for reviews, see Bottcher & Niehrs, 2005; Dorey & Amaya, 2010).

Analysis of FGF and FGFR expression in the developing mouse adrenal gland at 15.5 dpc detected FGFs 1, 2, and 9 and FGFRs 1 IIIc, 2 IIIb, 2 IIIc, and 3 IIIc in either predominantly the capsule (FGF2, 9) or the cortex (FGFR2 IIIb, IIIc, and FGFR3 IIIc) or both (FGFR1 IIIc), with FGFR2 IIIb and IIIc being expressed in subcapsular clusters of cells similar to Shh and β-catenin (Guasti, Sze, McKay, Grose, & King, 2013). FGF2 acts as a mitogen for adrenocortical cells both in culture and in gland regeneration experiments (Chu, Ho, & Dunn, 2009; Crickard, Ill, & Jaffe, 1981; Gospodarowicz, Ill, Hornsby, & Gill, 1977; Lepique et al., 2004) and has been shown to bind specifically to cells from the ZG (Basile & Holzwarth, 1994). The principal receptor for FGF2 is FGFR2 IIIc, and therefore, this observation is in line with the receptor expression analyses.

Embryos with a global *Fgfr2 IIIb* deletion have hypoplastic adrenal glands (Guasti et al., 2013; Revest et al., 2001), and deletion of both isoforms of FGFR2 from steroidogenic tissue recapitulates this phenotype and causes male-to-female sex reversal (Kim et al., 2007), implying that FGFR2 is not necessary for AGP formation but is required for the subsequent growth and development of the adrenal gland.

6.3. Fetal adrenal growth

Anencephaly is a common congenital abnormality caused by the failure of the neural tube to close. The severity of the defect is variable but generally results in the absence of the hypothalamus, although the anterior pituitary gland is often spared (Swaab, Boer, & Visser, 1978). Often, these fetuses exhibit adrenal hypoplasia, which can be detected from between 13 and 18 wpc (Benirschke, 1956; Gray & Abramovich, 1980; Mazzitelli, Vauthay, Grandi, Fuksman, & Rittler, 2002), indicating that early formation of the adrenal gland proceeds normally in the absence of hypothalamic input. The hypoplasia is mainly observed in the FZ, with a relatively normally sized DZ and medulla (Bocian-Sobkowska, Malendowicz, & Wozniak, 1997) and normal cellular proliferation (Staton, Grilliot, & Parker, 2004), and production of the FZ-derived steroid hormone DHEA is reduced in these fetuses (Easterling, Simmer, Dignam, Frankland, & Naftolin, 1966). Fetal ACTH is detectable by 50 dpc in adrenal development (Goto et al., 2006) and is thought to be responsible for androgen secretion from the FZ (Carr, Ohashi, MacDonald, & Simpson, 1981) and also for the short window of cortisol production from the DZ at this time (Goto et al., 2006). Anencephalic pituitary glands show a reduction in ACTH production by midgestation, indicating a requirement for continued input from the hypothalamus (Begeot, Dubois, & Dubois, 1977; Osamura, 1977), and this is supported by the observation that administration of ACTH to anencephalic fetuses *in utero* restores the size of the FZ (Honnebier & Swaab, 1973). Based on these data, it has been proposed that ACTH, or possibly another POMC peptide (Estivariz, Iturriza, McLean, Hope, & Lowry, 1982), is the primary regulator of adrenal development (Ishimoto & Jaffe, 2011). However, Pomc-null mice (Karpac et al., 2005) and *Mc2r*-null mice (Chida et al., 2007) have normal adrenal glands at birth, indicating that neither ACTH nor any other POMC peptide is required for fetal adrenal development in this species. It is possible that because the rodent placenta, unlike humans, expresses CYP17 and is able to make estrogen (Arensburg, Payne, & Orly, 1999; Durkee et al., 1992), whereas this is formed from adrenal androgens secreted from the human FZ under the control of ACTH, there is no need for ACTH signaling *in utero* in mice. This would indicate that the role of ACTH in humans is to maintain pregnancy through its actions on steroidogenesis rather than via its effects on growth.

It has also been hypothesized that ACTH itself is not mitogenic, but upregulates the expression of other growth factors such as IGF2 (Voutilainen & Miller, 1987) and FGF2 (Mesiano, Mellon, Gospodarowicz, Di Blasio, & Jaffe, 1991). Both these factors are potent mitogens for fetal adrenal cells *in vitro* (Crickard et al., 1981; Mesiano, Coulter, & Jaffe, 1993; Mesiano, Mellon, & Jaffe, 1993). IGF2 is not expressed in the adult cortex, but is proposed to be a major regulator of fetal adrenal growth (for review, see Mesiano & Jaffe, 1997). FGF2 has a more profound effect on the growth of the DZ than the FZ (Crickard et al., 1981).

6.4. Fetal zone involution and X zone regression

FZ involution takes place rapidly during the first 2 weeks following birth and is known to involve apoptosis of cells in the inner FZ (Jirasek, 1980; Spencer et al., 1999). Activin and inhibin are expressed in the FZ (Spencer, Rabinovici, Mesiano, Goldsmith, & Jaffe, 1992), and *in vitro* studies demonstrate that activin can inhibit FZ cell proliferation by triggering apoptosis, with no effect on DZ cells, indicating a possible role in involution (Spencer et al., 1999). The X zone of the mouse adrenal gland expresses high levels of activin receptors and Smad2, a downstream effector in the activin pathway, and activin injection into intact mouse adrenal glands *ex vivo*, or *in vitro* treatment of cell cultures, caused apoptosis specifically in the X zone cells (Beuschlein et al., 2003). These data indicate a potential equivalence of the FZ and X zones (see later).

7. MODELS OF DEVELOPMENT

7.1. SF-1 FAdE

As discussed previously, the primary driver of the development of the AGP appears to be SF-1. It is the first identified marker of these anlagen, and in its absence, the primordium undergoes apoptosis after specification. Analysis of the transcriptional regulation of SF-1 in the adrenal primordium revealed the presence of an adrenal gland-specific enhancer, referred to as the FAdE. The FAdE is active in mouse development from 10.5 dpc, the time at which the adrenal primordia and gonadal primordia separate, and in all cells of the adrenal primordium until 13.5 dpc. By 15.5 dpc, its activity was restricted to the inner cortex around the medulla and persisted in the X zone of the adult mouse until its regression (Zubair et al., 2006; Zubair, Parker, & Morohashi, 2008). This enhancer, which is located in the fourth intron (Zubair et al., 2006) and is conserved in mammals (Gardiner, Shima, Morohashi, & Swain, 2012), contains binding sites for Pbx complexes as well as SF-1 itself. Mutation of the Pbx binding sites in FAdE abolishes the expression of SF-1 in the adrenal primordium in transgenic mice (Zubair et al., 2006), and furthermore, as described earlier, Pbx1 is expressed early in the development of the urogenital ridge and its absence causes greatly reduced SF-1 expression in *Pbx1*-null mice (Schnabel et al., 2003). Mutation of the SF-1 sites does not impair the activity of the enhancer at 11.5 dpc, but by 17.5 dpc, activity was abolished (Zubair et al., 2006), suggesting that Pbx1 is required for initiation of SF-1 expression in the adrenal anlagen but its prolonged expression requires SF-1 to act in a feedforward autoregulatory mechanism.

The FAdE transgene was used in lineage-tracing experiments to drive expression of tamoxifen-inducible cre recombinase. If pregnant mothers were injected with tamoxifen at 10.5 dpc, the entire cortex was labeled at 14.5 dpc, but when injected at 12.5 and 13.5 dpc, there was a progressive decrease in labeling in cells of the outer cortex, and injecting at 14.5 dpc did not label the outer cortical cells at 15.5 dpc (Zubair et al., 2008). Interestingly, injecting at 10.5 dpc also revealed some weak labeling of the gonads, but this was not seen at later time points, suggesting that FAdE is repressed in the gonadal primordium. When similar experiments were performed but the adrenal glands were not studied until the animals were 2 months of age, those injected at 11.5 dpc were still labeled across the cortex, but those injected at 14.5 dpc retained no label at all (Zubair et al., 2008). This study demonstrates that mice adrenal glands possess fetal and DZs equivalent to those in the human adrenal gland and that the founder population of fetal adrenal cells can give rise to the DZ in early development of the cortex but loses this potential from around 14.5 dpc. Persistent labeling of the X zone by the FAdE transgene also demonstrates that the original FZ becomes the X zone.

These observations argue against earlier ideas that the primordium differentiates into the FZ while the DZ is formed from another group of migrating coelomic epithelial cells (Crowder, 1957; Uotila, 1940) or alternatively that only the inner portion of the primordium differentiates to form the FZ and the outer portion retains its primordial phenotype (Jirasek & Lojda, 1964). It is not known whether such a mechanism takes place in humans, but the very high conservation of the FAdE sequence across species would argue for its role in human adrenal development.

7.2. Shh/Gli1 lineages

An alternative view of adrenal development was provided by studies on Shh signaling in the adrenal cortex. Shh expression has been first detected at around 12.5 dpc in mouse adrenal glands; however, more sensitive lineage analyses using an Shh cre driver detected Shh expression as early as 11.5 dpc, in the adrenal glands but not in the gonads, suggesting that Shh expression is activated exclusively in cells of the adrenal primordium almost immediately after its separation from the AGP (King et al., 2009; Laufer et al., 2012). In this analysis, the number of labeled cells in the adrenal increased rapidly to 13% of all SF-1-positive cells by e12.5, mainly in the subcapsular region, and to 75% by 13.5 dpc, spread throughout the developing cortex. Analysis of the adrenal glands from these experiments shortly after birth showed that all cortical cells were labeled, including those of the ZG, ZF, and X zone (King et al., 2009), but not the capsule or medulla. When similar experiments were performed with an inducible *Shh* cre-T2 driver, labeled cells were seen at the periphery of the gland at birth following tamoxifen injection at 14.5 dpc, and after 28 days, columns were observed, which almost spanned the cortex to the medulla boundary. Adult mice were also injected and similar results were obtained with 7- and 13-day chases, with the labeled cells coexpressing CYP11B2 or CYP11B1 in increasing numbers. These data indicate that Shh-expressing cells are candidate stem/progenitor cells both in the fetal adrenal gland and for adult maintenance of the gland.

Deletion of *Shh* from the cortex causes adrenal hypoplasia but it does not act upon cortical cells; rather, it signals to mesenchymal cells in the capsule and in the periphery of the cortex. These are identified by their expression of Hh-responsive genes, such as *Gli1*, *Ptch1*, and *FoxD2*, and do not express the pan-steroidogenic markers SF-1 and CYP11A1. The lack of requirement for Hh signaling to cortical cells is evidenced by the fact that deletion of Smo from cortical cells using the *SF-1* cre driver has no effect on adrenal development (King et al., 2009). However, the adrenal hypoplasia was associated with a much thinner capsule with a greatly reduced proliferative capacity (Huang et al., 2010; King et al., 2009), suggesting that the effect of Shh was therefore chiefly via the capsule. Lineage-tracing experiments using inducible *Gli1* cre drivers demonstrated that the capsule and subcapsular mesenchymal cells were initially labeled by injection of tamoxifen at 14.5 or 15.5 dpc, but clusters of labeled cells, which expressed steroidogenic markers (King et al., 2009), were observed in the cortex (Huang et al., 2010; King et al., 2009) and expanded into columns with extended chases. Similar to the tracing using *Shh*, these columns eventually spanned the cortex to the medulla (Huang et al., 2010; King et al., 2009). At least some of these labeled cells were observed to express Shh (King et al., 2009). Postnatal induction of the *Gli* cre driver also labeled columns of cells entering the cortex (King et al., 2009). Taken together, these lineage-tracing studies provide direct confirmation of the centripetal migration hypothesis and that the stem/progenitor cell populations reside at the periphery of the gland. They imply that Shh signals to a mesenchymal population of cells that differentiate from a nonsteroidogenic to a steroidogenic phenotype, at least in part via an Shh-expressing intermediate (Fig. 7.4). These data also imply that the definitive cortex develops from the capsule and the adrenal cortex and is of both coelomic epithelial and mesenchymal origin and are therefore contradictory to the results from the FAdE experiment, which suggests that the cortex is entirely of SF-1-expressing coelomic epithelium origin.

One explanation for this conundrum has been to invoke a mechanism in which cells in the newly separated adrenal primordium enter the forming adrenal capsule whereupon their SF-1 expression is immediately extinguished before the onset of Shh expression. SF-1 expression is subsequently reactivated upon entry of these cells back into the cortex (Wood & Hammer, 2011). One candidate for a factor repressing SF-1 in the newly forming adrenal capsule is the helix–loop–helix transcription factor Pod1. Loss of *Pod1* in mice leads to the expansion of the SF-1-expressing domain in the testes, with increased differentiation of the steroidogenic Leydig cells and ectopic expression of SF-1 and SF-1 target genes (Cui et al., 2004). Pod1 can repress SF-1 expression through inhibiting binding of USF to an E box in its promoter (Cui et al., 2004). Pod1 is exclusively expressed in the capsule of the adrenal gland in adult mice (Kim & Hammer, 2007) and analysis of adrenal glands from *Pod1* null mice also demonstrated ectopic expression of SF-1 in some regions of the capsule (Kim & Hammer, 2007). It has therefore been hypothesized that Pod1 downregulates *Sf-1* expression in the fetal adrenal cells upon recruitment into the capsule. While this is an attractive hypothesis, other data do not support it. An SF-1 BAC transgene driving cre does not label the capsule (King et al., 2009) and the FAdE lineage tracing does not appear to label it either (Zubair et al., 2008). Furthermore, Pod1 expression in the developing adrenal gland is apparently detected within the cortex but not in the capsule at 16.5 dpc (Maezawa et al., 2012), at a time when SF-1 is undetectable in the capsule. The resolution of these apparently conflicting studies awaits further investigation.

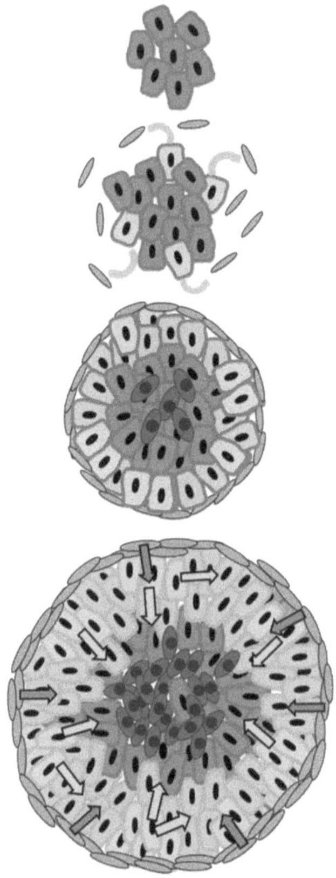

Figure 7.4 Lineages during adrenal development. SF-1 expression in the AP is driven by the FAdE from 10.5 dpc (dark cells). The AP grows and some cells (pink) express Shh from 11.5 dpc and signal to surrounding mesenchymal cells that express Gli1 and start to form the capsule (yellow). Some AP cells may enter the capsule and downregulate SF-1 at this point, although this is not proved. At around 14.5 dpc, the FAdE becomes inactive and SF-1 transcription is driven by the definitive enhancer from this point. Gli1-positive capsule cells enter the cortex and convert to an SF-1-positive identity in the steroidogenic lineage. At least some of these express Shh, and Shh-positive cells can differentiate into ZG and ZF cells of the definitive adrenal gland (light cells). Whether ZG and ZF cells can interconvert by transdifferentiation is not known at present. (See color plate.)

7.3. Stem/progenitor cells

The *Shh/Gli1* lineage-tracing studies referred to earlier provide genetic evidence for the long-held view that adrenocortical stem/progenitor cells are located in the capsule and subcapsular region of the gland. Analysis of proliferation and apoptosis in Shh mutant adrenal cortices revealed a reduction in PCNA staining in mutants compared to wild-type adrenal glands from 13.5 dpc that persisted through to the adult (Ching & Vilain, 2009), but no apparent increase in apoptosis (Ching & Vilain, 2009; Huang et al., 2010). These data suggest that Shh signaling to the capsule is not required for the production of a cortical survival signal from this compartment but may produce a proliferative signal. Interestingly, Huang et al. observed that the greatest effect on cell proliferation in Shh-null adrenal glands was observed in the capsule, indicating that Shh may act as a mitogen for the capsule (Huang et al., 2010). The remarkably thin capsule in these mutants may result from the lack of trophic support from the Shh signal, or Shh may alternatively be acting as a chemoattractant to promote the coalescence of the surrounding mesenchyme to form the capsule, or it may regulate the differentiation of the capsule into the steroidogenic lineages, with rapid depletion from the capsule during development as a consequence of its absence.

Thus, there is an Shh signal from the cortex to the capsule and potentially an Shh-dependent proliferative signal acting on the cortex from the capsule. Candidates for these signals include FGF2, exclusively expressed in the capsule of 15.5 dpc mouse adrenal glands and known to be a mitogen for adrenocortical cells, at least *in vitro* (see earlier), and Wnts. Deletion of *β-catenin* is extremely deleterious for adrenocortical cell proliferation, with an approximately 75% decrease in BrdU uptake in cells at 13.5 dpc (Kim et al., 2008), but the identity of the Wnts acting upon it is yet to be determined. Wnt4 is a candidate but is principally produced by cortical cells. Wnts exclusively expressed in the capsule of the rat adrenal gland include 2, 6, and 8b (Cavlan D and King P, unpublished results) and may represent Shh-dependent proliferative signals. It is as yet not known whether Shh signaling is required for FGF or canonical Wnt signaling in the subcapsular region. Interestingly, *Fgfr2 IIIb*-null mice have thicker capsules that display a disorganized arrangement of cells rather than the normal stratified arrangement, a phenotype also observed in *Mc2r*-null mice (Chida et al., 2007). Potentially, this indicates that signaling through FGFR2 IIIb or the MC2R regulates capsule growth, perhaps through regulation of the Shh signal in the subcapsular cells. This has not been investigated for *Mc2r*-mutant adrenal glands, but *Fgfr2 IIIb* mutant capsules have enhanced cell proliferation as well as upregulated Gli1 expression, suggesting that FGF signaling may in fact be a negative regulator of Shh expression. Further examination of the reciprocal signaling between the capsule and the subcapsular progenitor cells is warranted.

8. CONTROL OF POSTNATAL GROWTH, ZONATION, AND REMODELING

The adrenal cortex is a highly dynamic organ: there being a constant centripetal migration of cells during under normal conditions, the cortex rapidly responds to requirements for hormonal production by altering the relative sizes of the zones in response to alterations of HPA or RAS activity in a process known as adrenal remodeling. Furthermore, like the liver, the adrenal has the ability to regrow following injury. The stem/progenitor populations identified earlier are likely to be involved in these hypertrophic and hyperplastic processes and in the control of zonation, but genetic analyses have yet to be undertaken to follow the lineages recruited in these processes.

8.1. Normal growth

ACTH is often considered the principal regulator of adrenal growth. Under normal conditions, cell proliferation in the adrenal cortex displays a circadian rhythm with a peak observed at 4 am, principally in the outer ZF. This circadian rhythm follows that of ACTH, suggesting that this is the main stimulator of proliferation (Mitani et al., 2003; Miyamoto, Mitani, Mukai, Suematsu, & Ishimura, 1999). Furthermore, it is well established that hypophysectomy of experimental animals results in adrenocortical atrophy (Estivariz, Carino, Lowry, & Jackson, 1988; Estivariz, Morano, Carino, Jackson, & Lowry, 1988), while excess ACTH states such as Cushing's disease cause adrenocortical hyperplasia (Bertagna, Guignat, Groussin, & Bertherat, 2009). As discussed previously, *Pomc*-null mice are born with normal adrenal glands (Karpac et al., 2005), but it has been found that soon after birth, the gland atrophies, indicating a requirement for POMC peptides for adrenal maintenance and growth (Coll et al., 2004; Yaswen, Diehl, Brennan, & Hochgeschwender, 1999). Administration of ACTH restored the adrenal weight and function in these animals (Coll et al., 2004, 2006), but ACTH replacement does not restore adrenal size in hypophysectomized rats (Payet & Lehoux, 1980). Furthermore, ACTH has been shown to be antimitogenic in primary adrenal cells and mouse Y1 adrenocortical cells (Masui & Garren, 1971; Rocha, Forti, Lepique, & Armelin, 2003).

8.2. Zonation

The zonation of the adrenal cortex is established perinatally with the onset of CYP11B2 expression in the ZG in humans and rodents. The juxtaposition of the proposed capsular or subcapsular stem/progenitor cells with the ZG and ZF raises some interesting questions regarding the differentiation processes that occur during centripetal migration. In mice, where ZG and ZF cells are in direct contact, centripetal migration would predict the ability of ZG cells to transdifferentiate into ZF cells. However, in rats, these two populations are separated by the ZU (Mitani et al., 2003). This zone has been further characterized and shown to be subdivided in to two other zones, an outer Shh-expressing zone, named the outer ZU (oZU), and a smaller inner layer approximately one to three cells wide that does not express Shh, referred to as the inner ZU (iZU) (Guasti et al., 2011). This observation represents a conundrum for the maintenance of rat zonation. If, as in the mouse, the capsular cells delaminate and enter the cortex to become steroidogenic, then, at first glance, their initial identity would appear to be CYP11B2-positive ZG cells rather than Shh-positive cells. However, it is possible that the capsular cells first express Shh upon entering the cortex and these cells migrate through the CYP11B2-expressing cells to the oZU. Bidirectional differentiation of the Shh-positive progenitor cells of the oZU would then form the ZG and ZF, as has been proposed (Mitani et al., 1999, 2003). The iZU expresses high levels of steroidogenic enzymes, including CYP11A1, MC2R, and its accessory factor MRAP, and this might indicate that the iZU represents a differentiating TZ, which has lost Shh expression and is converting to a more steroidogenic phenotype, although it has yet to acquire CYP11B1 expression (Gorrigan, Guasti, King, Clark, & Chan, 2011). A pool of CYP11B1/CYP11B2-negative cells have been found in the human adrenal cortex (Nishimoto et al., 2010), suggesting the existence of the ZU in humans and the possibility of similar mechanisms of progenitor cell differentiation described for the rat.

The establishment of zonation in mice could also involve bidirectional differentiation of the Shh-expressing population. If this is not the case, and cells can transdifferentiate from a ZG to a ZF identity, the very sharp boundary between the ZG and the ZF must indicate that positional cues such as steep gradients of morphogens control these events. Transdifferentiation from ZG to ZF identity has not been formally demonstrated but is an interesting area for future study.

8.3. Remodeling

These considerations of stem/progenitor cell differentiation become all the more pertinent when remodeling is considered. Steroidal output responds rapidly to changes in HPA or RAAS activity, and when these are applied chronically in experimental models, profound changes in the size and steroidogenic expression of the zones are observed. For example, activation of the RAAS under low-sodium or high-potassium conditions causes a rapid expansion of the ZG to produce more CYP11B2-expressing cells and hence secrete more aldosterone, in rodent models (Romero et al., 2007), and adrenal remodeling with increased CYP11B2 expression has been reported in adrenal glands from patients with primary aldosteronism (Boulkroun et al., 2010, 2011). Conversely, suppression of the RAAS under high-sodium conditions, or by the administration of angiotensin-converting enzyme (ACE) inhibitors, causes a rapid shrinking of the ZG with consequent reduction in CYP11B2 expression and reduction in blood volume (Brennan, Chittka, Barker, & Vinson, 2008; Pignatelli et al., 2000). Similar changes occur in the ZF; HPA axis activation by chronic ACTH administration expands the ZF to secrete and express more CYP11B1 and thus secrete more glucocorticoid, but suppression by dexamethasone causes its contraction (Dallman, 1984). The processes involved in remodeling may involve a combination of enhanced differentiation of stem/progenitor populations, dedifferentiation of mature steroidogenic cells, cell proliferation, and apoptosis. For example, new ZG cells produced by RAAS activation may differentiate from capsular and/or subcapsular stem/progenitor populations, or ZF cells may transdifferentiate to become ZG cells. Loss of ZG cells following RAAS inhibition may involve their dedifferentiation back into Shh-expressing cell populations or apoptosis (McEwan, Lindop, & Kenyon, 1996). Activation of the RAAS or HPA axis causes enhanced proliferation in the ZU and the zone being remodeled, the ZG and ZF, respectively (McEwan, Lindop, & Kenyon, 1995; McEwan et al., 1996). As yet, the mechanisms controlling these processes are unclear, although migration of BrdU labeled cells from the ZU into the ZG in the adrenal glands of rats following RAAS activation (Mitani et al., 2003).

8.4. Regeneration following surgical injury

8.4.1 Enucleation

These studies, in which the adrenal capsule is slit and the gland squashed to remove the inner ZF and medulla, demonstrate the remarkable regeneration properties of the adrenal cortex that within 4 weeks can reestablish the ZG and ZF from the remaining capsule and adherent ZG cells (for review, see Bland, Desclozeaux, & Ingraham, 2003). This regeneration also occurs in humans (Feuerstein & Streeten, 1991). During adrenal regeneration in rats, CYP11B2 expression is lost in the remaining adherent ZG cells before an increase in proliferation in all cortical cells until CYP11B2 is reexpressed in the ZG, around 10 days postsurgery, and proliferation in this zone returns to normal levels. Proliferation remains increased throughout the ZF at this time point but returns to normal in the inner ZF by 28 days, remaining high in the outer ZF adjacent to the ZU (Ennen, Levay-Young, & Engeland, 2005). Shh or β-catenin expression has not been documented during these experiments, but Dlk1 (Pref-1, ZOG), which is coexpressed with Shh in the rat ZU (Halder et al., 1998), is rapidly downregulated and not reexpressed until the cortex has regenerated, consistent with its role as a negative regulator of cell differentiation in other systems (Sul, 2009). Whether Shh and/or β-catenin expression is similarly affected is not known at present. Interestingly, cell proliferation is observed in the adrenal capsule (Reiter & Pizzarello, 1966), which becomes initially thickened (Skelton, 1959), indicating a potential involvement of capsular cells in regeneration.

8.4.2 Unilateral adrenalectomy

Compensatory growth of the remaining adrenal gland following unilateral adrenalectomy is another paradigm of regeneration following injury. Unlike in adrenal enucleation, only growth of the ZF is observed following unilateral adrenalectomy to compensate for the loss of adrenal mass and resultant glucocorticoid deficit with proliferation being observed in the outer ZF (Engeland, Ennen, Elayaperumal, Durand, & Levay-Young, 2005). Cell proliferation is increased threefold in the outer ZF within 2 days of surgery, but, in contrast to enucleation, no increase in ZG proliferation is seen.

In all these remodeling paradigms, proliferating cells are located in and around the ZU, and this has led to the proposal that this is a stem cell zone in rats (Mitani et al., 1999, 2003). The expression of Shh in this region is highly suggestive of the involvement of this cell population in remodeling, and this merits further study. Interestingly, compensatory growth following adrenalectomy is not observed in *Sf-1*$^{+/-}$mice (Beuschlein et al., 2002).

8.5. ACTH and remodeling

Dax1 is transcriptionally inhibited by ACTH and activated by glucocorticoid (Gummow, Scheys, Cancelli, & Hammer, 2006; Ragazzon et al.,

2006), and given the inhibition Dax1 exerts over steroidogenesis, this potentially serves as a negative feedback loop to control glucocorticoid production in adrenocortical cells. However, as discussed earlier, Dax1 can function as a negative regulator of ES cell differentiation, and it is tempting to speculate, based on its subcapsular localization, at least in male mice, that it may play a similar role in controlling the mobilization of adrenal stem cell populations under the control of the HPA axis (Kim & Hammer, 2007). However, the role for ACTH in regulating proliferation during remodeling is controversial. Clearly, ACTH administration can provoke hyperplasia in the ZF, and the expression of MC2R and MRAP in the iZU (Gorrigan et al., 2011) may suggest a role for ACTH in the differentiation of these cells towards the ZF phenotype *in vivo*. This may involve downregulation of Dax1 expression to allow differentiation of the progenitor cells to become ZF cells. Alternatively, dexamethasone, which blocks the HPA axis and lowers POMC production, may upregulate Dax1 expression and inhibit the formation ZF cells. However, the data from remodeling experiments are conflicting. Dexamethasone administration has been shown to block ZF proliferation and compensatory growth following unilateral adrenalectomy in some studies (Engeland et al., 2005; Lowry, Silas, McLean, Linton, & Estivariz, 1983; Phillips, Crock, & Funder, 1985), but not others (Engeland, Shinsako, & Dallman, 1975; Grizzle & Dunlap, 1984). Hypophysectomy, the complete removal of the pituitary gland, blocks regeneration following enucleation (Estivariz, Carino, et al., 1988; Estivariz, Morano, et al., 1988) but does not completely block compensatory growth (Dallman, 1984; Dallman, Engeland, Holzwarth, & Scholz, 1980), and ACTH has even been shown to inhibit this process (Dallman et al., 1980). Nevertheless, hypophysectomy does reduce the overall extent of compensatory growth, suggesting the involvement of a pituitary factor. One such candidate for roles in both processes is N-POMC, which is derived from the secreted POMC cleavage product pro-γ-MSH. Neutralizing antibodies against pro-γ-MSH inhibit both adrenal regeneration and compensatory growth, but pro-γ-MSH has no mitogenic activity (Lowry et al., 1983). It was postulated that pro-γ-MSH was cleaved at the adrenal gland to provide local sources of mitogenic N-POMC, and, subsequently, a protease expressed specifically in the adrenal cortex, AsP, principally at the outer ZF boundary in rat adrenal glands, was isolated that cleaved pro-γ-MSH to N-POMC and whose expression was significantly increased during compensatory growth (Bicknell et al., 2001; Estivariz, Carino, et al., 1988; Estivariz, Morano, et al., 1988; Lowry et al., 1983). Interestingly, however, N-POMC cannot restore adrenal growth in *Pomc*-null mice (Coll et al., 2006).

These studies may suggest that ACTH is not involved in adrenal cell proliferation during adrenal remodeling, but it has been reported that the elevated ACTH levels that accompany enucleation are required for the dedifferentiation of the ZG cells that is an immediate response in the regenerating adrenal gland (Engeland & Levay-Young, 1999). Whether this works via inhibiting Dax1 expression has not been reported. ACTH does not affect proliferation in the ZG; the cell proliferation observed by placing animals on a low-sodium diet is inhibited by type 1 angiotensin receptor antagonists (McEwan et al., 1996), indicating that cell growth in this zone is controlled by the RAAS (Chatelain, Montel, Dickes-Coopman, Chatelain, & Deloof, 2003).

If N-POMC, rather than ACTH, provides the pituitary signal for compensatory growth, what is the nonpituitary signal observed? There is evidence for a neurally mediated signal consisting of afferent and efferent connections between the adrenal gland and the hypothalamus, with mechanical perturbation of the nerve-bearing adrenal pedicle causing increased proliferation in the contralateral adrenal gland (Dallman, Engeland, & Shinsako, 1976), and in support of this, compensatory growth is blocked by interruption of the nerve supply (Kleitman & Holzwarth, 1985). FGF2, recognized as a potent mitogen for adrenocortical cells *in vitro*, has been suggested to mediate this neural effect based on the observation that suramin, which is thought to block FGF2 binding to its receptors in the adrenal cortex, inhibits compensatory growth, although there is little supporting data (Basile & Holzwarth, 1994).

9. DISEASES OF THE ADRENAL CORTEX

Several genetic mutations that result in adrenal agenesis or aplasia have been referred to in previous sections, and often, these are not compatible with life and therefore do not manifest as causes of adrenal insufficiency in children or adults. A number of adrenal insufficiency syndromes have been identified and elucidating their genetic causes will be important for further understanding mechanisms of adrenal development, growth, and maintenance. These syndromes include those in which there is a relatively isolated defect in adrenal development; lack of trophic support from the pituitary gland; disorders of steroidogenesis; and occasionally, but less informatively, autoimmune destruction of adrenal tissue. Finally, adrenal cancer will be discussed. For a more extensive appreciation of syndromes including adrenal insufficiency, see Else and Hammer (2005) and Ferraz-de-Souza and Achermann (2008).

9.1. IMAGe

AHC has been considered in terms of the role of SF-1 and Dax1. However, it also forms a feature of the rare IMAGe (intrauterine growth restriction, metaphyseal dysplasia, AHC, and genital anomalies) syndrome (Vilain et al., 1999). Most affected individuals are born small and develop skeletal abnormalities at the ends of the long bones, scoliosis, or osteoporosis later in life. Males present with genital malformations including micropenis and undescended testicles and urethra anomalies, but the most clinically important condition of the syndrome is adrenal insufficiency, which shortly after birth causes salt wasting, hypoglycemia, and shock due to loss of both mineralocorticoid and glucocorticoid synthesis and can be life-threatening. This syndrome is not associated with mutations in either SF-1 or Dax1, but recently, a candidate gene has been reported, *CDKN1C*, which is paternally imprinted and encodes the cell cycle regulator p57Kip2 (Arboleda et al., 2012). Mutations were detected in the PCNA-binding domain of p57Kip2 and interaction between the two proteins is disrupted. *p57Kip2*-null mice, however, display adrenal hyperplasia, reflecting its role as a cyclin-dependent kinase inhibitor (Pateras, Apostolopoulou, Niforou, Kotsinas, & Gorgoulis, 2009). p57Kip2 is expressed in the cortex with highest levels of expression adjacent to the capsule in both humans (Arboleda et al., 2012) and rats (Kobayashi et al., 2006). Its expression is upregulated by ACTH treatment in the ZG in rats, and it is tempting to speculate that this is to prevent proliferation in this zone. The discrepancy between these observations suggests that the loss of PCNA binding is insufficient to create a loss-of-function mutation of p57Kip2, and further consequences of this mutation that causes gain of function in IMAGe, such as the potential impact upon its ubiquitination and stability (Arboleda et al., 2012), remain to be elucidated.

9.2. Familial glucocorticoid deficiency

In contrast with AHC, familial glucocorticoid deficiency (FGD) is a defect of glucocorticoid production with no effect on the production of mineralocorticoids from the ZG (Meimaridou et al., 2013). Patients typically present in infancy and childhood with a low cortisol and raised ACTH. Low cortisol levels manifest in hypoglycemia, failure to thrive, and recurrent infections, which can be fatal if untreated by hormone replacement. Sometimes, ACTH levels are high enough to cause hyperpigmentation via its actions on MC1R (Meimaridou et al., 2013). Several mechanisms have come to light over the recent years.

9.2.1 *MC2R2*

Mutations of the ACTH receptor *MC2R* were the first identified cause of FGD and account for approximately 25% of known cases. More than 40 have now been identified. Varied aspects of the receptor, from ACTH binding to signaling, are blocked by inactivating mutations targeting amino acids throughout the GPCR, prohibiting adrenal cortisol production in response to pituitary signals. Untreated patients have adrenal glands with severely atrophied ZF and ZR at postmortem (Clark & Weber, 1998).

9.2.2 *MRAP*

MRAP is a single transmembrane protein that binds MC2R via its transmembrane domain and exports it from the endoplasmic reticulum to the plasma membrane (Metherell et al., 2005). Mutations in this protein account for a further 20% of FGD cases and include those that result in loss of the transmembrane domain and hence sequestering of the receptor in the ER and mutations that retain trafficking function but do not allow the formation of a signaling-competent receptor (Chung, Chan, Metherell, & Clark, 2010; Hughes et al., 2010). MRAP and MC2R expression patterns overlap in rat fetal and adult adrenal glands (Gorrigan et al., 2011), displaying a gradient of expression from a maximum in the ZU and fading to the inner ZF. The ability of the ZU to respond to ACTH signals suggests that these cells are primed to convert to a ZF identity and may well be the cells that proliferate with a circadian rhythm discussed earlier.

9.2.3 *StAR/CYP11A1*

Almost half of all cases of FGD are caused by an inability to respond to ACTH. Further causes of FGD have been identified that implicate steroidogenesis, with mutations in StAR (Metherell et al., 2009) and CYP11A1 (Parajes et al., 2011; Sahakitrungruang, Tee, Blackett, & Miller, 2011). These are unexpected findings because mutations in these genes had previously been found in another defect of steroidogenesis, lipoid congenital adrenal hyperplasia (LCAH), which, because of the central role of these genes in steroidogenesis, involves severe defects in glucocorticoid, mineralocorticoid, and sex hormone synthesis, with accumulation of cholesterol in the cells causing cell death (King, Bhangoo, & Stocco, 2011). The mutations identified in FGD appear to have a less severe phenotype than do the ones in LCAH and selectively affect steroidogenesis in the ZF. This is likely to be because the ZF cells have a more steroidogenic phenotype than ZG cells, being lipid-rich and containing mitochondria with tubulovesicular cristae. Mutations in these steroidogenic enzymes account for a further 5% of cases of FGD.

9.2.4 *MCM4*

An unusual form of FGD was observed in a cohort of Irish travelers in whom the condition developed relatively late in childhood following a period of normal function. The affected individuals also presented with chromosomal breakage, natural killer cell deficiency, and short stature (Gineau et al., 2012; Hughes et al., 2012). Genetic analysis of the cohort for a gene responsible for FGD or NK cell deficiency identified the same candidate, minichromosome maintenance-deficient 4 homologue (*MCM4*). This is a component of the heterohexameric prereplication complex along with MCM2, 3, 4, 6, and 7 that forms at the origin of replication to initiate DNA synthesis during cell division. The identified mutation was a missense mutation in the splice acceptor site of exon 2, which was predicted to lead to a one nucleotide shift in this site, thus generating a frameshift that caused a prematurely terminated protein consisting of only 27 amino acids. Given the essential role of this factor in cells, as evidenced by the fact that *Mcm4*-null mice that died before implantation (Shima et al., 2007), and the serious foreshortening of the protein predicted by the mutation identified, it was very surprising that this homozygous mutation was compatible with life. However, although full-length MCM4 could not be detected in lymphocytes from affected individuals, shorter proteins lacking the N-terminus were detected, potentially being translated from in-frame ATGs located in exon 2. The N-terminal 130 amino acids are dispensable for activity of the protein (Masai et al., 2006), and the mutant protein appears to be able to retain function in the majority of tissues in these patients. It is unclear why the adrenal gland should be particularly affected by this mutation, but the study of the adrenal glands in a mouse model that reduces MCM4 to the lowest levels compatible with life demonstrated that their steroidogenic cells are increasingly replaced by nonsteroidogenic spindle-shaped cells as the animals age, suggesting a progressive loss of steroidogenic capacity in keeping with the FGD phenotype (Hughes et al., 2012).

9.2.5 NNT

The most recent causative mutations reported for FGD have been found in nicotinamide nucleotide transhydrogenase (NNT) (Meimaridou et al., 2012). NNT is located in the inner mitochondrial membrane where it reduces $NADP^+$ to NADPH, which is in turn used to maintain high reduced glutathione levels. This allows glutathione peroxidases to detoxify harmful reactive oxygen species (ROS) such as superoxide and hydrogen peroxide, which are produced during oxidative phosphorylation and steroidogenesis. Twenty-one mutations have been found spread throughout the gene in 15 families and knockdown of NNT levels in the human adrenocortical carcinoma cell line H295R lowered reduced glutathione levels as well as increasing levels of ROS and activating apoptosis. Comparison of the adrenal glands from substrains of C57Bl6/J, one of which contains a spontaneously occurring *Nnt* mutation, showed that although the appearance of the glands and the steroidogenic gene expression is similar, increased apoptosis was observed in the ZF and corticosterone levels were reduced both basally and upon ACTH stimulation in the *Nnt* mutant substrain, indicating an impairment of steroidogenesis. As ROS are produced as a by-product of cellular metabolism, it is again perhaps surprising that the adrenal ZF is particularly susceptible to mutations in one of the antioxidant defense systems. Increased ROS levels are known to be deleterious and steroidogenic cells are exposed to high levels because of the use of molecular oxygen in steroidogenesis and the production of free radicals during CYP reactions (Hornsby, 1989; Hornsby & Crivello, 1983). Again, because of the relatively high levels of steroidogenesis in the ZF compared to the ZG and other steroidogenic tissues, and the critical importance of glucocorticoids for health, impairment of these cells may be noticed before effects on other tissues. Interestingly, the C57Bl6/J strain harboring the *Nnt* mutation develops glucose intolerance, and it may be necessary to monitor FGD patients with *NNT* mutations for extra-adrenal complications.

These causes of FGD still only explain just over 60% of all cases, and the involvement of cell cycle and antioxidant defense genes indicates that the adrenal cortex is particularly susceptible to the impairment of seemingly common systems. It is anticipated that the discovery of the causes of the remaining cases of FGD will give further insight into the control of adrenal steroidogenesis and ZF maintenance.

9.3. Triple-A syndrome

Triple-A syndrome, also known as Allgrove syndrome, is an autosomal recessive disease characterized by adrenal insufficiency together with achalasia of the cardia and alacrima (Huebner et al., 2000). The gene responsible, *AAAS*, encodes a nuclear pore protein named ALADIN (Handschug et al., 2001; Tullio-Pelet et al., 2000) and mutations in this gene cause ALADIN to sequester in the cytoplasm (Cronshaw & Matunis, 2003) and defective nuclear transport of DNA repair enzymes and antioxidant defense proteins (Hirano, Furiya, Asai, Yasui, & Ueno, 2006; Storr et al., 2009). Fibroblasts from these patients are susceptible to oxidative stress, and the pathogenesis of AAAS for the adrenal gland may be similar to that observed for Nnt, adding further support to the importance of controlling antioxidant levels for adrenal viability.

9.4. Adrenal cancer

Adrenocortical tumors are common, with the majority, referred to as incidentalomas, being asymptomatic and revealed during investigations for other disorders or at autopsy. These benign, nonmetastatic, adenomas usually remain undiagnosed unless they are very large or hypersecrete hormones such as cortisol-producing adenomas (CPAs) in Cushing's syndrome and aldosterone-producing adenomas (APAs) in Conn's syndrome. Adrenocortical carcinomas, by contrast, are rare (between 1 and 12 per million adults; Grumbach et al., 2003) and malignant with a poor prognosis and a median survival time of less than 12 months from detection (Sidhu, Sywak, Robinson, & Delbridge, 2004), but they can also be secreting or nonsecreting. It is as yet unclear whether adenomas can progress to become carcinomas or whether adenomas and carcinomas are diseases with different etiologies.

Steroid-secreting adrenocortical carcinomas have increased expression levels of steroidogenic genes, *CYP11B2* in APAs and *CYP11B1* in CPAs, as well as increased SF-1 levels (Bassett et al., 2005; Boulkroun et al., 2013). Analysis of adrenalectomies from patients with APAs indicates that both the tumor and peritumoral tissues express more CYP11B2, indicating a remodeling of the ZG, although transcriptomic analyses of the peritumoral tissue do not support a mechanism suggesting that the remodeling is an initial step towards APA formation. APAs often have the histological appearance and transcriptional profile of ZF cells rather than ZG cells, but whether they are ZG cells that have acquired ZF characteristics or ZF cells that have activated markers of ZG expression, such as CYP11B2 and Dab2, is currently controversial but interesting for the etiology of tumor formation (Azizan et al., 2012; Boulkroun et al., 2013). Somatic and germline mutations in

KCNJ5, a gene encoding a potassium channel; *GIRK4* (Boulkroun et al., 2012, 2013; Choi et al., 2011); and somatic mutations in two ATPases, *ATP1A1* encoding the α1-subunit of the Na^+/K^+ ATPase and *ATP2B3*, encoding the plasma membrane calcium-transporting ATPase 3, have been identified in approximately 7% of APAs (Beuschlein et al., 2013). These proteins are involved in maintaining cell membrane potential, and mutations lead to excess aldosterone production as a result of membrane depolarization and calcium influx, which activates CYP11B2 expression and aldosterone synthesis.

Other inherited forms of adrenocortical carcinoma include Li–Fraumeni syndrome, MEN-1, and Beckwith–Wiedemann syndrome (BWS). Li–Fraumeni syndrome is an autosomal dominant syndrome of cancers including osteosarcomas, soft tissue sarcomas, breast cancer, and adrenocortical carcinoma (Li & Fraumeni, 1969), as a consequence of germline mutations in *TP53*, which encodes the p53 tumor suppressor protein (Varley et al., 1999). The frequency of childhood adrenocortical carcinoma is highest in southern Brazil and associated with a germline *TP53* mutation that does not appear to cause other cancers in the Li–Fraumeni syndrome (Ribeiro et al., 2001). Interestingly, these tumors display very high chromosomal damage including frequent amplification of the 9q33–q34 locus that contains the SF-1 gene (Figueiredo et al., 1999), and overexpression of SF-1 in mice causes hyperplasia and adrenocortical cancer (Doghman et al., 2007), although the tumors have a gonadal phenotype (see later). *TP53* mutations have also been observed in sporadic adrenocortical carcinomas but not adenomas, suggesting that *TP53* mutation could be an important event in malignant metastatic adrenal cancer.

BWS is a pediatric overgrowth syndrome with features comprising macrosomia, abdominal wall defects, macroglossia, gigantism, and childhood tumors. In 5% of cases, these include adrenocortical carcinoma (Elliott, Bayly, Cole, Temple, & Maher, 1994). BWS is principally caused by misregulation of an imprinted region on the short arm of chromosome 11 at 11p15 that includes the maternally expressed *IGF2* and paternally expressed *CDKN1C* and *H19* genes. Inheritance of two copies of the paternal 11p15 allele (paternal uniparental isodisomy), leading to overexpression of IGF2 and loss of expression of p57Kip2, is observed in 10–20% of BWS, and mutations of *CDKN1C* account for 5–10% of cases (Jacob, Robinson, & Lefebvre, 2013). Increased IGF2 levels are commonly observed in adrenocortical carcinoma, but not adenomas, and there is a positive correlation between levels of expression, malignancy, and recurrence rates (Boulle, Logie, Gicquel, Perin, & Le Bouc, 1998; Gicquel et al., 2001). *TP53* mutations can also lead to the upregulation of IGF2 as seen in Li–Fraumeni syndrome (Ribeiro & Figueiredo, 2004). Several lines of evidence have indicated the importance of IGFs in adrenal growth and differentiation; IGF2 levels are high during fetal adrenal development and IGF2 transgenic mice have large adrenal glands with hyperplastic ZFs, and other components of the IGF system, namely, the receptor IGFR1 and the binding protein IGFBP2, but not IGF1, are also upregulated in ACC (Boulle et al., 1998; Weber, Auernhammer, Kiess, & Engelhardt, 1997). *Tp53* mutant mice (Else et al., 2009) or those in which IGF2 is overexpressed (Drelon et al., 2012; Heaton et al., 2012) do not develop adrenal tumors, however, indicating these events in themselves are not sufficient for neoplasia.

9.4.1 Wnt/β-catenin

As described earlier, canonical Wnt signaling plays an important role in adrenal development, and it also appears to be involved in adrenal adenoma and carcinoma. Increased cytoplasmic and/or nuclear β-catenin is a frequent observation in both adenomas and carcinomas (Bonnet et al., 2011; Tissier et al., 2005), and mutations in β-catenin are often observed in adrenocortical tumors (Tadjine, Lampron, Ouadi, & Bourdeau, 2008). These mutations cluster in exon 3 of *CTNNB1* that encodes amino acids that are required for targeting the protein for destruction, in particular serines and threonines that are phosphorylated by the destruction complex or amino acids adjacent to these phosphorylated residues, with mutation of serine 45 particularly common in adrenocortical tumors. These serve as activating mutations of β-catenin as it is no longer degraded and can accumulate in the cytoplasm and nucleus and activate TCF/LEF-mediated transcription. Biallelic inactivating mutations in APC have also been observed in patients with adrenal tumors and familial adenomatous polyposis, and together, these data indicate that Wnt/β-catenin pathway activation is an important event in adrenocortical tumor etiology. Histological studies indicate that there is greater β-catenin accumulation in adrenocortical carcinomas than adenomas and nuclear localization of β-catenin is associated with a more aggressive form of cancer and a poorer prognosis (Gaujoux et al., 2011). Pathway mutations are observed more frequently in larger, nonsecreting adenomas, suggesting that they produce less differentiated and more aggressive forms of tumor (Bonnet et al., 2011). Adrenal-specific expression of an active form of β-catenin (Δcat; Berthon et al., 2010) in mice has recently been shown to cause adrenal hyperplasia, confirming a role for this pathway in tumor initiation, but only a third of these animals developed malignant tumors, implying that other events were required to trigger this transformation. Although activation of the Wnt/β-catenin pathway appears to be correlated with adrenal tumor formation, β-catenin-activating mutations and APC-inactivating mutations are only found in a minority of carcinomas (26%; Gaujoux et al., 2010). Mutations in Axin2 have recently been described in a small number of both adrenal adenomas and carcinomas, but alterations in other pathway components such as downregulation of extracellular inhibitors of the pathway or overexpression of Wnts could underlie this phenomenon. One potential factor is IGF2, which is known to activate β-catenin signaling *in vitro* by inhibiting GSK3β phosphorylation (Cross, Alessi, Cohen, Andjelkovich, & Hemmings, 1995) and promoting direct phosphorylation of β-catenin on serine 552 via AKT that inhibits degradation (Fang et al., 2007). When β-catenin was activated and IGF2 overexpressed together in mouse adrenal glands, hyperplasia was observed earlier and progressed more frequently to adenoma than was seen with active β-catenin alone and in one case developed into a carcinoma (Heaton et al., 2012). These data may indicate that activation of the Wnt/β-catenin pathway may be a common event in adrenal cancer and that adenomas do progress to carcinomas if activating mutations in other pathways accumulate.

9.4.2 PKA pathway

Another pathway that is frequently activated in adrenal tumors is the protein kinase A pathway. Inactivating mutations of the *PRKAR1A* gene that encodes the type 1α regulatory subunit of PKA are found in 10% of adrenocortical adenomas that tend to be small and pigmented and are classed as primary pigmented nodular adrenocortical disease (PPNAD). PPNAD may be associated with ACTH-independent Cushing's syndrome and is also a common feature of the autosomal dominant multiple endocrine neoplasia syndrome known as Carney complex. 80% of Carney complex patients with Cushing's syndrome have mutations in *PRKAR1A* (Kirschner et al., 2000). LOH of the locus containing *PRKAR1A*, 17q22–24, is frequently observed in adrenal cancer, reported to be observed in 23% of adenomas and 53% of carcinomas (Bertherat et al., 2003). Mutations of the gene are less frequent, however, observed in 10% of adenomas and not seen in carcinomas. The adrenal gland-specific knockout of *Prkar1a* in mice (Sahut-Barnola et al., 2010) recapitulates the main features of PPNAD, ACTH-independent Cushing's syndrome, and adrenocortical hyperplasia.

Definitively, demonstrating that loss of activity of the R1α-subunit of PKA results in PPNAD and Carney complex, these studies give insight into the role of the PKA pathway in adrenal development. The hyperplasia observed in these animals appears to arise from the medulla/ZF boundary, with a population of eosinophilic, hypertrophic cells expanding outward from the medulla to the ZG from 5 to 18 months. This cell population resembles the X zone and expresses the X zone marker 20αHSD as well as markers of the ZF. CYP17, which is normally only expressed in the fetal rodent adrenal gland, is also expressed in these cells, and, as a consequence, the mutant adrenal glands make cortisol as well as corticosterone. The mutant adrenal glands express higher levels of inhibin, which has also been observed in human PPNAD nodules, and inhibin is more highly expressed in human fetal than adult adrenal glands (Voutilainen, Eramaa, & Ritvos, 1991). It has been proposed that activins are involved in fetal and X zone apoptosis (see earlier), and one consequence of dysregulated PKA activity may be enhanced expression of inhibin that by competitively inhibiting activin signaling causes the persistence of the fetal/X zone cells leading to adrenal hyperplasia.

Other examples of disorders resulting from PKA pathway dysregulation include McCune–Albright syndrome (MAS), a pediatric disorder not only classified by peripheral precocious puberty, polyostotic fibrous dysplasia, and café-au-lait pigmentation but also associated with other endocrine disorders such as ACTH-independent macronodular hyperplasia (Zacharin, 2007). MAS is caused by a mutation in *GNAS1*, the gene that encodes the stimulatory G protein subunit $G_{\alpha s}$ that activates adenylate cyclase and cAMP-dependent PKA signaling following GPCR stimulation, as is the case in MC2R activation by ACTH. The mutation results in constitutive activation of adenylate cyclase and the PKA pathway. This mutation has been identified in cortisol-secreting adrenal adenomas but not so far in carcinomas. Other mutations in the PKA pathway associated with adrenocortical tumors include *PDE11A* and *PDE8B*, genes encoding phosphodiesterases that negatively regulate the pathway and have been observed not only in hyperplasia but also in 19% of adenomas and 16% of carcinomas (Libe et al., 2008; Vezzosi et al., 2012). Like AKT, PKA phosphorylates and inactivates GSK3β (Fang et al., 2000) and phosphorylates β-catenin (Hino, Tanji, Nakayama, & Kikuchi, 2005) to activate β-catenin-dependent transcription, and β-catenin accumulation has been reported in PPNAD (Gaujoux et al., 2008). This may again indicate that β-catenin pathway activation is a requirement for adrenocortical cancer.

9.4.3 Gata4, Gata6, and adrenocortical hyperplasia

An interesting feature shared by some of the mouse models of adrenocortical disease and cancer described earlier is the development of spindle-shaped cell hyperplasia at the periphery of the gland as the animal ages, such as in Mcm4-depleted mice (Hughes et al., 2012), *Prkar1a*-null mice (Sahut-Barnola et al., 2010), β-catenin-overexpressing Δcat mice (Berthon et al., 2010), SF-1-overexpressing mice (Doghman et al., 2007), and IGF2-overexpressing mice (Drelon et al., 2012). The phenotype of this spindle-shaped cell hyperplasia is very similar to that observed in certain inbred mouse strains that develop adrenocortical tumors following prepubertal gonadectomy. These strains include DBA/2J, CE/J, C3H, Balb/c, IQI/Jic, and NU/J mice (Bernichtein, Alevizaki, & Huhtaniemi, 2008; Bernichtein, Petretto, et al., 2008; Bielinska et al., 2006; Kim, Kubota, Kiuchi, Doi, & Saegusa, 1997). Shortly after gonadectomy, densely packed nests of small, spindle-shaped, basophilic cells appear just below the adrenal capsule, referred to as A cells (for reviews, see Parviainen et al., 2007; Vuorenoja et al., 2007). These cells are nonsteroidogenic, expressing neither SF-1 nor other steroidogenic enzymes, but as they invade the steroidogenic cortex in a centripetal manner similar to that observed for normal adrenal cell migration, they become interspersed with larger, rounder, eosinophilic steroidogenic cells, referred to as B cells (Parviainen et al., 2007). Human and rodent adult steroidogenic cells express the zinc finger transcription factor Gata6, but both A and B cells express Gata4 (Parviainen et al., 2007), which is normally expressed only in the fetal adrenal glands, and other markers of gonadal cells such as Amhr2 (Krachulec et al., 2012). The steroidogenic B cells express CYP17, required for sex steroid as well as cortisol synthesis, gremlin (Krachulec et al., 2012), and the gonadotropin receptor LHCGR, hallmarks of gonadal-like cells (Bernichtein, Alevizaki, & Huhtaniemi, 2008; Bernichtein, Petretto, et al., 2008; Parviainen et al., 2007; Vuorenoja et al., 2007). Neither A nor B cells express MC2R or other adrenal steroidogenic markers such as CYP11B1 or CYP11B2, indicating that these cells have adopted a gonadal rather than adrenal identity, and the B cells can secrete sex steroids (Bernichtein, Alevizaki, & Huhtaniemi, 2008; Bernichtein, Petretto, et al., 2008). Gonadectomy interrupts the feedback of the hypothalamic–pituitary–gonadal (HPG) axis by gonadal sex steroids, and it was likely that the absence of these or the resulting increase in pituitary gonadotropins was responsible for the development of the hyperplastic phenotype. Xenografts of CHO cells engineered to express human chorionic gonadotropin in NU/J mice also caused the development of spindle-cell hyperplasia, suggesting that it is the elevated gonadotropins that drive the hyperplasia (Bielinska et al., 2005).

Several steroidogenic enzyme promoters contain consensus GATA binding sites, including StAR, 17β HSD, CYP17, and CYP19 as well as SF-1, with which Gata factors can act in synergy to activate transcription of these promoters *in vitro* (for reviews, see Parviainen et al., 2007; Tremblay & Viger, 2003), indicating that in the adult adrenal gland, Gata6 can regulate steroidogenic gene expression. Several investigations have demonstrated that the switch in expression from Gata6 to Gata4 is causative in the development of hyperplasia. First, mice in which Gata6 is conditionally deleted from steroidogenic cells, as well as displaying a thin cortex with a thickened capsule, also develop Gata4-positive gonadal-like spindle-cell hyperplasia that is enhanced by gonadectomy, with the accumulation of LHCGR- and CYP17-positive B cells (Pihlajoki et al., 2013). *Gata4*-haploinsufficient B6D2F1 mice are resistant to gonadectomy-induced adrenal hyperplasia, and *Amhr2* cre-mediated Gata4 deletion in these mice produces only small nests of A cells following gonadectomy that do not extend far into the cortex and do not develop into B cells, indicating that Gata4 expression is necessary for both gonadectomy-induced adrenal neoplasia and its progression and not just a consequence of it (Krachulec et al., 2012). Furthermore, overexpression of Gata4 in mice under the control of the *Cyp21* promoter causes spindle-cell hyperplasia in females, which appears in males and is increased in females following gonadectomy, leading to adenoma formation (Chrusciel et al., 2013).

Two other mouse models that develop adrenal tumors following gonadectomy are the *inhibin* α-null mouse (Matzuk, Finegold, Su, Hsueh, & Bradley, 1992) and the *Inhα/Tag* mouse in which SV40 large T antigen is expressed under the control of the *inhibin* α promoter (Kananen et al., 1996, 1997). These tumors are also Gata4-positive and express gonadal markers and secrete sex steroids, but unlike in the models mentioned earlier, they appear to form at the medulla boundary, although spindle-shaped cells have been observed at the periphery in some cases, and it has been proposed that the tumors arise from these subcapsular cells (Looyenga & Hammer, 2006). The *Prkar1a* mouse, as well as developing central tumors composed of large SF-1-positive steroidogenic cells, also displays small subcapsular spindle-shaped cells (Sahut-Barnola et al., 2010). Furthermore, this is also seen in the Δcat constitutively active β catenin mouse model (Berthon et al., 2010). Again, two kinds of hyperplastic cells, SF-1-positive steroidogenic cells found in the inner zones of the cortex and in the central medulla region, which go on to form a carcinoma as the mice age, and basophilic, SF-1-negative and Gata4-positive, spindle-shaped cells at the periphery of the gland, which by 10 months have traversed the cortex to reach the medulla. Interestingly, these cells do not express β-catenin, suggesting that β-catenin activation has indirect effects on other cell populations that give rise to these cells. In the Mcm4 mouse model, spindle-shaped cells had infiltrated the cortex from the subcapsular region by 3 months of age and by 12 months had spanned the cortex from the capsule to the medulla (Hughes et al., 2012). These were phenotypically equivalent to A cells, expressing Gata4 but no steroidogenic markers or LHCGR. They were, however, both β-catenin- and Gli1-positive, and interestingly, the capsule was markedly thinned adjacent to the areas of infiltrating hyperplastic cells (Fig. 7.5). Loss of Mcms in mouse models is associated with stem cell defects (Chuang, Wallace, Abratte, Southard, & Schimenti, 2010; Pruitt, Bailey, & Freeland, 2007), and the apparently specific impingement of the MCM4 defect on adrenal function may be a consequence of its effect on the growth and differentiation of these putative capsular stem/progenitor cells.

Taken together, these data may indicate that the central tumors arise from fetal/X zone cells, whereas the subcapsular spindle-shaped cells may result from a misspecification of capsular or subcapsular mesenchymal stem cells that, rather than differentiating to enter the steroidogenic lineage as Gata6-positive cells, instead express Gata4 and retain Gli1 expression. In support of this hypothesis is the observation that Gli1 expression was also elevated in adrenal glands with hyperplastic spindle-shaped cell infiltration in *Gata6*-null mice (Pihlajoki et al., 2013). Furthermore, overexpression of

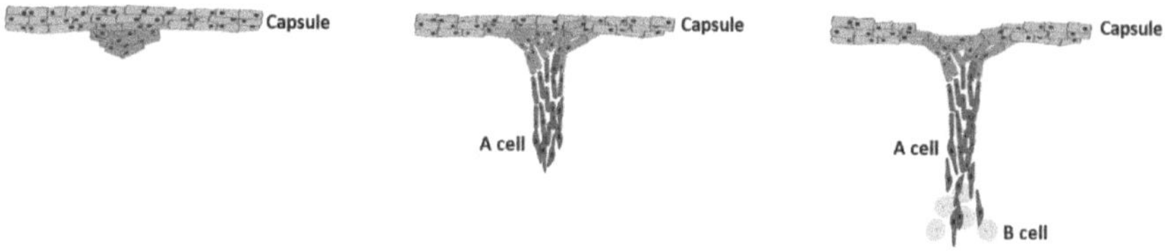

Figure 7.5 Spindle-cell hyperplasia. Spindle-cell hyperplasia develops in several mouse models. Spindle-shaped basophilic cells are observed near the capsule and extend into the cortex over time. These A cells are nonsteroidogenic but express Gata4, unlike the surrounding steroidogenic cells. They have also been shown to express Gli1 and Pod1, suggesting that they are capsular cells that have not differentiated down the correct adrenal lineage. A cells have also been shown to express Amhr2, which, like Gata4, is a marker of gonadal cells. In some models, B cells appear following gonadectomy as a consequence of the increase in LH. These are steroidogenic but resemble gonadal rather than adrenal cells, expressing CYP17, LHCGR, and gremlin. (See color plate.)

IGF2 in mouse adrenal glands also led to the production of subcapsular Gata4-positive spindle-shaped cells that expressed Gli1 and another capsule marker, Pod1 (Drelon et al., 2012). Interestingly, even though the *Cyp21* promoter was used to overexpress Gata4 in the adrenal gland (Chrusciel et al., 2013), no CYP21-positive cells and spindle-shaped cells were observed. This may indicate that the neoplastic cells are derived from differentiating stem cells that transiently activate CYP21 expression before it is extinguished by the Gata4 overexpression. This view is supported by the formation of subcapsular spindle-cell hyperplasia in the *Prkar1a* conditional-null, β-catenin, and IGF2-overexpressing mouse models in which the genetic recombinations are controlled by *ark1b7* cre. This cre is only expressed at high levels in the ZF (Lambert-Langlais et al., 2009; Sahut-Barnola et al., 2010), again suggesting that the subcapsular spindle-shaped cell hyperplasia is likely be a non-cell-autonomous effect of these modifications. Indeed, the spindle-shaped cells do not express β-catenin in the Δcat mouse (Berthon et al., 2010).

Are these observations relevant for human disease? It is interesting to note that adrenal tumorigenesis in humans sometimes occurs following disruption of the HPG axis, for example, in postmenopausal women (Mijnhout, Danner, van de Goot, & van Dam, 2004; Pabon et al., 1996) and men with testicular atrophy (Romberger & Wong, 1989). These patients have marked elevation in gonadotropins, a situation equivalent to that in the gonadectomized animals. Furthermore, subcapsular spindle-cell tumors are reported in some of these cases (Fidler, 1977), and Gata4 and LHCGR expression has been detected as well as downregulation of Gata6 (Barbosa, Giacaglia, Martin, Mendonca, & Lin, 2004; Kiiveri et al., 2005; Saner-Amigh et al., 2006). Interestingly, male patients with congenital adrenal hyperplasia (CAH) have a high incidence of testicular tumors, called adrenal rests, which have features of adrenal steroidogenic cells, expressing MC2R, CYP21, and CYP11B1 (Claahsen-van der Grinten, Otten, Stikkelbroeck, Sweep, & Hermus, 2009; Claahsen-van der Grinten, Stikkelbroeck, Sweep, Hermus, & Otten, 2006). CAH is predominantly caused by mutations in the cortisol pathway synthetic enzymes CYP21 or CYP11B1, such that the HPA axis negative feedback is impaired resulting in elevated levels of ACTH (Maitra & Shirwalkar, 2003). Perhaps, excess ACTH or gonadotropins can direct the differentiation of stem cells in the gonads or adrenal glands down an adrenal or gonadal pathway, respectively, possibly by upregulating very low levels of MC2R or LHR expression in these cells. If the spindle-shaped cells described earlier do arise from the adrenal capsular stem/progenitor cells but adopt the phenotype of gonadal stromal cells, it raises the intriguing possibility that the stem/progenitor cells have the ability to convert to a gonadal as well as an adrenal identity. It certainly appears likely that the elevated LH levels following gonadectomy are required for the development of B cells, with perhaps a feedforward mechanism with Gata4 upregulating LHCGR expression, although this has not been demonstrated. An alternative explanation is that the segregation of the adrenal and gonadal primordia is not a completely efficient process and these inappropriately localized cells can respond to tropic stimuli. ACTH-responsive CYP11B1- and CYP21-positive cells have been observed in mouse testes from 13.5 dpc through to adulthood, perhaps in support of this idea (Val, Jeays-Ward, & Swain, 2006).

10. SUMMARY

Our understanding of the control of adrenal development has increased over the last few years with the increasing use of animal models and lineage-tracing techniques, but there is still much left to be discovered. We have yet to definitively identify adrenocortical stem/progenitor cells and it is likely that these cell populations have important roles in adult homeostasis and disease, another important area for study. The constant remodeling of the zones of the adrenal cortex presumably requires the precise control of the growth of these stem/progenitor populations and their differentiation into the appropriate mature steroidogenic cells. We do not know whether these are principally in the capsule or in the subcapsular cortex or whether both populations can play a role. Nor do we know the relative contributions of Shh and Wnt/β-catenin signaling or how these may be controlled by other signals, including those from the RAAS and HPA axes. The mechanisms that control zonation, such as positional cues during centripetal migration, are also undetermined at present. Understanding these processes and the mechanisms controlling them will help to elucidate the causes of adrenal insufficiency and cancer, conditions that are not optimally managed at present. Ultimately, identification of the stem cells and understanding their origins and how the balance between self-renewal and differentiation is maintained may allow targeted cell replacement approaches to treat adrenal insufficiency disorders that require lifelong hormone replacement therapies. Identifying novel pathways to target for treatment of adrenocortical carcinoma is a particularly important therapeutic goal.

It is less than 20 years since the discovery that deletion of Sf-1 caused adrenal agenesis, a seminal publication that ushered in an era of investigation of the mechanisms controlling adrenocortical development and function. It can be hoped that at the current rate of progress, it will not be too long before these and other questions have been answered.

REFERENCES

Achermann, J. C., Ito, M., Hindmarsh, P. C., & Jameson, J. L. (1999). A mutation in the gene encoding steroidogenic factor-1 causes XY sex reversal and adrenal failure in humans. *Nature Genetics*, *22*(2), 125–126.

Achermann, J. C., Ozisik, G., Ito, M., Orun, U. A., Harmanci, K., Gurakan, B., et al. (2002). Gonadal determination and adrenal development are regulated by the orphan nuclear receptor steroidogenic factor-1, in a dose-dependent manner. *The Journal of Clinical Endocrinology and Metabolism*, *87*(4), 1829–1833.

Addison, T. (1855). *On the constitutional and local effects of disease of the supra-renal capsules*. Baltimore: Williams & Wilkins.

Anderson, D. J., Carnahan, J. F., Michelsohn, A., & Patterson, P. H. (1991). Antibody markers identify a common progenitor to sympathetic neurons and chromaffin cells in vivo and reveal the timing of commitment to neuronal differentiation in the sympathoadrenal lineage. *The Journal of Neuroscience: The Official Journal of the Society for Neuroscience*, *11*(11), 3507–3519.

Arboleda, V. A., Lee, H., Parnaik, R., Fleming, A., Banerjee, A., Ferraz-de-Souza, B., et al. (2012). Mutations in the PCNA-binding domain of CDKN1C cause IMAGe syndrome. *Nature Genetics*, *44*(7), 788–792.

Arensburg, J., Payne, A. H., & Orly, J. (1999). Expression of steroidogenic genes in maternal and extraembryonic cells during early pregnancy in mice. *Endocrinology*, *140*(11), 5220–5232.

Arnold, J. (1886). Ein Beitrag zur feineren Struktur und dem Chemismus der Nebennieren. *Virchows Archiv für pathologische Anatomie und Physiologie und für klinische Medizin*, *35*, 64–107.

Aumo, L., Rusten, M., Mellgren, G., Bakke, M., & Lewis, A. E. (2010). Functional roles of protein kinase A (PKA) and exchange protein directly activated by 3′,5′-cyclic adenosine 5′-monophosphate (cAMP) 2 (EPAC2) in cAMP-mediated actions in adrenocortical cells. *Endocrinology*, *151*(5), 2151–2161.

Azizan, E. A., Lam, B. Y., Newhouse, S. J., Zhou, J., Kuc, R. E., Clarke, J., et al. (2012). Microarray, qPCR, and KCNJ5 sequencing of aldosterone-producing adenomas reveal differences in genotype and phenotype between zona glomerulosa- and zona fasciculata-like tumors. *The Journal of Clinical Endocrinology and Metabolism*, *97*(5), E819–E829.

Babu, P. S., Bavers, D. L., Beuschlein, F., Shah, S., Jeffs, B., Jameson, J. L., et al. (2002). Interaction between Dax-1 and steroidogenic factor-1 in vivo: Increased adrenal responsiveness to ACTH in the absence of Dax-1. *Endocrinology*, *143*(2), 665–673.

Baird, P. N., Groves, N., Haber, D. A., Housman, D. E., & Cowell, J. K. (1992). Identification of mutations in the WT1 gene in tumours from patients with the WAGR syndrome. *Oncogene*, 7(11), 2141–2149.

Baker, D. D., & Baillif, R. N. (1939). Rôle of capsule in suprarenal regeneration studied with aid of colchicine. *Experimental Biology and Medicine*, *40*, 117–121.

Bamforth, S. D., Braganca, J., Eloranta, J. J., Murdoch, J. N., Marques, F. I., Kranc, K. R., et al. (2001). Cardiac malformations, adrenal agenesis, neural crest defects and exencephaly in mice lacking Cited2, a new Tfap2 co-activator. *Nature Genetics*, *29*(4), 469–474.

Bamforth, S. D., Braganca, J., Farthing, C. R., Schneider, J. E., Broadbent, C., Michell, A. C., et al. (2004). Cited2 controls left-right patterning and heart development through a Nodal-Pitx2c pathway. *Nature Genetics, 36*(11), 1189–1196.

Barbosa, A. S., Giacaglia, L. R., Martin, R. M., Mendonca, B. B., & Lin, C. J. (2004). Assessment of the role of transcript for GATA-4 as a marker of unfavorable outcome in human adrenocortical neoplasms. *BMC Endocrine Disorders, 4*(1), 3.

Basile, D. P., & Holzwarth, M. A. (1994). Basic fibroblast growth factor receptor in the rat adrenal cortex: Effects of suramin and unilateral adrenalectomy on receptor numbers. *Endocrinology, 134*(6), 2482–2489.

Bassett, M. H., Mayhew, B., Rehman, K., White, P. C., Mantero, F., Arnaldi, G., et al. (2005). Expression profiles for steroidogenic enzymes in adrenocortical disease. *The Journal of Clinical Endocrinology and Metabolism, 90*(9), 5446–5455.

Begeot, M., Dubois, M. P., & Dubois, P. M. (1977). Growth hormone and ACTH in the pituitary of normal and anencephalic human fetuses: Immunocytochemical evidence for hypothalamic influences during development. *Neuroendocrinology, 24*(3–4), 208–220.

Benirschke, K. (1956). Adrenals in anencephaly and hydrocephaly. *Obstetrics and Gynecology, 8*(4), 412–425.

Bernichtein, S., Alevizaki, M., & Huhtaniemi, I. (2008). Is the adrenal cortex a target for gonadotropins? *Trends in Endocrinology and Metabolism: TEM, 19*(7), 231–238.

Bernichtein, S., Petretto, E., Jamieson, S., Goel, A., Aitman, T. J., Mangion, J. M., et al. (2008). Adrenal gland tumorigenesis after gonadectomy in mice is a complex genetic trait driven by epistatic loci. *Endocrinology, 149*(2), 651–661.

Bertagna, X., Guignat, L., Groussin, L., & Bertherat, J. (2009). Cushing's disease. *Best Practice & Research. Clinical endocrinology & Metabolism, 23*(5), 607–623.

Bertherat, J., Groussin, L., Sandrini, F., Matyakhina, L., Bei, T., Stergiopoulos, S., et al. (2003). Molecular and functional analysis of PRKAR1A and its locus (17q22-24) in sporadic adrenocortical tumors: 17q losses, somatic mutations, and protein kinase A expression and activity. *Cancer Research, 63*(17), 5308–5319.

Bertholet, J. Y. (1980). Proliferative activity and cell migration in the adrenal cortex of fetal and neonatal rats: An autoradiographic study. *The Journal of Endocrinology, 87*(1), 1–9.

Berthon, A., Martinez, A., Bertherat, J., & Val, P. (2012). Wnt/beta-catenin signalling in adrenal physiology and tumour development. *Molecular and Cellular Endocrinology, 351*(1), 87–95.

Berthon, A., Sahut-Barnola, I., Lambert-Langlais, S., de Joussineau, C., Damon-Soubeyrand, C., Louiset, E., et al. (2010). Constitutive beta-catenin activation induces adrenal hyperplasia and promotes adrenal cancer development. *Human Molecular Genetics, 19*(8), 1561–1576.

Beuschlein, F., Boulkroun, S., Osswald, A., Wieland, T., Nielsen, H. N., Lichtenauer, U. D., et al. (2013). Somatic mutations in ATP1A1 and ATP2B3 lead to aldosterone-producing adenomas and secondary hypertension. *Nature Genetics, 45*(4), 440–444, 444e441-442.

Beuschlein, F., Looyenga, B. D., Bleasdale, S. E., Mutch, C., Bavers, D. L., Parlow, A. F., et al. (2003). Activin induces x-zone apoptosis that inhibits luteinizing hormone-dependent adrenocortical tumor formation in inhibin-deficient mice. *Molecular and Cellular Biology, 23*(11), 3951–3964.

Beuschlein, F., Mutch, C., Bavers, D. L., Ulrich-Lai, Y. M., Engeland, W. C., Keegan, C., et al. (2002). Steroidogenic factor-1 is essential for compensatory adrenal growth following unilateral adrenalectomy. *Endocrinology, 143*(8), 3122–3135.

Bhattacharya, S., Michels, C. L., Leung, M. K., Arany, Z. P., Kung, A. L., & Livingston, D. M. (1999). Functional role of p35srj, a novel p300/CBP binding protein, during transactivation by HIF-1. *Genes & Development, 13*(1), 64–75.

Bicknell, A. B., Lomthaisong, K., Woods, R. J., Hutchinson, E. G., Bennett, H. P., Gladwell, R. T., et al. (2001). Characterization of a serine protease that cleaves pro-gamma-melanotropin at the adrenal to stimulate growth. *Cell*, *105*(7), 903–912.

Bielinska, M., Genova, E., Boime, I., Parviainen, H., Kiiveri, S., Leppaluoto, J., et al. (2005). Gonadotropin-induced adrenocortical neoplasia in NU/J nude mice. *Endocrinology*, *146*(9), 3975–3984.

Bielinska, M., Kiiveri, S., Parviainen, H., Mannisto, S., Heikinheimo, M., & Wilson, D. B. (2006). Gonadectomy-induced adrenocortical neoplasia in the domestic ferret (Mustela putorius furo) and laboratory mouse. *Veterinary Pathology*, *43*(2), 97–117.

Bijlsma, M. F., Spek, C. A., Zivkovic, D., van de Water, S., Rezaee, F., & Peppelenbosch, M. P. (2006). Repression of smoothened by patched-dependent (pro-)vitamin D3 secretion. *PLoS Biology*, *4*(8), e232.

Bird, I. M., Hanley, N. A., Word, R. A., Mathis, J. M., McCarthy, J. L., Mason, J. I., et al. (1993). Human NCI-H295 adrenocortical carcinoma cells: A model for angiotensin-II-responsive aldosterone secretion. *Endocrinology*, *133*(4), 1555–1561.

Bitgood, M. J., & McMahon, A. P. (1995). Hedgehog and Bmp genes are coexpressed at many diverse sites of cell–cell interaction in the mouse embryo. *Developmental Biology*, *172*(1), 126–138.

Bland, M. L., Desclozeaux, M., & Ingraham, H. A. (2003). Tissue growth and remodeling of the embryonic and adult adrenal gland. *Annals of the New York Academy of Sciences*, *995*, 59–72.

Bland, M. L., Fowkes, R. C., & Ingraham, H. A. (2004). Differential requirement for steroidogenic factor-1 gene dosage in adrenal development versus endocrine function. *Molecular Endocrinology*, *18*(4), 941–952.

Bland, M. L., Jamieson, C. A., Akana, S. F., Bornstein, S. R., Eisenhofer, G., Dallman, M. F., et al. (2000). Haploinsufficiency of steroidogenic factor-1 in mice disrupts adrenal development leading to an impaired stress response. *Proceedings of the National Academy of Sciences of the United States of America*, *97*(26), 14488–14493.

Bocian-Sobkowska, J., Malendowicz, L. K., & Wozniak, W. (1997). Comparative stereological study on zonation and cellular composition of adrenal glands of normal and anencephalic human fetuses. I. Zonation of the gland. *Histology and Histopathology*, *12*(2), 311–317.

Bonnet, S., Gaujoux, S., Launay, P., Baudry, C., Chokri, I., Ragazzon, B., et al. (2011). Wnt/beta-catenin pathway activation in adrenocortical adenomas is frequently due to somatic CTNNB1-activating mutations, which are associated with larger and non-secreting tumors: A study in cortisol-secreting and -nonsecreting tumors. *The Journal of Clinical Endocrinology and Metabolism*, *96*(2), E419–E426.

Bose, J., Grotewold, L., & Ruther, U. (2002). Pallister-Hall syndrome phenotype in mice mutant for Gli3. *Human Molecular Genetics*, *11*(9), 1129–1135.

Bottcher, R. T., & Niehrs, C. (2005). Fibroblast growth factor signaling during early vertebrate development. *Endocrine Reviews*, *26*(1), 63–77.

Boulkroun, S., Beuschlein, F., Rossi, G. P., Golib-Dzib, J. F., Fischer, E., Amar, L., et al. (2012). Prevalence, clinical, and molecular correlates of KCNJ5 mutations in primary aldosteronism. *Hypertension*, *59*(3), 592–598.

Boulkroun, S., Golib Dzib, J. F., Samson-Couterie, B., Rosa, F. L., Rickard, A. J., Meatchi, T., et al. (2013). KCNJ5 mutations in aldosterone producing adenoma and relationship with adrenal cortex remodeling. *Molecular and Cellular Endocrinology*, *371*(1–2), 221–227.

Boulkroun, S., Samson-Couterie, B., Dzib, J. F., Lefebvre, H., Louiset, E., Amar, L., et al. (2010). Adrenal cortex remodeling and functional zona glomerulosa hyperplasia in primary aldosteronism. *Hypertension*, *56*(5), 885–892.

Boulkroun, S., Samson-Couterie, B., Golib-Dzib, J. F., Amar, L., Plouin, P. F., Sibony, M., et al. (2011). Aldosterone-producing adenoma formation in the adrenal cortex involves expression of stem/progenitor cell markers. *Endocrinology*, *152*(12), 4753–4763.

Boulle, N., Logie, A., Gicquel, C., Perin, L., & Le Bouc, Y. (1998). Increased levels of insulin-like growth factor II (IGF-II) and IGF-binding protein-2 are associated with malignancy in sporadic adrenocortical tumors. *The Journal of Clinical Endocrinology and Metabolism*, *83*(5), 1713–1720.

Braganca, J., Eloranta, J. J., Bamforth, S. D., Ibbitt, J. C., Hurst, H. C., & Bhattacharya, S. (2003). Physical and functional interactions among AP-2 transcription factors, p300/CREB-binding protein, and CITED2. *The Journal of Biological Chemistry*, *278*(18), 16021–16029.

Brennan, C. H., Chittka, A., Barker, S., & Vinson, G. P. (2008). Eph receptors and zonation in the rat adrenal cortex. *The Journal of Endocrinology*, *198*(1), 185–191.

Brown-Sequard, E. (1856). Recherches experimentales sur le physiologie et la pathologie des capsules surrénales. *Archives Generales de Medecine*, *8*, 572–598.

Carr, B. R., Ohashi, M., MacDonald, P. C., & Simpson, E. R. (1981). Human anencephalic adrenal tissue: Low density lipoprotein metabolism and cholesterol synthesis. *The Journal of Clinical Endocrinology and Metabolism*, *53*(2), 406–411.

Carr, B. R., & Simpson, E. R. (1981). Lipoprotein utilization and cholesterol synthesis by the human fetal adrenal gland. *Endocrine Reviews*, *2*(3), 306–326.

Carrasco, G. A., & Van de Kar, L. D. (2003). Neuroendocrine pharmacology of stress. *European Journal of Pharmacology*, *463*(1–3), 235–272.

Cater, D. B., & Lever, J. D. (1954). The zona intermedia of the adrenal cortex: A correlation of possible functional significance with development, morphology and histochemistry. *Journal of Anatomy*, *88*(4), 437–454.

Chatelain, D., Montel, V., Dickes-Coopman, A., Chatelain, A., & Deloof, S. (2003). Trophic and steroidogenic effects of water deprivation on the adrenal gland of the adult female rat. *Regulatory Peptides*, *110*(3), 249–255.

Chemaitilly, W., Goldenberg, A., Baujat, G., Thibaud, E., Cormier-Daire, V., & Abadie, V. (2003). Adrenal insufficiency and abnormal genitalia in a 46XX female with Smith-Lemli-Opitz syndrome. *Hormone Research*, *59*(5), 254–256 [Case Reports].

Chida, D., Nakagawa, S., Nagai, S., Sagara, H., Katsumata, H., Imaki, T., et al. (2007). Melanocortin 2 receptor is required for adrenal gland development, steroidogenesis, and neonatal gluconeogenesis. *Proceedings of the National Academy of Sciences of the United States of America*, *104*(46), 18205–18210.

Ching, S., & Vilain, E. (2009). Targeted disruption of Sonic Hedgehog in the mouse adrenal leads to adrenocortical hypoplasia. *Genesis*, *47*(9), 628–637.

Choi, M., Scholl, U. I., Yue, P., Bjorklund, P., Zhao, B., Nelson-Williams, C., et al. (2011). K+ channel mutations in adrenal aldosterone-producing adenomas and hereditary hypertension. *Science*, *331*(6018), 768–772.

Chrousos, G. P. (1995). The hypothalamic-pituitary-adrenal axis and immune-mediated inflammation. *The New England Journal of Medicine*, *332*(20), 1351–1362.

Chrusciel, M., Vuorenoja, S., Mohanty, B., Rivero-Muller, A., Li, X., Toppari, J., et al. (2013). Transgenic GATA-4 expression induces adrenocortical tumorigenesis in C57Bl/6 mice. *Journal of Cell Science*, *126*(Pt. 8), 1845–1857.

Chu, Y., Ho, W. J., & Dunn, J. C. (2009). Basic fibroblast growth factor delivery enhances adrenal cortical cellular regeneration. *Tissue Engineering. Part A*, *15*(8), 2093–2101.

Chuang, C. H., Wallace, M. D., Abratte, C., Southard, T., & Schimenti, J. C. (2010). Incremental genetic perturbations to MCM2-7 expression and subcellular distribution reveal exquisite sensitivity of mice to DNA replication stress. *PLoS Genetics*, *6*(9), e1001110.

Chung, T. T., Chan, L. F., Metherell, L. A., & Clark, A. J. (2010). Phenotypic characteristics of familial glucocorticoid deficiency (FGD) type 1 and 2. *Clinical Endocrinology, 72*(5), 589–594.

Claahsen-van der Grinten, H. L., Otten, B. J., Stikkelbroeck, M. M., Sweep, F. C., & Hermus, A. R. (2009). Testicular adrenal rest tumours in congenital adrenal hyperplasia. *Best practice & Research. Clinical Endocrinology & Metabolism, 23*(2), 209–220.

Claahsen-van der Grinten, H. L., Stikkelbroeck, N. M., Sweep, C. G., Hermus, A. R., & Otten, B. J. (2006). Fertility in patients with congenital adrenal hyperplasia. *Journal of Pediatric Endocrinology & Metabolism: JPEM, 19*(5), 677–685.

Clark, A. J., & Weber, A. (1998). Adrenocorticotropin insensitivity syndromes. *Endocrine Reviews, 19*(6), 828–843.

Clipsham, R., & McCabe, E. R. (2003). DAX1 and its network partners: Exploring complexity in development. *Molecular Genetics and Metabolism, 80*(1–2), 81–120.

Coll, A. P., Challis, B. G., Yeo, G. S., Snell, K., Piper, S. J., Halsall, D., et al. (2004). The effects of proopiomelanocortin deficiency on murine adrenal development and responsiveness to adrenocorticotropin. *Endocrinology, 145*(10), 4721–4727.

Coll, A. P., Fassnacht, M., Klammer, S., Hahner, S., Schulte, D. M., Piper, S., et al. (2006). Peripheral administration of the N-terminal pro-opiomelanocortin fragment 1–28 to $Pomc^{-/-}$ mice reduces food intake and weight but does not affect adrenal growth or corticosterone production. *The Journal of Endocrinology, 190*(2), 515–525.

Collier, J. G., Robinson, B. F., & Vane, J. R. (1973). Reduction of pressor effects of angiotensin I in man by synthetic nonapeptide (B.P.P. 9a or SQ 20,881) which inhibits converting enzyme. *Lancet, 1*(7794), 72–74.

Combes, A. N., Spiller, C. M., Harley, V. R., Sinclair, A. H., Dunwoodie, S. L., Wilhelm, D., et al. (2010). Gonadal defects in Cited2-mutant mice indicate a role for SF1 in both testis and ovary differentiation. *The International Journal of Developmental Biology, 54*(4), 683–689.

Coulter, C. L., & Jaffe, R. B. (1998). Functional maturation of the primate fetal adrenal in vivo: 3. Specific zonal localization and developmental regulation of CYP21A2 (P450c21) and CYP11B1/CYP11B2 (P450c11/aldosterone synthase) lead to integrated concept of zonal and temporal steroid biosynthesis. *Endocrinology, 139*(12), 5144–5150.

Crawford, P. A., Sadovsky, Y., & Milbrandt, J. (1997). Nuclear receptor steroidogenic factor 1 directs embryonic stem cells toward the steroidogenic lineage. *Molecular and Cellular Biology, 17*(7), 3997–4006.

Crickard, K., Ill, C. R., & Jaffe, R. B. (1981). Control of proliferation of human fetal adrenal cells in vitro. *The Journal of Clinical Endocrinology and Metabolism, 53*(4), 790–796.

Cronshaw, J. M., & Matunis, M. J. (2003). The nuclear pore complex protein ALADIN is mislocalized in triple A syndrome. *Proceedings of the National Academy of Sciences of the United States of America, 100*(10), 5823–5827.

Cross, D. A., Alessi, D. R., Cohen, P., Andjelkovich, M., & Hemmings, B. A. (1995). Inhibition of glycogen synthase kinase-3 by insulin mediated by protein kinase B. *Nature, 378*(6559), 785–789.

Crowder, R. E. (1957). The development of the adrenal gland in man, with special reference to origin and ultimate location of cell types and evidence in favor of the "cell migration" theory. *Contributions to Embryology, 36*, 193–210.

Cui, S., Ross, A., Stallings, N., Parker, K. L., Capel, B., & Quaggin, S. E. (2004). Disrupted gonadogenesis and male-to-female sex reversal in Pod1 knockout mice. *Development, 131*(16), 4095–4105.

Dallman, M. F. (1984). Control of adrenocortical growth in vivo. *Endocrine Research, 10*(3–4), 213–242.

Dallman, M. F., Engeland, W. C., Holzwarth, M. A., & Scholz, P. M. (1980). Adrenocorticotropin inhibits compensatory adrenal growth after unilateral adrenalectomy. *Endocrinology, 107*(5), 1397–1404.
Dallman, M. F., Engeland, W. C., & Shinsako, J. (1976). Compensatory adrenal growth: A neurally mediated reflex. *The American Journal of Physiology, 231*(2), 408–414.
de Kloet, E. R. (1995). Steroids, stability and stress. *Frontiers in Neuroendocrinology, 16*(4), 416–425.
Dobbie, J. W., & Symington, T. (1966). The human adrenal gland with special reference to the vasculature. *The Journal of Endocrinology, 34*(4), 479–489.
Doghman, M., Karpova, T., Rodrigues, G. A., Arhatte, M., De Moura, J., Cavalli, L. R., et al. (2007). Increased steroidogenic factor-1 dosage triggers adrenocortical cell proliferation and cancer. *Molecular Endocrinology, 21*(12), 2968–2987.
Domoto, D. T., Boyd, J. E., Mulrow, P. J., & Kashgarian, M. (1973). The ultrastructure of the adrenal zona glomerulosa of rats on patassium-supplemented or sodium-depleted diets. *The American Journal of Pathology, 72*(3), 433–446.
Dorey, K., & Amaya, E. (2010). FGF signalling: Diverse roles during early vertebrate embryogenesis. *Development, 137*(22), 3731–3742.
Drelon, C., Berthon, A., Ragazzon, B., Tissier, F., Bandiera, R., Sahut-Barnola, I., et al. (2012). Analysis of the role of Igf2 in adrenal tumour development in transgenic mouse models. *PLoS one*, 7(8), e44171.
Dubourg, C., Bendavid, C., Pasquier, L., Henry, C., Odent, S., & David, V. (2007). Holoprosencephaly. *Orphanet Journal of Rare Diseases, 2*, 8.
Durkee, T. J., McLean, M. P., Hales, D. B., Payne, A. H., Waterman, M. R., Khan, I., et al. (1992). P450(17 alpha) and P450SCC gene expression and regulation in the rat placenta. *Endocrinology, 130*(3), 1309–1317.
Easterling, W. E., Jr., Simmer, H. H., Dignam, W. J., Frankland, M. V., & Naftolin, F. (1966). Neutral C19-steroids and steroid sulfates in human pregnancy. II. Dehydroepiandrosterone sulfate, 16-alpha-hydroxydehydroepiandrosterone, and 16-alpha-hydroxydehydroepiandrosterone sulfate in maternal and fetal blood of pregnancies with anencephalic and normal fetuses. *Steroids, 8*(2), 157–178.
Ehrhart-Bornstein, M., Breidert, M., Guadanucci, P., Wozniak, W., Bocian-Sobkowska, J., Malendowicz, L. K., et al. (1997). 17 alpha-Hydroxylase and chromogranin A in 6th week human fetal adrenals. *Hormone and Metabolic Research, 29*(1), 30–32.
El-Khairi, R., & Achermann, J. C. (2012). Steroidogenic factor-1 and human disease. *Seminars in Reproductive Medicine, 30*(5), 374–381.
El-Khairi, R., Martinez-Aguayo, A., Ferraz-de-Souza, B., Lin, L., & Achermann, J. C. (2011). Role of DAX-1 (NR0B1) and steroidogenic factor-1 (NR5A1) in human adrenal function. *Endocrine Development, 20*, 38–46.
Elliott, M., Bayly, R., Cole, T., Temple, I. K., & Maher, E. R. (1994). Clinical features and natural history of Beckwith-Wiedemann syndrome: Presentation of 74 new cases. *Clinical Genetics, 46*(2), 168–174.
Else, T., & Hammer, G. D. (2005). Genetic analysis of adrenal absence: Agenesis and aplasia. *Trends in Endocrinology and Metabolism: TEM, 16*(10), 458–468.
Else, T., Trovato, A., Kim, A. C., Wu, Y., Ferguson, D. O., Kuick, R. D., et al. (2009). Genetic p53 deficiency partially rescues the adrenocortical dysplasia phenotype at the expense of increased tumorigenesis. *Cancer Cell, 15*(6), 465–476.
El Wakil, A., & Lalli, E. (2011). The Wnt/beta-catenin pathway in adrenocortical development and cancer. *Molecular and Cellular Endocrinology, 332*(1–2), 32–37.
Endoh, A., Kristiansen, S. B., Casson, P. R., Buster, J. E., & Hornsby, P. J. (1996). The zona reticularis is the site of biosynthesis of dehydroepiandrosterone and dehydroepiandrosterone sulfate in the adult human adrenal cortex resulting from its low expression of 3

beta-hydroxysteroid dehydrogenase. *The Journal of Clinical Endocrinology and Metabolism, 81*(10), 3558–3565.
Engeland, W. C., Ennen, W. B., Elayaperumal, A., Durand, D. A., & Levay-Young, B. K. (2005). Zone-specific cell proliferation during compensatory adrenal growth in rats. *American Journal of Physiology. Endocrinology and Metabolism, 288*(2), E298–E306.
Engeland, W. C., & Levay-Young, B. K. (1999). Changes in the glomerulosa cell phenotype during adrenal regeneration in rats. *The American Journal of Physiology, 276*(5 Pt. 2), R1374–R1382.
Engeland, W. C., Shinsako, J., & Dallman, M. F. (1975). Corticosteroids and ACTH are not required for compensatory adrenal growth. *The American Journal of Physiology, 229*(5), 1461–1464.
Ennen, W. B., Levay-Young, B. K., & Engeland, W. C. (2005). Zone-specific cell proliferation during adrenocortical regeneration after enucleation in rats. *American Journal of Physiology. Endocrinology and Metabolism, 289*(5), E883–E891.
Ernsberger, U., Esposito, L., Partimo, S., Huber, K., Franke, A., Bixby, J. L., et al. (2005). Expression of neuronal markers suggests heterogeneity of chick sympathoadrenal cells prior to invasion of the adrenal anlagen. *Cell and Tissue Research, 319*(1), 1–13.
Essafi, A., Webb, A., Berry, R. L., Slight, J., Burn, S. F., Spraggon, L., et al. (2011). A wt1-controlled chromatin switching mechanism underpins tissue-specific wnt4 activation and repression. *Developmental Cell, 21*(3), 559–574.
Estivariz, F. E., Carino, M., Lowry, P. J., & Jackson, S. (1988). Further evidence that N-terminal pro-opiomelanocortin peptides are involved in adrenal mitogenesis. *The Journal of Endocrinology, 116*(2), 201–206.
Estivariz, F. E., Iturriza, F., McLean, C., Hope, J., & Lowry, P. J. (1982). Stimulation of adrenal mitogenesis by N-terminal proopiocortin peptides. *Nature, 297*(5865), 419–422.
Estivariz, F. E., Morano, M. I., Carino, M., Jackson, S., & Lowry, P. J. (1988). Adrenal regeneration in the rat is mediated by mitogenic N-terminal pro-opiomelanocortin peptides generated by changes in precursor processing in the anterior pituitary. *The Journal of Endocrinology, 116*(2), 207–216.
Eustachio, B. (1563–1564). Opuscula anatomicae. Venice: V. Lucchino.
Fang, D., Hawke, D., Zheng, Y., Xia, Y., Meisenhelder, J., Nika, H., et al. (2007). Phosphorylation of beta-catenin by AKT promotes beta-catenin transcriptional activity. *The Journal of Biological Chemistry, 282*(15), 11221–11229.
Fang, X., Yu, S. X., Lu, Y., Bast, R. C., Jr., Woodgett, J. R., & Mills, G. B. (2000). Phosphorylation and inactivation of glycogen synthase kinase 3 by protein kinase A. *Proceedings of the National Academy of Sciences of the United States of America, 97*(22), 11960–11965.
Ferraz-de-Souza, B., & Achermann, J. C. (2008). Disorders of adrenal development. *Endocrine Development, 13*, 19–32.
Ferraz-de-Souza, B., Lin, L., & Achermann, J. C. (2011). Steroidogenic factor-1 (SF-1, NR5A1) and human disease. *Molecular and Cellular Endocrinology, 336*(1–2), 198–205.
Ferraz-de-Souza, B., Martin, F., Mallet, D., Hudson-Davies, R. E., Cogram, P., Lin, L., et al. (2009). CBP/p300-interacting transactivator, with Glu/Asp-rich C-terminal domain, 2, and pre-B-cell leukemia transcription factor 1 in human adrenal development and disease. *The Journal of Clinical Endocrinology and Metabolism, 94*(2), 678–683.
Feuerstein, B., & Streeten, D. H. (1991). Recovery of adrenal function after failure resulting from traumatic bilateral adrenal hemorrhages. *Annals of Internal Medicine, 115*(10), 785–786.
Fidler, W. J. (1977). Ovarian thecal metaplasia in adrenal glands. *American Journal of Clinical Pathology, 67*(4), 318–323.
Figueiredo, B. C., Stratakis, C. A., Sandrini, R., DeLacerda, L., Pianovsky, M. A., Giatzakis, C., et al. (1999). Comparative genomic hybridization analysis of adrenocortical tumors of childhood. *The Journal of Clinical Endocrinology and Metabolism, 84*(3), 1116–1121.

Finotto, S., Krieglstein, K., Schober, A., Deimling, F., Lindner, K., Bruhl, B., et al. (1999). Analysis of mice carrying targeted mutations of the glucocorticoid receptor gene argues against an essential role of glucocorticoid signalling for generating adrenal chromaffin cells. *Development, 126*(13), 2935–2944.

Ford, J. K., & Young, R. W. (1963). Cell proliferation and displacement in the adrenal cortex of young rats injected with tritiated thymidine. *The Anatomical Record, 146*, 125–137.

Franchimont, D. (2004). Overview of the actions of glucocorticoids on the immune response: A good model to characterize new pathways of immunosuppression for new treatment strategies. *Annals of the New York Academy of Sciences, 1024*, 124–137.

Fukami, M. H., Haavik, J., & Flatmark, T. (1990). Phenylalanine as substrate for tyrosine hydroxylase in bovine adrenal chromaffin cells. *The Biochemical Journal, 268*(2), 525–528.

Fuller, P. J., & Young, M. J. (2005). Mechanisms of mineralocorticoid action. *Hypertension, 46*(6), 1227–1235.

Gardiner, J. R., Shima, Y., Morohashi, K., & Swain, A. (2012). SF-1 expression during adrenal development and tumourigenesis. *Molecular and Cellular Endocrinology, 351*(1), 12–18.

Gaujoux, S., Grabar, S., Fassnacht, M., Ragazzon, B., Launay, P., Libe, R., et al. (2011). beta-Catenin activation is associated with specific clinical and pathologic characteristics and a poor outcome in adrenocortical carcinoma. *Clinical Cancer Research: An Official Journal of the American Association for Cancer Research, 17*(2), 328–336.

Gaujoux, S., Pinson, S., Gimenez-Roqueplo, A. P., Amar, L., Ragazzon, B., Launay, P., et al. (2010). Inactivation of the APC gene is constant in adrenocortical tumors from patients with familial adenomatous polyposis but not frequent in sporadic adrenocortical cancers. *Clinical Cancer Research: An Official Journal of the American Association for Cancer Research, 16*(21), 5133–5141.

Gaujoux, S., Tissier, F., Groussin, L., Libe, R., Ragazzon, B., Launay, P., et al. (2008). Wnt/beta-catenin and 3′,5′-cyclic adenosine 5′-monophosphate/protein kinase A signaling pathways alterations and somatic beta-catenin gene mutations in the progression of adrenocortical tumors. *The Journal of Clinical Endocrinology and Metabolism, 93*(10), 4135–4140.

Gicquel, C., Bertagna, X., Gaston, V., Coste, J., Louvel, A., Baudin, E., et al. (2001). Molecular markers and long-term recurrences in a large cohort of patients with sporadic adrenocortical tumors. *Cancer Research, 61*(18), 6762–6767.

Gineau, L., Cognet, C., Kara, N., Lach, F. P., Dunne, J., Veturi, U., et al. (2012). Partial MCM4 deficiency in patients with growth retardation, adrenal insufficiency, and natural killer cell deficiency. *The Journal of Clinical Investigation, 122*(3), 821–832.

Glenn, D. J., & Maurer, R. A. (1999). MRG1 binds to the LIM domain of Lhx2 and may function as a coactivator to stimulate glycoprotein hormone alpha-subunit gene expression. *The Journal of Biological Chemistry, 274*(51), 36159–36167.

Gondo, S., Yanase, T., Okabe, T., Tanaka, T., Morinaga, H., Nomura, M., et al. (2004). SF-1/Ad4BP transforms primary long-term cultured bone marrow cells into ACTH-responsive steroidogenic cells. *Genes to Cells: Devoted to Molecular & Cellular Mechanisms, 9*(12), 1239–1247.

Gorrigan, R. J., Guasti, L., King, P., Clark, A. J., & Chan, L. F. (2011). Localisation of the melanocortin-2-receptor and its accessory proteins in the developing and adult adrenal gland. *Journal of Molecular Endocrinology, 46*(3), 227–232.

Gospodarowicz, D., Ill, C. R., Hornsby, P. J., & Gill, G. N. (1977). Control of bovine adrenal cortical cell proliferation by fibroblast growth factor. Lack of effect of epidermal growth factor. *Endocrinology, 100*(4), 1080–1089.

Goto, M., Piper Hanley, K., Marcos, J., Wood, P. J., Wright, S., Postle, A. D., et al. (2006). In humans, early cortisol biosynthesis provides a mechanism to safeguard female sexual development. *The Journal of Clinical Investigation, 116*(4), 953–960.

Gottschau, M. (1883). Struktur und embryonale Entwicklung der Nebennieren bei Saugethieren. *Archiv fur Anatomie und Physiologie, 9*, 412–458.

Gray, E. S., & Abramovich, D. R. (1980). Morphologic features of the anencephalic adrenal gland in early pregnancy. *American Journal of Obstetrics and Gynecology, 137*(4), 491–495.

Grizzle, W. E., & Dunlap, N. E. (1984). Aldosterone blocks adrenal compensatory hypertrophy in the rat. *The American Journal of Physiology, 246*(4 Pt. 1), E306–E310.

Grumbach, M. M., Biller, B. M., Braunstein, G. D., Campbell, K. K., Carney, J. A., Godley, P. A., et al. (2003). Management of the clinically inapparent adrenal mass ("incidentaloma"). *Annals of Internal Medicine, 138*(5), 424–429.

Guasti, L., Paul, A., Laufer, E., & King, P. (2011). Localization of Sonic hedgehog secreting and receiving cells in the developing and adult rat adrenal cortex. *Molecular and Cellular Endocrinology, 336*(1–2), 117–122.

Guasti, L., Sze, W. C., McKay, T., Grose, R., & King, P. J. (2013). FGF signalling through Fgfr2 isoform IIIb regulates adrenal cortex development. *Molecular and Cellular Endocrinology, 371*(1–2), 182–188.

Gummow, B. M., Scheys, J. O., Cancelli, V. R., & Hammer, G. D. (2006). Reciprocal regulation of a glucocorticoid receptor-steroidogenic factor-1 transcription complex on the Dax-1 promoter by glucocorticoids and adrenocorticotropic hormone in the adrenal cortex. *Molecular Endocrinology, 20*(11), 2711–2723.

Gummow, B. M., Winnay, J. N., & Hammer, G. D. (2003). Convergence of Wnt signaling and steroidogenic factor-1 (SF-1) on transcription of the rat inhibin alpha gene. *The Journal of Biological Chemistry, 278*(29), 26572–26579.

Haber, D. A., & Housman, D. E. (1992). The genetics of Wilms' tumor. *Advances in Cancer Research, 59*, 41–68.

Halder, S. K., Takemori, H., Hatano, O., Nonaka, Y., Wada, A., & Okamoto, M. (1998). Cloning of a membrane-spanning protein with epidermal growth factor-like repeat motifs from adrenal glomerulosa cells. *Endocrinology, 139*(7), 3316–3328.

Hall, J. G., Pallister, P. D., Clarren, S. K., Beckwith, J. B., Wiglesworth, F. W., Fraser, F. C., et al. (1980). Congenital hypothalamic hamartoblastoma, hypopituitarism, imperforate anus and postaxial polydactyly—A new syndrome? Part I: Clinical, causal, and pathogenetic considerations. *American Journal of Medical Genetics*, 7(1), 47–74.

Handschug, K., Sperling, S., Yoon, S. J., Hennig, S., Clark, A. J., & Huebner, A. (2001). Triple A syndrome is caused by mutations in AAAS, a new WD-repeat protein gene. *Human Molecular Genetics, 10*(3), 283–290.

Hanley, N. A., Ball, S. G., Clement-Jones, M., Hagan, D. M., Strachan, T., Lindsay, S., et al. (1999). Expression of steroidogenic factor 1 and Wilms' tumour 1 during early human gonadal development and sex determination. *Mechanisms of Development, 87*(1–2), 175–180.

Hanley, N. A., Rainey, W. E., Wilson, D. I., Ball, S. G., & Parker, K. L. (2001). Expression profiles of SF-1, DAX1, and CYP17 in the human fetal adrenal gland: Potential interactions in gene regulation. *Molecular Endocrinology, 15*(1), 57–68.

Hatano, O., Takakusu, A., Nomura, M., & Morohashi, K. (1996). Identical origin of adrenal cortex and gonad revealed by expression profiles of Ad4BP/SF-1. *Genes to Cells: Devoted to Molecular & Cellular Mechanisms, 1*(7), 663–671.

Hatini, V., Huh, S. O., Herzlinger, D., Soares, V. C., & Lai, E. (1996). Essential role of stromal mesenchyme in kidney morphogenesis revealed by targeted disruption of Winged Helix transcription factor BF-2. *Genes & Development, 10*(12), 1467–1478.

Heaton, J. H., Wood, M. A., Kim, A. C., Lima, L. O., Barlaskar, F. M., Almeida, M. Q., et al. (2012). Progression to adrenocortical tumorigenesis in mice and humans through insulin-like growth factor 2 and beta-catenin. *The American Journal of Pathology, 181*(3), 1017–1033.

Heikkila, M., Peltoketo, H., Leppaluoto, J., Ilves, M., Vuolteenaho, O., & Vainio, S. (2002). Wnt-4 deficiency alters mouse adrenal cortex function, reducing aldosterone production. *Endocrinology, 143*(11), 4358–4365.

Hino, S., Tanji, C., Nakayama, K. I., & Kikuchi, A. (2005). Phosphorylation of beta-catenin by cyclic AMP-dependent protein kinase stabilizes beta-catenin through inhibition of its ubiquitination. *Molecular and Cellular Biology, 25*(20), 9063–9072.

Hirano, M., Furiya, Y., Asai, H., Yasui, A., & Ueno, S. (2006). ALADINI482S causes selective failure of nuclear protein import and hypersensitivity to oxidative stress in triple A syndrome. *Proceedings of the National Academy of Sciences of the United States of America, 103*(7), 2298–2303.

Honnebier, W. J., & Swaab, D. F. (1973). The influence of anencephaly upon intrauterine growth of fetus and placenta and upon gestation length. *The Journal of Obstetrics and Gynaecology of the British Commonwealth, 80*(7), 577–588.

Hornsby, P. J. (1989). Steroid and xenobiotic effects on the adrenal cortex: Mediation by oxidative and other mechanisms. *Free Radical Biology & Medicine, 6*(1), 103–115.

Hornsby, P. J., & Crivello, J. F. (1983). The role of lipid peroxidation and biological antioxidants in the function of the adrenal cortex. Part 2. *Molecular and Cellular Endocrinology, 30*(2), 123–147.

Hu, J., Zhang, Z., Shen, W. J., & Azhar, S. (2010). Cellular cholesterol delivery, intracellular processing and utilization for biosynthesis of steroid hormones. *Nutrition & Metabolism,* 7, 47.

Huang, C. C., Miyagawa, S., Matsumaru, D., Parker, K. L., & Yao, H. H. (2010). Progenitor cell expansion and organ size of mouse adrenal is regulated by sonic hedgehog. *Endocrinology, 151*(3), 1119–1128.

Huebner, A., Yoon, S. J., Ozkinay, F., Hilscher, C., Lee, H., Clark, A. J., et al. (2000). Triple A syndrome—Clinical aspects and molecular genetics. *Endocrine Research, 26*(4), 751–759.

Hughes, C. R., Chung, T. T., Habeb, A. M., Kelestimur, F., Clark, A. J., & Metherell, L. A. (2010). Missense mutations in the melanocortin 2 receptor accessory protein that lead to late onset familial glucocorticoid deficiency type 2. *The Journal of Clinical Endocrinology and Metabolism, 95*(7), 3497–3501.

Hughes, C. R., Guasti, L., Meimaridou, E., Chuang, C. H., Schimenti, J. C., King, P. J., et al. (2012). MCM4 mutation causes adrenal failure, short stature, and natural killer cell deficiency in humans. *The Journal of Clinical Investigation, 122*(3), 814–820.

Hui, X. G., Akahira, J., Suzuki, T., Nio, M., Nakamura, Y., Suzuki, H., et al. (2009). Development of the human adrenal zona reticularis: Morphometric and immunohistochemical studies from birth to adolescence. *The Journal of Endocrinology, 203*(2), 241–252.

Iannaccone, P., Morley, S., Skimina, T., Mullins, J., & Landini, G. (2003). Cord-like mosaic patches in the adrenal cortex are fractal: Implications for growth and development. *FASEB Journal: Official Publication of the Federation of American Societies for Experimental Biology, 17*(1), 41–43.

Ikeda, Y., Shen, W. H., Ingraham, H. A., & Parker, K. L. (1994). Developmental expression of mouse steroidogenic factor-1, an essential regulator of the steroid hydroxylases. *Molecular Endocrinology, 8*(5), 654–662.

Ikeda, Y., Swain, A., Weber, T. J., Hentges, K. E., Zanaria, E., Lalli, E., et al. (1996). Steroidogenic factor 1 and Dax-1 colocalize in multiple cell lineages: Potential links in endocrine development. *Molecular Endocrinology, 10*(10), 1261–1272.

Ikeda, Y., Takeda, Y., Shikayama, T., Mukai, T., Hisano, S., & Morohashi, K. I. (2001). Comparative localization of Dax-1 and Ad4BP/SF-1 during development of the hypothalamic-pituitary-gonadal axis suggests their closely related and distinct functions. *Developmental Dynamics: An Official Publication of the American Association of Anatomists, 220*(4), 363–376.

Ingle, D. J., & Higgins, J. R. (1939). The extent of regeneration of the enucleated adrenal gland in the rat as influenced by the amount of capsule left at operation. *Endocrinology, 24*, 379–382.

Ishimoto, H., & Jaffe, R. B. (2011). Development and function of the human fetal adrenal cortex: A key component in the feto-placental unit. *Endocrine Reviews, 32*(3), 317–355.

Itaranta, P., Viiri, K., Kaartinen, V., & Vainio, S. (2009). Lumbo-sacral neural crest derivatives fate mapped with the aid of Wnt-1 promoter integrate but are not essential to kidney development. *Differentiation; Research in Biological Diversity, 77*(2), 199–208.

Iyer, A. K., & McCabe, E. R. (2004). Molecular mechanisms of DAX1 action. *Molecular Genetics and Metabolism, 83*(1–2), 60–73.

Jacob, K., Robinson, W., & Lefebvre, L. (2013). Beckwith-Wiedemann and Silver-Russell syndromes: Opposite developmental imbalances in imprinted regulators of placental function and embryonic growth. *Clinical Genetics, 84*(4), 326–334.

Jeays-Ward, K., Dandonneau, M., & Swain, A. (2004). Wnt4 is required for proper male as well as female sexual development. *Developmental Biology, 276*(2), 431–440.

Jirasek, J. E. (1980). *Human fetal endocrines.* (pp. 69–82). London: Martinus Nijhoff.

Jirasek, J. E., & Lojda, Z. (1964). A Histochemical contribution to the development of the adrenal cortex in human embryos and fetuses. *Acta Histochemica, 18*, 65–67.

Johnson, J. A., & Davis, J. O. (1973). Angiotensin. II. Important role in the maintenance of arterial blood pressure. *Science, 179*(4076), 906–907.

Jones, I. C. (1948). Variation in the mouse adrenal cortex with special reference to the zona reticularis and to brown degeneration, together with a discussion of the cell migration theory. *The Quarterly Journal of Microscopical Science, 89*(Pt. 1), 53–74.

Jozan, S., Aziza, J., Chatelin, S., Evra, C., Courtade-Saidi, M., Parant, O., et al. (2007). Human fetal chromaffin cells: A potential tool for cell pain therapy. *Experimental Neurology, 205*(2), 525–535.

Kananen, K., Markkula, M., Mikola, M., Rainio, E. M., McNeilly, A., & Huhtaniemi, I. (1996). Gonadectomy permits adrenocortical tumorigenesis in mice transgenic for the mouse inhibin alpha-subunit promoter/simian virus 40T-antigen fusion gene: Evidence for negative autoregulation of the inhibin alpha-subunit gene. *Molecular Endocrinology, 10*(12), 1667.

Kananen, K., Rilianawati, Paukku, T., Markkula, M., Rainio, E. M., & Huhtaniemi, I. (1997). Suppression of gonadotropins inhibits gonadal tumorigenesis in mice transgenic for the mouse inhibin alpha-subunit promoter/simian virus 40 T-antigen fusion gene. *Endocrinology, 138*(8), 3521–3531.

Karpac, J., Ostwald, D., Bui, S., Hunnewell, P., Shankar, M., & Hochgeschwender, U. (2005). Development, maintenance, and function of the adrenal gland in early postnatal proopiomelanocortin-null mutant mice. *Endocrinology, 146*(6), 2555–2562.

Kataoka, Y., Ikehara, Y., & Hattori, T. (1996). Cell proliferation and renewal of mouse adrenal cortex. *Journal of Anatomy, 188*(Pt. 2), 375–381.

Katoh-Fukui, Y., Owaki, A., Toyama, Y., Kusaka, M., Shinohara, Y., Maekawa, M., et al. (2005). Mouse Polycomb M33 is required for splenic vascular and adrenal gland formation through regulating Ad4BP/SF1 expression. *Blood, 106*(5), 1612–1620.

Katoh-Fukui, Y., Tsuchiya, R., Shiroishi, T., Nakahara, Y., Hashimoto, N., Noguchi, K., et al. (1998). Male-to-female sex reversal in M33 mutant mice. *Nature, 393*(6686), 688–692.

Keene, M. F. (1927). Observations on the development of the human suprarenal gland. *Journal of Anatomy, 61*(Pt. 3), 302–324.

Kelley, R. I., & Hennekam, R. C. (2000). The Smith-Lemli-Opitz syndrome. *Journal of Medical Genetics, 37*(5), 321–335.

Kiefer, S. M., Ohlemiller, K. K., Yang, J., McDill, B. W., Kohlhase, J., & Rauchman, M. (2003). Expression of a truncated Sall1 transcriptional repressor is responsible for Townes-Brocks syndrome birth defects. *Human Molecular Genetics*, *12*(17), 2221–2227.

Kiiveri, S., Liu, J., Arola, J., Heikkila, P., Kuulasmaa, T., Lehtonen, E., et al. (2005). Transcription factors GATA-6, SF-1, and cell proliferation in human adrenocortical tumors. *Molecular and Cellular Endocrinology*, *233*(1–2), 47–56.

Kim, A. C., Barlaskar, F. M., Heaton, J. H., Else, T., Kelly, V. R., Krill, K. T., et al. (2009). In search of adrenocortical stem and progenitor cells. *Endocrine Reviews*, *30*(3), 241–263.

Kim, Y., Bingham, N., Sekido, R., Parker, K. L., Lovell-Badge, R., & Capel, B. (2007). Fibroblast growth factor receptor 2 regulates proliferation and Sertoli differentiation during male sex determination. *Proceedings of the National Academy of Sciences of the United States of America*, *104*(42), 16558–16563.

Kim, A. C., & Hammer, G. D. (2007). Adrenocortical cells with stem/progenitor cell properties: Recent advances. *Molecular and Cellular Endocrinology*, *265–266*, 10–16.

Kim, J. S., Kubota, H., Kiuchi, Y., Doi, K., & Saegusa, J. (1997). Subcapsular cell hyperplasia and mast cell infiltration in the adrenal cortex of mice: Comparative study in 7 inbred strains. *Experimental Animals/Japanese Association for Laboratory Animal Science*, *46*(4), 303–306.

Kim, A. C., Reuter, A. L., Zubair, M., Else, T., Serecky, K., Bingham, N. C., et al. (2008). Targeted disruption of beta-catenin in Sf1-expressing cells impairs development and maintenance of the adrenal cortex. *Development*, *135*(15), 2593–2602.

King, S. R., Bhangoo, A., & Stocco, D. M. (2011). Functional and physiological consequences of StAR deficiency: Role in lipoid congenital adrenal hyperplasia. *Endocrine Development*, *20*, 47–53.

King, P. J., Guasti, L., & Laufer, E. (2008). Hedgehog signalling in endocrine development and disease. *The Journal of Endocrinology*, *198*(3), 439–450.

King, P., Paul, A., & Laufer, E. (2009). Shh signaling regulates adrenocortical development and identifies progenitors of steroidogenic lineages. *Proceedings of the National Academy of Sciences of the United States of America*, *106*(50), 21185–21190.

Kirschner, L. S., Carney, J. A., Pack, S. D., Taymans, S. E., Giatzakis, C., Cho, Y. S., et al. (2000). Mutations of the gene encoding the protein kinase A type I-alpha regulatory subunit in patients with the Carney complex. *Nature Genetics*, *26*(1), 89–92.

Kleitman, N., & Holzwarth, M. A. (1985). Compensatory adrenal cortical growth is inhibited by sympathectomy. *The American Journal of Physiology*, *248*(2 Pt. 1), E261–E263.

Kobayashi, H., Kambe, F., Imai, T., Hibi, Y., Kikumori, T., Ohmori, S., et al. (2006). Differential expression of cyclin-dependent kinase inhibitors, p27Kip1 and p57Kip2, by corticotropin in rat adrenal cortex. *The Journal of Endocrinology*, *189*(3), 671–679.

Kohlhase, J., Wischermann, A., Reichenbach, H., Froster, U., & Engel, W. (1998). Mutations in the SALL1 putative transcription factor gene cause Townes-Brocks syndrome. *Nature Genetics*, *18*(1), 81–83.

Kölliker, A. (1852). *Handbuch der Braunschweig Gewebelehre der Menschen*. Leipzig: Wilhelm Engelmann.

Krachulec, J., Vetter, M., Schrade, A., Lobs, A. K., Bielinska, M., Cochran, R., et al. (2012). GATA4 is a critical regulator of gonadectomy-induced adrenocortical tumorigenesis in mice. *Endocrinology*, *153*(6), 2599–2611.

Kreidberg, J. A., Sariola, H., Loring, J. M., Maeda, M., Pelletier, J., Housman, D., et al. (1993). WT-1 is required for early kidney development. *Cell*, *74*(4), 679–691.

Kume, T., Deng, K., & Hogan, B. L. (2000). Minimal phenotype of mice homozygous for a null mutation in the forkhead/winged helix gene, Mf2. *Molecular and Cellular Biology*, *20*(4), 1419–1425.

Lambert-Langlais, S., Val, P., Guyot, S., Ragazzon, B., Sahut-Barnola, I., De Haze, A., et al. (2009). A transgenic mouse line with specific Cre recombinase expression in the adrenal cortex. *Molecular and Cellular Endocrinology, 300*(1–2), 197–204.

Laufer, E., Kesper, D., Vortkamp, A., & King, P. (2012). Sonic hedgehog signaling during adrenal development. *Molecular and Cellular Endocrinology, 351*(1), 19–27.

Lepique, A. P., Moraes, M. S., Rocha, K. M., Eichler, C. B., Hajj, G. N., Schwindt, T. T., et al. (2004). c-Myc protein is stabilized by fibroblast growth factor 2 and destabilized by ACTH to control cell cycle in mouse Y1 adrenocortical cells. *Journal of Molecular Endocrinology, 33*(3), 623–638.

Li, F. P., & Fraumeni, J. F., Jr. (1969). Soft-tissue sarcomas, breast cancer, and other neoplasms. A familial syndrome? *Annals of Internal Medicine, 71*(4), 747–752.

Libe, R., Fratticci, A., Coste, J., Tissier, F., Horvath, A., Ragazzon, B., et al. (2008). Phosphodiesterase 11A (PDE11A) and genetic predisposition to adrenocortical tumors. *Clinical Cancer Research: An Official Journal of the American Association for Cancer Research, 14*(12), 4016–4024.

Lichtenauer, U. D., Duchniewicz, M., Kolanczyk, M., Hoeflich, A., Hahner, S., Else, T., et al. (2007). Pre-B-cell transcription factor 1 and steroidogenic factor 1 synergistically regulate adrenocortical growth and steroidogenesis. *Endocrinology, 148*(2), 693–704.

Lombardo, L. D., & Cortesini, C. (1988). The origin and differentiation of adrenocortical cells in rats with portacaval shunt. A structural and ultrastructural study. *Histology and Histopathology, 3*(2), 163–171.

Looyenga, B. D., & Hammer, G. D. (2006). Origin and identity of adrenocortical tumors in inhibin knockout mice: Implications for cellular plasticity in the adrenal cortex. *Molecular Endocrinology, 20*(11), 2848–2863.

Lowry, P. J., Silas, L., McLean, C., Linton, E. A., & Estivariz, F. E. (1983). Pro-gamma-melanocyte-stimulating hormone cleavage in adrenal gland undergoing compensatory growth. *Nature, 306*(5938), 70–73.

Luo, X., Ikeda, Y., & Parker, K. L. (1994). A cell-specific nuclear receptor is essential for adrenal and gonadal development and sexual differentiation. *Cell,* 77(4), 481–490.

Ma, Y., Chai, L., Cortez, S. C., Stopa, E. G., Steinhoff, M. M., Ford, D., et al. (2002). SALL1 expression in the human pituitary-adrenal/gonadal axis. *The Journal of Endocrinology, 173*(3), 437–448.

Maezawa, Y., Binnie, M., Li, C., Thorner, P., Hui, C. C., Alman, B., et al. (2012). A new Cre driver mouse line, Tcf21/Pod1-Cre, targets metanephric mesenchyme. *PLoS one,* 7(7), e40547.

Maitra, A., & Shirwalkar, H. (2003). Congenital adrenal hyperplasia: Biochemical and molecular perspectives. *Indian Journal of Experimental Biology, 41*(7), 701–709.

Mandel, H., Shemer, R., Borochowitz, Z. U., Okopnik, M., Knopf, C., Indelman, M., et al. (2008). SERKAL syndrome: An autosomal-recessive disorder caused by a loss-of-function mutation in WNT4. *American Journal of Human Genetics, 82*(1), 39–47.

Masai, H., Taniyama, C., Ogino, K., Matsui, E., Kakusho, N., Matsumoto, S., et al. (2006). Phosphorylation of MCM4 by Cdc7 kinase facilitates its interaction with Cdc45 on the chromatin. *The Journal of Biological Chemistry, 281*(51), 39249–39261.

Masui, H., & Garren, L. D. (1971). Inhibition of replication in functional mouse adrenal tumor cells by adrenocorticotropic hormone mediated by adenosine 3′:5′-cyclic monophosphate. *Proceedings of the National Academy of Sciences of the United States of America, 68*(12), 3206–3210.

Matzuk, M. M., Finegold, M. J., Su, J. G., Hsueh, A. J., & Bradley, A. (1992). Alpha-inhibin is a tumour-suppressor gene with gonadal specificity in mice. *Nature, 360*(6402), 313–319.

Mazzitelli, N., Vauthay, L., Grandi, C., Fuksman, R., & Rittler, M. (2002). Reviewing old concepts at the start of a new millenium: Growth restriction, adrenal hypoplasia, and thymomegaly in human anencephaly. *Teratology, 66*(3), 105–114.

McEwan, P. E., Lindop, G. B., & Kenyon, C. J. (1995). In vivo studies of the control of DNA synthesis in the rat adrenal cortex and medulla. *Endocrine Research*, *21*(1–2), 91–102.

McEwan, P. E., Lindop, G. B., & Kenyon, C. J. (1996). Control of cell proliferation in the rat adrenal gland in vivo by the renin-angiotensin system. *The American Journal of Physiology*, *271*(1 Pt. 1), E192–E198.

McNutt, N. S., & Jones, A. L. (1970). Observations on the ultrastructure of cytodifferentiation in the human fetal adrenal cortex. *Laboratory Investigation; A Journal of Technical Methods and Pathology*, *22*(6), 513–527.

Meimaridou, E., Hughes, C. R., Kowalczyk, J., Guasti, L., Chapple, J. P., King, P. J., et al. (2013). Familial glucocorticoid deficiency: New genes and mechanisms. *Molecular and Cellular Endocrinology*, *371*(1–2), 195–200.

Meimaridou, E., Kowalczyk, J., Guasti, L., Hughes, C. R., Wagner, F., Frommolt, P., et al. (2012). Mutations in NNT encoding nicotinamide nucleotide transhydrogenase cause familial glucocorticoid deficiency. *Nature Genetics*, *44*(7), 740–742.

Mesiano, S., Coulter, C. L., & Jaffe, R. B. (1993). Localization of cytochrome P450 cholesterol side-chain cleavage, cytochrome P450 17 alpha-hydroxylase/17, 20-lyase, and 3 beta-hydroxysteroid dehydrogenase isomerase steroidogenic enzymes in human and rhesus monkey fetal adrenal glands: Reappraisal of functional zonation. *The Journal of Clinical Endocrinology and Metabolism*, 77(5), 1184–1189.

Mesiano, S., & Jaffe, R. B. (1997). Role of growth factors in the developmental regulation of the human fetal adrenal cortex. *Steroids*, *62*(1), 62–72.

Mesiano, S., Mellon, S. H., Gospodarowicz, D., Di Blasio, A. M., & Jaffe, R. B. (1991). Basic fibroblast growth factor expression is regulated by corticotropin in the human fetal adrenal: A model for adrenal growth regulation. *Proceedings of the National Academy of Sciences of the United States of America*, *88*(12), 5428–5432.

Mesiano, S., Mellon, S. H., & Jaffe, R. B. (1993). Mitogenic action, regulation, and localization of insulin-like growth factors in the human fetal adrenal gland. *The Journal of Clinical Endocrinology and Metabolism*, *76*(4), 968–976.

Metherell, L. A., Chapple, J. P., Cooray, S., David, A., Becker, C., Ruschendorf, F., et al. (2005). Mutations in MRAP, encoding a new interacting partner of the ACTH receptor, cause familial glucocorticoid deficiency type 2. *Nature Genetics*, *37*(2), 166–170.

Metherell, L. A., Naville, D., Halaby, G., Begeot, M., Huebner, A., Nurnberg, G., et al. (2009). Nonclassic lipoid congenital adrenal hyperplasia masquerading as familial glucocorticoid deficiency. *The Journal of Clinical Endocrinology and Metabolism*, *94*(10), 3865–3871.

Mijnhout, G. S., Danner, S. A., van de Goot, F. R., & van Dam, E. W. (2004). Macronodular adrenocortical hyperplasia in a postmenopausal woman. *The Netherlands Journal of Medicine*, *62*(11), 454–455 [Case Reports].

Miller, E. D. (1981). The role of the renin-angiotensin-aldosterone system in circulatory control and hypertension. *British Journal of Anaesthesia*, *53*(7), 711–718.

Ming, J. E., Kaupas, M. E., Roessler, E., Brunner, H. G., Golabi, M., Tekin, M., et al. (2002). Mutations in PATCHED-1, the receptor for SONIC HEDGEHOG, are associated with holoprosencephaly. *Human Genetics*, *110*(4), 297–301.

Mitani, F., Mukai, K., Miyamoto, H., Suematsu, M., & Ishimura, Y. (1999). Development of functional zonation in the rat adrenal cortex. *Endocrinology*, *140*(7), 3342–3353.

Mitani, F., Mukai, K., Miyamoto, H., Suematsu, M., & Ishimura, Y. (2003). The undifferentiated cell zone is a stem cell zone in adult rat adrenal cortex. *Biochimica et Biophysica Acta*, *1619*(3), 317–324.

Mitani, F., Suzuki, H., Hata, J., Ogishima, T., Shimada, H., & Ishimura, Y. (1994). A novel cell layer without corticosteroid-synthesizing enzymes in rat adrenal cortex: Histochemical detection and possible physiological role. *Endocrinology*, *135*(1), 431–438.

Miyamoto, H., Mitani, F., Mukai, K., Suematsu, M., & Ishimura, Y. (1999). Studies on cytogenesis in adult rat adrenal cortex: Circadian and zonal variations and their modulation by adrenocorticotropic hormone. *Journal of Biochemistry*, *126*(6), 1175–1183.

Mizusaki, H., Kawabe, K., Mukai, T., Ariyoshi, E., Kasahara, M., Yoshioka, H., et al. (2003). Dax-1 (dosage-sensitive sex reversal-adrenal hypoplasia congenita critical region on the X chromosome, gene 1) gene transcription is regulated by wnt4 in the female developing gonad. *Molecular Endocrinology*, *17*(4), 507–519.

Moore, A. W., McInnes, L., Kreidberg, J., Hastie, N. D., & Schedl, A. (1999). YAC complementation shows a requirement for Wt1 in the development of epicardium, adrenal gland and throughout nephrogenesis. *Development*, *126*(9), 1845–1857.

Morley, S. D., Viard, I., Chung, B. C., Ikeda, Y., Parker, K. L., & Mullins, J. J. (1996). Variegated expression of a mouse steroid 21-hydroxylase/beta-galactosidase transgene suggests centripetal migration of adrenocortical cells. *Molecular Endocrinology*, *10*(5), 585–598.

Mukai, T., Kusaka, M., Kawabe, K., Goto, K., Nawata, H., Fujieda, K., et al. (2002). Sexually dimorphic expression of Dax-1 in the adrenal cortex. *Genes to Cells: Devoted to Molecular & Cellular Mechanisms*, 7(7), 717–729.

Muscatelli, F., Strom, T. M., Walker, A. P., Zanaria, E., Recan, D., Meindl, A., et al. (1994). Mutations in the DAX-1 gene give rise to both X-linked adrenal hypoplasia congenita and hypogonadotropic hypogonadism. *Nature*, *372*(6507), 672–676.

Nachtigal, M. W., Hirokawa, Y., Enyeart-VanHouten, D. L., Flanagan, J. N., Hammer, G. D., & Ingraham, H. A. (1998). Wilms' tumor 1 and Dax-1 modulate the orphan nuclear receptor SF-1 in sex-specific gene expression. *Cell*, *93*(3), 445–454.

Nakamura, Y., Gang, H. X., Suzuki, T., Sasano, H., & Rainey, W. E. (2009). Adrenal changes associated with adrenarche. *Reviews in Endocrine & Metabolic Disorders*, *10*(1), 19–26.

Nanni, L., Ming, J. E., Bocian, M., Steinhaus, K., Bianchi, D. W., Die-Smulders, C., et al. (1999). The mutational spectrum of the sonic hedgehog gene in holoprosencephaly: SHH mutations cause a significant proportion of autosomal dominant holoprosencephaly. *Human Molecular Genetics*, *8*(13), 2479–2488.

Nishimoto, K., Nakagawa, K., Li, D., Kosaka, T., Oya, M., Mikami, S., et al. (2010). Adrenocortical zonation in humans under normal and pathological conditions. *The Journal of Clinical Endocrinology and Metabolism*, *95*(5), 2296–2305.

Nishinakamura, R., Matsumoto, Y., Nakao, K., Nakamura, K., Sato, A., Copeland, N. G., et al. (2001). Murine homolog of SALL1 is essential for ureteric bud invasion in kidney development. *Development*, *128*(16), 3105–3115.

Nogueira, E. F., Bollag, W. B., & Rainey, W. E. (2009). Angiotensin II regulation of adrenocortical gene transcription. *Molecular and Cellular Endocrinology*, *302*(2), 230–236.

Nussdorfer, G. G. (1986). Cytophysiology of the adrenal cortex. *International Review of Cytology*, *98*, 1–405.

Oparil, S., Sanders, C. A., & Haber, E. (1970). In-vivo and in-vitro conversion of angiotensin I to angiotensin II in dog blood. *Circulation Research*, *26*(5), 591–599.

Ornitz, D. M., & Itoh, N. (2001). Fibroblast growth factors. *Genome Biology*, *2*(3), Reviews 3005.1–Reviews 3005.12.

Osamura, R. Y. (1977). Functional prenatal development of anencephalic and normal anterior pituitary glands. In human and experimental animals studied by peroxidase-labeled antibody method. *Acta Pathologica Japonica*, *27*(4), 495–509.

Pabon, J. E., Li, X., Lei, Z. M., Sanfilippo, J. S., Yussman, M. A., & Rao, C. V. (1996). Novel presence of luteinizing hormone/chorionic gonadotropin receptors in human adrenal glands. *The Journal of Clinical Endocrinology and Metabolism*, *81*(6), 2397–2400.

Pan, Y., Bai, C. B., Joyner, A. L., & Wang, B. (2006). Sonic hedgehog signaling regulates Gli2 transcriptional activity by suppressing its processing and degradation. *Molecular and Cellular Biology*, *26*(9), 3365–3377.

Parajes, S., Kamrath, C., Rose, I. T., Taylor, A. E., Mooij, C. F., Dhir, V., et al. (2011). A novel entity of clinically isolated adrenal insufficiency caused by a partially inactivating mutation of the gene encoding for P450 side chain cleavage enzyme (CYP11A1). *The Journal of Clinical Endocrinology and Metabolism*, *96*(11), E1798–E1806.

Parker, T. L., Kesse, W. K., Mohamed, A. A., & Afework, M. (1993). The innervation of the mammalian adrenal gland. *Journal of Anatomy*, *183*(Pt. 2), 265–276.

Parker, K. L., Rice, D. A., Lala, D. S., Ikeda, Y., Luo, X., Wong, M., et al. (2002). Steroidogenic factor 1: An essential mediator of endocrine development. *Recent Progress in Hormone Research*, *57*, 19–36.

Parviainen, H., Kiiveri, S., Bielinska, M., Rahman, N., Huhtaniemi, I. T., Wilson, D. B., et al. (2007). GATA transcription factors in adrenal development and tumors. *Molecular and Cellular Endocrinology*, *265–266*, 17–22.

Pateras, I. S., Apostolopoulou, K., Niforou, K., Kotsinas, A., & Gorgoulis, V. G. (2009). p57KIP2: "Kip"ing the cell under control. *Molecular Cancer Research: MCR*, 7(12), 1902–1919.

Paul, A., & Laufer, E. (2011). Endogenous biotin as a marker of adrenocortical cells with steroidogenic potential. *Molecular and Cellular Endocrinology*, *336*(1–2), 133–140.

Payet, N., & Lehoux, J. G. (1980). A comparative study of the role of vasopressin and ACTH in the regulation of growth and function of rat adrenal glands. *Journal of Steroid Biochemistry*, *12*, 461–467.

Payne, A. H., & Hales, D. B. (2004). Overview of steroidogenic enzymes in the pathway from cholesterol to active steroid hormones. *Endocrine Reviews*, *25*(6), 947–970.

Pezzi, V., Clyne, C. D., Ando, S., Mathis, J. M., & Rainey, W. E. (1997). Ca(2+)-regulated expression of aldosterone synthase is mediated by calmodulin and calmodulin-dependent protein kinases. *Endocrinology*, *138*(2), 835–838.

Phillips, R., Crock, C., & Funder, J. (1985). Effects of mineralocorticoids and glucocorticoids on compensatory adrenal growth in rats. *The American Journal of Physiology*, *248*(4 Pt. 1), E450–E456.

Pignatelli, D., Maia, M., Bento, M. J., Sousa, S., Azevedo, M. E., Magalhaes, M. M., et al. (2000). Captopril effects on the rat adrenal cortex. *Endocrine Research*, *26*(4), 965–972.

Pihlajoki, M., Gretzinger, E., Cochran, R., Kyronlahti, A., Schrade, A., Hiller, T., et al. (2013). Conditional mutagenesis of Gata6 in SF1-positive cells causes gonadal-like differentiation in the adrenal cortex of mice. *Endocrinology*, *154*(5), 1754–1767.

Pitetti, J. L., Calvel, P., Romero, Y., Conne, B., Truong, V., Papaioannou, M. D., et al. (2013). Insulin and IGF1 receptors are essential for XX and XY gonadal differentiation and adrenal development in mice. *PLoS Genetics*, *9*(1), e1003160.

Pritchard, L. E., & White, A. (2007). Neuropeptide processing and its impact on melanocortin pathways. *Endocrinology*, *148*(9), 4201–4207.

Pruitt, S. C., Bailey, K. J., & Freeland, A. (2007). Reduced Mcm2 expression results in severe stem/progenitor cell deficiency and cancer. *Stem Cells*, *25*(12), 3121–3132.

Ragazzon, B., Lefrancois-Martinez, A. M., Val, P., Sahut-Barnola, I., Tournaire, C., Chambon, C., et al. (2006). Adrenocorticotropin-dependent changes in SF-1/DAX-1 ratio influence steroidogenic genes expression in a novel model of glucocorticoid-producing adrenocortical cell lines derived from targeted tumorigenesis. *Endocrinology*, *147*(4), 1805–1818.

Reissmann, E., Ernsberger, U., Francis-West, P. H., Rueger, D., Brickell, P. M., & Rohrer, H. (1996). Involvement of bone morphogenetic protein-4 and bone morphogenetic protein-7 in the differentiation of the adrenergic phenotype in developing sympathetic neurons. *Development*, *122*(7), 2079–2088.

Reiter, R. J., & Pizzarello, D. J. (1966). Radioautographic study of cellular replacement in the adrenal cortex of male rats. *Texas Reports on Biology and Medicine*, *24*(2), 189–194.

Revest, J. M., Spencer-Dene, B., Kerr, K., De Moerlooze, L., Rosewell, I., & Dickson, C. (2001). Fibroblast growth factor receptor 2-IIIb acts upstream of Shh and Fgf4 and is required for limb bud maintenance but not for the induction of Fgf8, Fgf10, Msx1, or Bmp4. *Developmental Biology*, *231*(1), 47–62.

Ribeiro, R. C., & Figueiredo, B. (2004). Childhood adrenocortical tumours. *European Journal of Cancer*, *40*(8), 1117–1126.

Ribeiro, R. C., Sandrini, F., Figueiredo, B., Zambetti, G. P., Michalkiewicz, E., Lafferty, A. R., et al. (2001). An inherited p53 mutation that contributes in a tissue-specific manner to pediatric adrenal cortical carcinoma. *Proceedings of the National Academy of Sciences of the United States of America*, *98*(16), 9330–9335.

Riobo, N. A. (2012). Cholesterol and its derivatives in Sonic Hedgehog signaling and cancer. *Current Opinion in Pharmacology*, *12*(6), 736–741.

Rocha, K. M., Forti, F. L., Lepique, A. P., & Armelin, H. A. (2003). Deconstructing the molecular mechanisms of cell cycle control in a mouse adrenocortical cell line: Roles of ACTH. *Microscopy Research and Technique*, *61*(3), 268–274.

Roessler, E., Du, Y. Z., Mullor, J. L., Casas, E., Allen, W. P., Gillessen-Kaesbach, G., et al. (2003). Loss-of-function mutations in the human GLI2 gene are associated with pituitary anomalies and holoprosencephaly-like features. *Proceedings of the National Academy of Sciences of the United States of America*, *100*(23), 13424–13429.

Romberger, C. F., & Wong, T. W. (1989). Thecal metaplasia in the adrenal gland of a man with acquired bilateral testicular atrophy. *Archives of Pathology & Laboratory Medicine*, *113*(9), 1071–1075 [Case Reports].

Romero, D. G., Yanes, L. L., de Rodriguez, A. F., Plonczynski, M. W., Welsh, B. L., Reckelhoff, J. F., et al. (2007). Disabled-2 is expressed in adrenal zona glomerulosa and is involved in aldosterone secretion. *Endocrinology*, *148*(6), 2644–2652.

Sadovsky, Y., Crawford, P. A., Woodson, K. G., Polish, J. A., Clements, M. A., Tourtellotte, L. M., et al. (1995). Mice deficient in the orphan receptor steroidogenic factor 1 lack adrenal glands and gonads but express P450 side-chain-cleavage enzyme in the placenta and have normal embryonic serum levels of corticosteroids. *Proceedings of the National Academy of Sciences of the United States of America*, *92*(24), 10939–10943.

Sahakitrungruang, T., Tee, M. K., Blackett, P. R., & Miller, W. L. (2011). Partial defect in the cholesterol side-chain cleavage enzyme P450scc (CYP11A1) resembling nonclassic congenital lipoid adrenal hyperplasia. *The Journal of Clinical Endocrinology and Metabolism*, *96*(3), 792–798.

Sahut-Barnola, I., de Joussineau, C., Val, P., Lambert-Langlais, S., Damon, C., Lefrancois-Martinez, A. M., et al. (2010). Cushing's syndrome and fetal features resurgence in adrenal cortex-specific Prkar1a knockout mice. *PLoS Genetics*, *6*(6), e1000980.

Sakaue, M., & Hoffman, B. B. (1991). Glucocorticoids induce transcription and expression of the alpha 1B adrenergic receptor gene in DTT1 MF-2 smooth muscle cells. *The Journal of Clinical Investigation*, *88*(2), 385–389.

Salmon, T. N., & Zwemer, R. L. (1941). A study of the life history of cortico-adrenal cells of the rat by means of trypan blue injections. *The Anatomical Record*, *80*(4), 421–429.

Saner-Amigh, K., Mayhew, B. A., Mantero, F., Schiavi, F., White, P. C., Rao, C. V., et al. (2006). Elevated expression of luteinizing hormone receptor in aldosterone-producing adenomas. *The Journal of Clinical Endocrinology and Metabolism*, *91*(3), 1136–1142 [Research Support, N.I.H., Extramural].

Sapolsky, R. M., Romero, L. M., & Munck, A. U. (2000). How do glucocorticoids influence stress responses? Integrating permissive, suppressive, stimulatory, and preparative actions. *Endocrine Reviews*, *21*(1), 55–89.

Scheys, J. O., Heaton, J. H., & Hammer, G. D. (2011). Evidence of adrenal failure in aging Dax1-deficient mice. *Endocrinology*, *152*(9), 3430–3439.

Schinner, S., & Bornstein, S. R. (2005). Cortical-chromaffin cell interactions in the adrenal gland. *Endocrine Pathology*, *16*(2), 91–98.
Schnabel, C. A., Selleri, L., & Cleary, M. L. (2003). Pbx1 is essential for adrenal development and urogenital differentiation. *Genesis*, *37*(3), 123–130.
Scornik, O. A., & Paladini, A. C. (1964). Angiotensin blood levels in hemorrhagic hypotension and other related Conditions. *The American Journal of Physiology*, *206*, 553–556.
Selleri, L., Depew, M. J., Jacobs, Y., Chanda, S. K., Tsang, K. Y., Cheah, K. S., et al. (2001). Requirement for Pbx1 in skeletal patterning and programming chondrocyte proliferation and differentiation. *Development*, *128*(18), 3543–3557.
Sewer, M. B., Dammer, E. B., & Jagarlapudi, S. (2007). Transcriptional regulation of adrenocortical steroidogenic gene expression. *Drug Metabolism Reviews*, *39*(2–3), 371–388.
Shah, N. M., Groves, A. K., & Anderson, D. J. (1996). Alternative neural crest cell fates are instructively promoted by TGFbeta superfamily members. *Cell*, *85*(3), 331–343.
Shigematsu, K., Niwa, M., Kurihara, M., Yamashita, K., Kawai, K., & Tsuchiyama, H. (1989). Receptor autoradiographic localization of insulin-like growth factor-I (IGF-I) binding sites in human fetal and adult adrenal glands. *Life Sciences*, *45*(5), 383–389.
Shima, N., Alcaraz, A., Liachko, I., Buske, T. R., Andrews, C. A., Munroe, R. J., et al. (2007). A viable allele of Mcm4 causes chromosome instability and mammary adenocarcinomas in mice. *Nature Genetics*, *39*(1), 93–98.
Sidhu, S., Sywak, M., Robinson, B., & Delbridge, L. (2004). Adrenocortical cancer: Recent clinical and molecular advances. *Current Opinion in Oncology*, *16*(1), 13–18.
Skelton, F. R. (1959). Adrenal regeneration and adrenal-regeneration hypertension. *Physiological Reviews*, *39*(1), 162–182.
So, P. L., & Danielian, P. S. (1999). Cloning and expression analysis of a mouse gene related to Drosophila odd-skipped. *Mechanisms of Development*, *84*(1–2), 157–160.
Spencer, S. J., Mesiano, S., Lee, J. Y., & Jaffe, R. B. (1999). Proliferation and apoptosis in the human adrenal cortex during the fetal and perinatal periods: Implications for growth and remodeling. *The Journal of Clinical Endocrinology and Metabolism*, *84*(3), 1110–1115.
Spencer, S. J., Rabinovici, J., Mesiano, S., Goldsmith, P. C., & Jaffe, R. B. (1992). Activin and inhibin in the human adrenal gland. Regulation and differential effects in fetal and adult cells. *The Journal of Clinical Investigation*, *90*(1), 142–149.
Stark, K., Vainio, S., Vassileva, G., & McMahon, A. P. (1994). Epithelial transformation of metanephric mesenchyme in the developing kidney regulated by Wnt-4. *Nature*, *372*(6507), 679–683.
Staton, B. A., Grilliot, M., & Parker, C. R., Jr. (2004). Effects of anencephaly on adrenocortical cell proliferation. *Endocrine Research*, *30*(4), 527.
Stone, D. M., Hynes, M., Armanini, M., Swanson, T. A., Gu, Q., Johnson, R. L., et al. (1996). The tumour-suppressor gene patched encodes a candidate receptor for Sonic hedgehog. *Nature*, *384*(6605), 129–134.
Storr, H. L., Kind, B., Parfitt, D. A., Chapple, J. P., Lorenz, M., Koehler, K., et al. (2009). Deficiency of ferritin heavy-chain nuclear import in triple a syndrome implies nuclear oxidative damage as the primary disease mechanism. *Molecular Endocrinology*, *23*(12), 2086–2094.
Sucheston, M. E., & Cannon, M. S. (1968). Development of zonular patterns in the human adrenal gland. *Journal of Morphology*, *126*(4), 477–491.
Sucheston, M. E., & Cannon, M. S. (1972). The transient-zone in the human and mouse adrenal gland. *The Ohio Journal of Science*, *72*, 120–126.
Sul, H. S. (2009). Minireview: Pref-1: Role in adipogenesis and mesenchymal cell fate. *Molecular Endocrinology*, *23*(11), 1717–1725.
Sun, C., Nakatake, Y., Akagi, T., Ura, H., Matsuda, T., Nishiyama, A., et al. (2009). Dax1 binds to Oct3/4 and inhibits its transcriptional activity in embryonic stem cells. *Molecular and Cellular Biology*, *29*(16), 4574–4583.

Swaab, D. F., Boer, G. J., & Visser, M. (1978). The fetal brain and intrauterine growth. *Postgraduate Medical Journal, 54*(Suppl. 1), 63–73.

Swain, A., Zanaria, E., Hacker, A., Lovell-Badge, R., & Camerino, G. (1996). Mouse Dax1 expression is consistent with a role in sex determination as well as in adrenal and hypothalamus function. *Nature Genetics, 12*(4), 404–409.

Tadjine, M., Lampron, A., Ouadi, L., & Bourdeau, I. (2008). Frequent mutations of beta-catenin gene in sporadic secreting adrenocortical adenomas. *Clinical Endocrinology, 68*(2), 264–270.

Taki, T. M., & Nickerson, P. A. (1985). Differentiation and proliferation of adrenocortical cells during the early stages of regeneration. *Laboratory Investigation; A Journal of Technical Methods and Pathology, 53*(1), 91–100.

Tien, E. S., Davis, J. W., & Vanden Heuvel, J. P. (2004). Identification of the CREB-binding protein/p300-interacting protein CITED2 as a peroxisome proliferator-activated receptor alpha coregulator. *The Journal of Biological Chemistry, 279*(23), 24053–24063.

Tissier, F., Cavard, C., Groussin, L., Perlemoine, K., Fumey, G., Hagnere, A. M., et al. (2005). Mutations of beta-catenin in adrenocortical tumors: Activation of the Wnt signaling pathway is a frequent event in both benign and malignant adrenocortical tumors. *Cancer Research, 65*(17), 7622–7627.

Tobian, L., Tomboulian, A., & Janecek, J. (1959). The effect of high perfusion pressures on the granulation of juxtaglomerular cells in an isolated kidney. *The Journal of Clinical Investigation, 38*(4), 605–610.

Tremblay, J. J., & Viger, R. S. (2003). Novel roles for GATA transcription factors in the regulation of steroidogenesis. *The Journal of Steroid Biochemistry and Molecular Biology, 85*(2–5), 291–298.

Tullio-Pelet, A., Salomon, R., Hadj-Rabia, S., Mugnier, C., de Laet, M. H., Chaouachi, B., et al. (2000). Mutant WD-repeat protein in triple-A syndrome. *Nature Genetics, 26*(3), 332–335.

Turner, N., & Grose, R. (2010). Fibroblast growth factor signalling: From development to cancer. *Nature Reviews. Cancer, 10*(2), 116–129.

Uotila, U. U. (1940). The early embryological development of the fetal and permanent adrenal cortex in man. *The Anatomical Record, 76*(2), 183–203.

Usongo, M., & Farookhi, R. (2012). beta-catenin/Tcf-signaling appears to establish the murine ovarian surface epithelium (OSE) and remains active in selected postnatal OSE cells. *BMC Developmental Biology, 12*, 17.

Vainio, S., Heikkila, M., Kispert, A., Chin, N., & McMahon, A. P. (1999). Female development in mammals is regulated by Wnt-4 signalling. *Nature, 397*(6718), 405–409.

Val, P., Jeays-Ward, K., & Swain, A. (2006). Identification of a novel population of adrenal-like cells in the mammalian testis. *Developmental Biology, 299*(1), 250–256.

Val, P., Martinez-Barbera, J. P., & Swain, A. (2007). Adrenal development is initiated by Cited2 and Wt1 through modulation of Sf-1 dosage. *Development, 134*(12), 2349–2358.

Van de Kar, L. D., & Blair, M. L. (1999). Forebrain pathways mediating stress-induced hormone secretion. *Frontiers in Neuroendocrinology, 20*(1), 1–48.

Varley, J. M., McGown, G., Thorncroft, M., James, L. A., Margison, G. P., Forster, G., et al. (1999). Are there low-penetrance TP53 Alleles? Evidence from childhood adrenocortical tumors. *American Journal of Human Genetics, 65*(4), 995–1006.

Vezzosi, D., Libe, R., Baudry, C., Rizk-Rabin, M., Horvath, A., Levy, I., et al. (2012). Phosphodiesterase 11A (PDE11A) gene defects in patients with ACTH-independent macronodular adrenal hyperplasia (AIMAH): Functional variants may contribute to genetic susceptibility of bilateral adrenal tumors. *The Journal of Clinical Endocrinology and Metabolism, 97*(11), E2063–E2069.

Vilain, E., Le Merrer, M., Lecointre, C., Desangles, F., Kay, M. A., Maroteaux, P., et al. (1999). IMAGe, a new clinical association of intrauterine growth retardation,

metaphyseal dysplasia, adrenal hypoplasia congenita, and genital anomalies. *The Journal of Clinical Endocrinology and Metabolism, 84*(12), 4335–4340.
Vinson, G. P. (2003). Adrenocortical zonation and ACTH. *Microscopy Research and Technique, 61*(3), 227–239.
Vokes, S. A., Ji, H., McCuine, S., Tenzen, T., Giles, S., Zhong, S., et al. (2007). Genomic characterization of Gli-activator targets in sonic hedgehog-mediated neural patterning. *Development, 134*(10), 1977–1989.
Voutilainen, R., Eramaa, M., & Ritvos, O. (1991). Hormonally regulated inhibin gene expression in human fetal and adult adrenals. *The Journal of Clinical Endocrinology and Metabolism, 73*(5), 1026–1030.
Voutilainen, R., & Miller, W. L. (1987). Coordinate tropic hormone regulation of mRNAs for insulin-like growth factor II and the cholesterol side-chain-cleavage enzyme, P450scc [corrected], in human steroidogenic tissues. *Proceedings of the National Academy of Sciences of the United States of America, 84*(6), 1590–1594.
Vuorenoja, S., Rivero-Muller, A., Kiiveri, S., Bielinska, M., Heikinheimo, M., Wilson, D. B., et al. (2007). Adrenocortical tumorigenesis, luteinizing hormone receptor and transcription factors GATA-4 and GATA-6. *Molecular and Cellular Endocrinology, 269*(1–2), 38–45.
Wang, Q., Lan, Y., Cho, E. S., Maltby, K. M., & Jiang, R. (2005). Odd-skipped related 1 (Odd 1) is an essential regulator of heart and urogenital development. *Developmental Biology, 288*(2), 582–594.
Waterman, M. R., & Bischof, L. J. (1996). Mechanisms of ACTH(cAMP)-dependent transcription of adrenal steroid hydroxylases. *Endocrine Research, 22*(4), 615–620.
Weber, M. M., Auernhammer, C. J., Kiess, W., & Engelhardt, D. (1997). Insulin-like growth factor receptors in normal and tumorous adult human adrenocortical glands. *European Journal of Endocrinology/European Federation of Endocrine Societies, 136*(3), 296–303.
Wehling, M., Spes, C. H., Win, N., Janson, C. P., Schmidt, B. M., Theisen, K., et al. (1998). Rapid cardiovascular action of aldosterone in man. *The Journal of Clinical Endocrinology and Metabolism, 83*(10), 3517–3522.
Wilburn, L. A., Goldsmith, P. C., Chang, K. J., & Jaffe, R. B. (1986). Ontogeny of enkephalin and catecholamine-synthesizing enzymes in the primate fetal adrenal medulla. *The Journal of Clinical Endocrinology and Metabolism, 63*(4), 974–980.
Wojcikiewicz, R. J., & Nahorski, S. R. (1993). Modulation of signalling initiated by phosphoinositidase-C-linked receptors. *The Journal of Experimental Biology, 184*, 145–159.
Wong, D. L. (2003). Why is the adrenal adrenergic? *Endocrine Pathology, 14*(1), 25–36.
Wood, M. A., & Hammer, G. D. (2011). Adrenocortical stem and progenitor cells: Unifying model of two proposed origins. *Molecular and Cellular Endocrinology, 336*(1–2), 206–212.
Wurtman, R. J., & Axelrod, J. (1965). Adrenaline synthesis: Control by the pituitary gland and adrenal glucocorticoids. *Science, 150*(3702), 1464–1465.
Xu, B., Yang, W. H., Gerin, I., Hu, C. D., Hammer, G. D., & Koenig, R. J. (2009). Dax-1 and steroid receptor RNA activator (SRA) function as transcriptional coactivators for steroidogenic factor 1 in steroidogenesis. *Molecular and Cellular Biology, 29*(7), 1719–1734.
Yamamoto, M., Yanai, R., & Arishima, K. (2004). Study of migration of neural crest cells to adrenal medulla by three-dimensional reconstruction. *The Journal of Veterinary Medical Science/The Japanese Society of Veterinary Science, 66*(6), 635–641.
Yaswen, L., Diehl, N., Brennan, M. B., & Hochgeschwender, U. (1999). Obesity in the mouse model of pro-opiomelanocortin deficiency responds to peripheral melanocortin. *Nature Medicine, 5*(9), 1066–1070.
Yu, R. N., Ito, M., Saunders, T. L., Camper, S. A., & Jameson, J. L. (1998). Role of Ahch in gonadal development and gametogenesis. *Nature Genetics, 20*(4), 353–357.

Zacharin, M. (2007). The spectrum of McCune Albright syndrome. *Pediatric Endocrinology Reviews: PER*, *4*(Suppl. 4), 412–418.

Zanaria, E., Muscatelli, F., Bardoni, B., Strom, T. M., Guioli, S., Guo, W., et al. (1994). An unusual member of the nuclear hormone receptor superfamily responsible for X-linked adrenal hypoplasia congenita. *Nature*, *372*(6507), 635–641.

Zubair, M., Ishihara, S., Oka, S., Okumura, K., & Morohashi, K. (2006). Two-step regulation of Ad4BP/SF-1 gene transcription during fetal adrenal development: Initiation by a Hox-Pbx1-Prep1 complex and maintenance via autoregulation by Ad4BP/SF-1. *Molecular and Cellular Biology*, *26*(11), 4111–4121.

Zubair, M., Oka, S., Parker, K. L., & Morohashi, K. (2009). Transgenic expression of Ad4BP/SF-1 in fetal adrenal progenitor cells leads to ectopic adrenal formation. *Molecular Endocrinology*, *23*(10), 1657–1667.

Zubair, M., Parker, K. L., & Morohashi, K. (2008). Developmental links between the fetal and adult zones of the adrenal cortex revealed by lineage tracing. *Molecular and Cellular Biology*, *28*(23), 7030–7040.

Zwemer, R. L., Wotton, R. M., & Norkus, M. G. (1938). A study of corticoadrenal cells. *The Anatomical Record*, *72*(2), 249–263.

INDEX

Note: Page numbers followed by "*f*" indicate figures and "*t*" indicate tables.

N

O

P

R

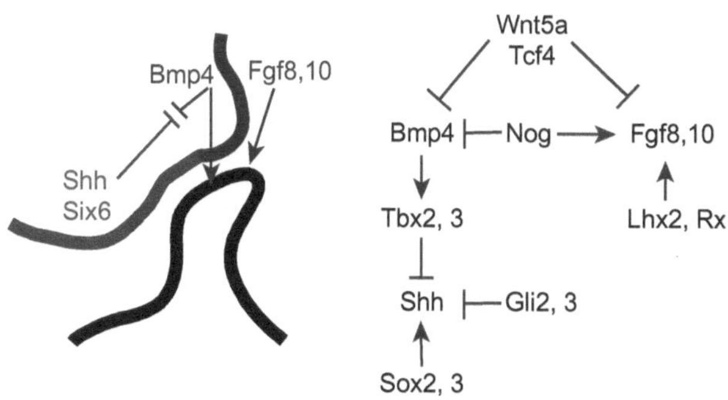

Shannon W. Davis *et al.*, Figure 1.1 Signaling pathways initiating in the organizing center regulate anterior pituitary gland growth and shape. The oral ectoderm invaginates at the roof of the mouth to produce Rathke's pouch (black). The overlying neural ectoderm is defined molecularly by expression domain of *Shh* and *Six6* and a more dorsal region expressing *Bmp* and *Fgf*. These regions exert inhibitory effects on each other, creating a balance that induces normal patterning and growth of Rathke's pouch. The interaction between stimulatory and inhibitory transcription factors and signaling molecules is diagrammed (right).

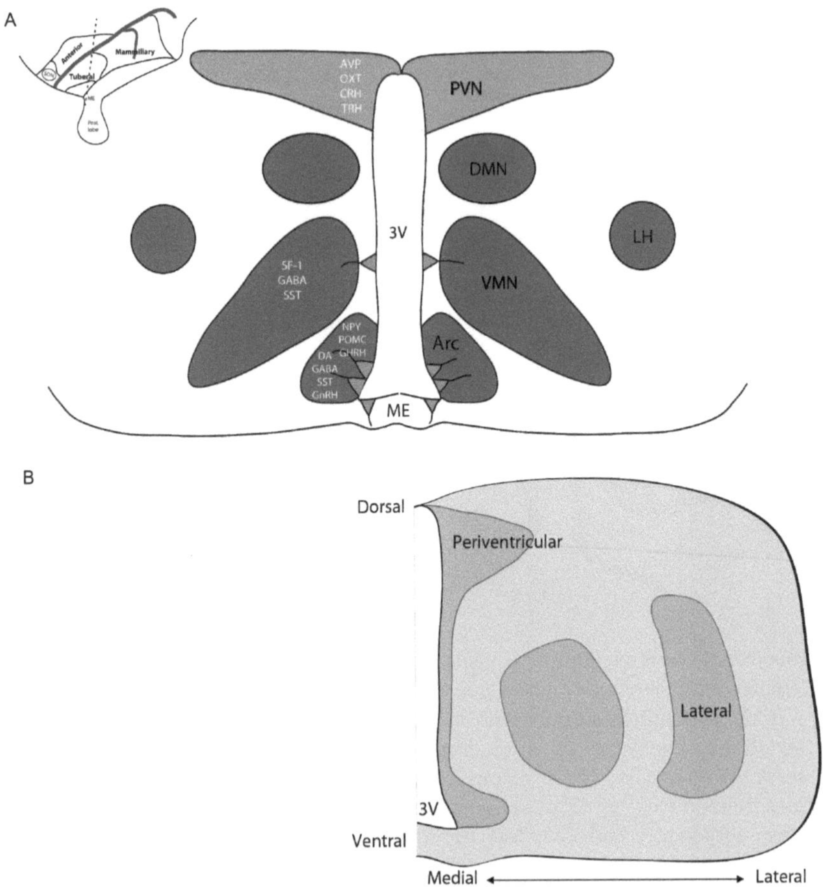

Caroline Alayne Pearson and Marysia Placzek, Figure 2.1 Organization of the medial hypothalamus. (A) Inset: divisions of adult hypothalamus; SON marks preoptic area. Main panel shows positions of medial hypothalamic nuclei relative to third ventricle and median eminence; plane of sections shown by dotted line in inset. Red shows tuberal nuclei, names on right hand side; orange shows anterior nuclei. Neuronal effectors (see text for details) shown on left-hand side. Tanycytes (green) are located in the ventral two-thirds of the ventricular zone (VZ). β-Tanycytes line the median eminence. (B) Mediolateral divisions of hypothalamus. SON, supraoptic nucleus; LH, lateral nucleus; PVN, paraventricular nucleus; DMN, dorsomedial nucleus; VMN, ventromedial nucleus; ARC, arcuate nucleus; ME, median eminence.

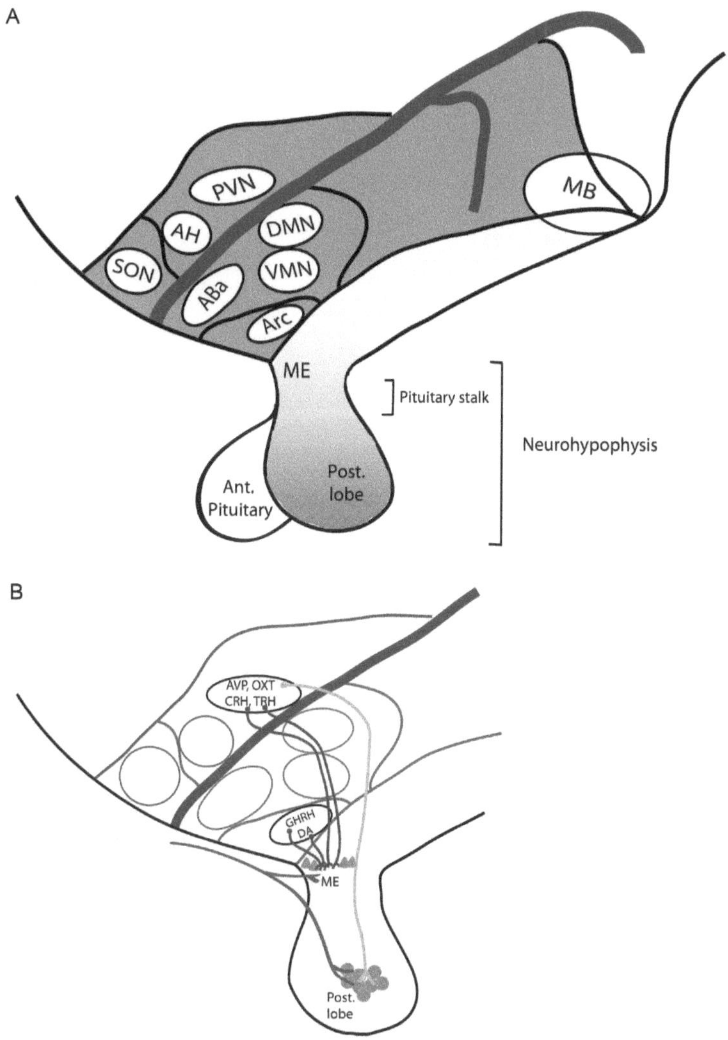

Caroline Alayne Pearson and Marysia Placzek, Figure 2.2 Structure of the adult neurohypophysis; side view. (A) Positions of hypothalamic nuclei and components of neurohypophysis. We include ME, pituitary stalk, and posterior lobe in the definition of the neurohypophysis. Originally named for being a "growth or attachment underneath," the neurohypophysis has been variously defined as the posterior lobe or the posterior lobe and pituitary stalk. Both exclude the ME from its definition. However, other studies include the ME as an integral part of the neurohypophysis (e.g., Wittkowski et al., 1999). The similar embryonic origins and features of cells in the ME and posterior lobe lead us to use this broader definition. (B) Cells and cellular components of hypothalamic-neurohypophyseal interfaces. Magnocellular and parvocellular axons project to the neurohypophysis, terminating in the posterior lobe (magnocellular neurons, turquoise) or median eminence (parvocellular neurons, blue). Axon termini are in intimate contact with blood capillaries (red) and glial cells (green) composed of tanycytes in the ME and pituicytes in the posterior lobe. SON, supraoptic nucleus; AH, anterior hypothalamic nucleus; PVN, paraventricular nucleus; Aba, anterobasal nucleus; VMN, ventromedial nucleus; Arc, arcuate nucleus; MB, mammillary bodies; ME, median eminence; post. lobe, posterior lobe of the pituitary; OXT, oxytocin; AVP, arginine vasopressin; CRH, corticotropin-releasing hormone; TRH, thyrotropin-releasing hormone; GHRH, growth hormone-releasing hormone; DA, dopamine.

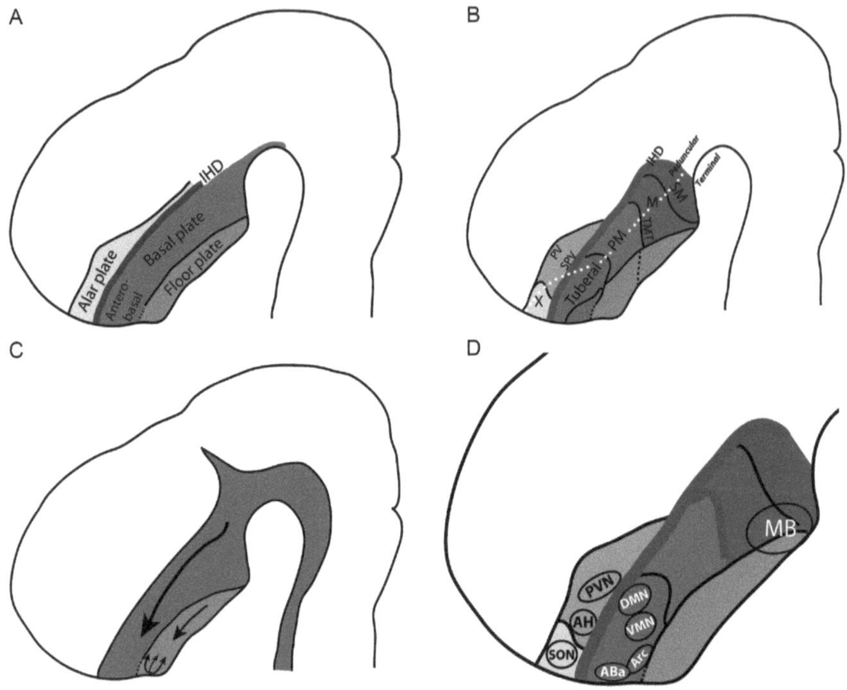

Caroline Alayne Pearson and Marysia Placzek, Figure 2.3 Combinatorial patterns of gene expression identify developing hypothalamic regions that prefigure adult nuclei. (A) The developing hypothalamus can be broadly divided along the dorsoventral axis into alar plate, basal plate, and floor plate. The alar and basal plates are separated by the intrahypothalamic diagonal (IHD). Each region can be identified by expression of a specific gene or combination of genes. Floor plate: *Six3* (high), *Shh* (transient), *Tbx1, 2,* and *3, Nkx2.1*. Basal plate: *Six3* (low), *Nkx2.1, Shh*. Alar/basal plate boundary: *Nkx2.2*. IHD: *Arx3, Gad67*. Alar plate: *Pax6, Sim1* (excluding rostral-most). (B) The developing hypothalamus can be further subdivided along the anterior–posterior (future rostrocaudal) axis by additional markers that define specific territories in the floor, basal, and alar plates. The alar plate is divided into two regions, an unnamed region (X) that expresses Nkx6.2 and the paraventricular and subparaventricular regions that express Pax6 and Sim1. Within the basal plate, there is a high degree of subregional heterogeneity in the early embryo, identified through particular markers. Tuberal: *Pomc/SF-1*. Premamillary: *Lef1*. Tuberomammillary terminal: *Arx*. Mammillary: *Foxb1*. Supramammillary: *Irx3*. The posterior IHD expresses *Lhx6*, whereas the anterior IHD expresses *Lhx8*. Note that anterior, tuberal, premammillary, and mammillary areas have constituent parts in the terminal hypothalamus and associated retro counterparts in the more caudal peduncular hypothalamus that are not included in this figure (see Morales-Delgado et al., 2011). Note also that cartoon should be treated as a "working model." Most of the transcription factors shown show highly dynamic profiles. (C) Fate-mapping studies show that populations of hypothalamic cells migrate along the rostrocaudal axis. Basal plate cells from a more caudal position undergo a rostral migration to intermingle with cells of the hypothalamic floor plate. A small subpopulation of rostral floor plate cells migrate caudolaterally to surround midline cells and form a collar (see Fig. 2.5). (D) Schematic shows general position of nascent hypothalamic nuclei and indicates their mapping onto embryonic regional territories. SON, supraoptic nucleus; AH, anterior hypothalamus; PVN, paraventricular nucleus; ABa, anterior basal nucleus; DMN, dorsomedial nucleus; VMN, ventromedial nucleus; Arc, arcuate nucleus; MB, mammillary bodies.

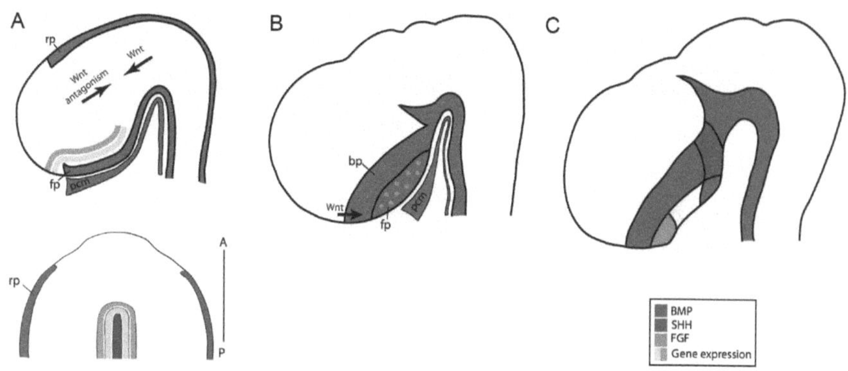

Caroline Alayne Pearson and Marysia Placzek, Figure 2.4 Dynamic expression of secreted signaling ligands at three successive times, all prior to formation of the infundibulum/neurohypophysis. (A) Side (top) or ventral (bottom) views illustrating early patterning events. Wnt antagonists promote anterior neural/hypothalamic fate by repressing Wnt activity (black arrows). Shh from the prechordal mesoderm/nascent hypothalamic floor plate is opposed by BMPs from the telencephalic roof plate. Shh from the prechordal mesoderm and hypothalamic floor plate establishes arcs of gene expression around the hypothalamic ventral midline. (B) The onset of BMP activity in the prechordal mesoderm leads to the downregulation of *Shh* and the upregulation of *Bmp* and *Fgf* in the hypothalamic floor plate. *Shh* is secondarily induced in the basal plate. Rostrally derived Wnt signals restrict the *Fgf/Bmp* expression domain. (C) *Fgf* and *Bmp* expression domains resolve, concomitant with the relative backward shift of the prechordal mesoderm. *Fgfs* become restricted to the tuberal region of the hypothalamic floor, where they will play a role in the development of the infundibulum/neurohypophysis. *Bmps* become restricted to the supramammillary hypothalamus. The *Shh*-expressing basal plate surrounds the hypothalamic floor plate. rp, roof plate; fp, floor plate; pcm, prechordal mesoderm; bp, basal plate.

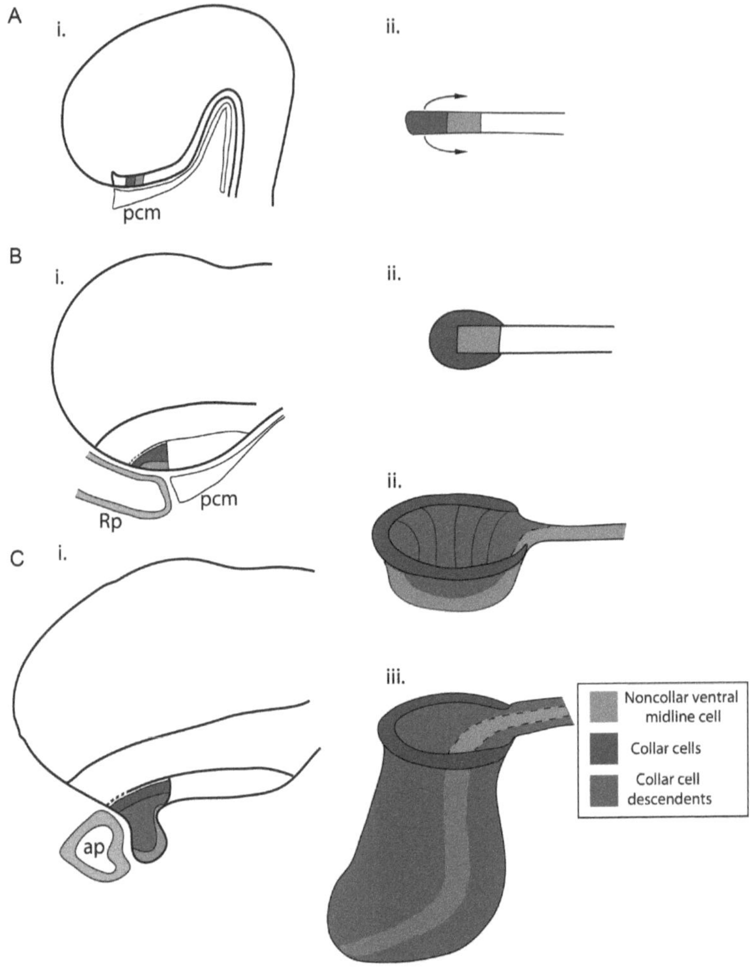

Caroline Alayne Pearson and Marysia Placzek, Figure 2.5 Multiple cell types form the infundibulum. The infundibulum is derived from at least three populations of cells termed collar cells, collar cell descendants, and noncollar rostral ventral midline cells. Collar cell descendants undergo a ventral migration, sculpting the infundibular walls. (A) Collar and noncollar rostral ventral midline cells are derived from two populations of floor plate cells directly adjacent to each other on the anterior–posterior axis. Lateral and dorsal views of the floor plate in the early hypothalamus (stage 10 in chick) indicate these two regions (collar cells in dark green and noncollar rostral ventral midline cells in light green) overlying the prechordal mesoderm. (i) is the lateral view and (ii) is the dorsal view. (B) Collar cells migrate caudally to surround the noncollar rostral ventral midline cells rostrally and laterally. During this time, the prechordal mesoderm retracts and part of the oral ectoderm, termed Rathke's pouch (in blue), comes to underlie collar cells and noncollar rostral ventral midline cells. Rathke's pouch is the presumptive anterior pituitary. (i) shows the lateral view and (ii) the dorsal view. (C) Collar cells show neural stem cell-like properties. Some appear to self-renew; others proliferate and give rise to collar cell descendants. These migrate ventrally toward the anterior pituitary (blue) and sculpt the forming infundibulum (medium green). Additionally, noncollar rostral ventral midline cells migrate ventrally. Collar cells remain as a band around the developing infundibulum and continue to give rise to descendants into late embryogenesis. (i) shows the lateral view; (ii) and (iii) show a three-dimensional view of the sculpting of the infundibulum (between HH st 24 and 32 in chick). pcm, prechordal mesoderm; Rp, Rathke's pouch; ap, anterior pituitary.

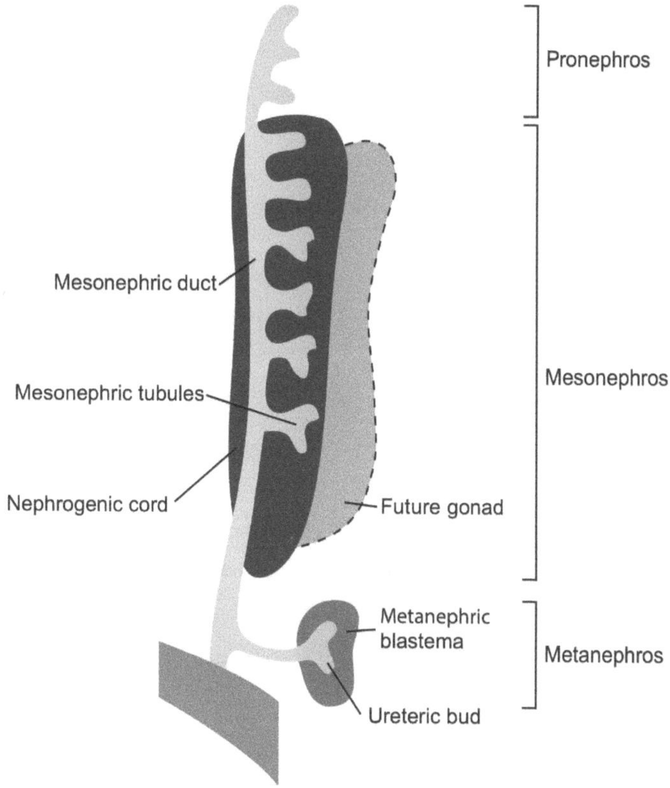

Dagmar Wilhelm *et al.*, Figure 3.1 Schematic representation of a urogenital ridge showing position of the pronephros, mesonephros, and metanephros. The gonads develop from a thickening on the ventromedial surface of the mesonephros (shown in cyan).

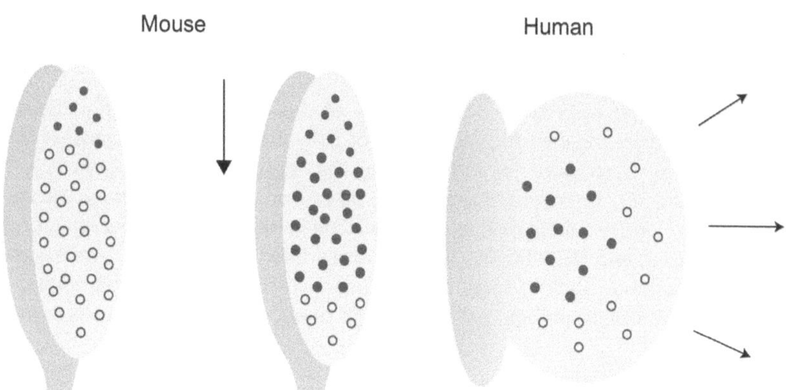

Dagmar Wilhelm *et al.*, Figure 3.2 Comparison of germ cell meiosis in mouse and human ovaries. In mouse, germ cells (indicated by the red circles) undergo meiosis (closed circles) in an anterior-to-posterior wave in response to retinoic acid (RA) produced by the mesonephros (shown in yellow). In humans, the ovary produces RA and germ cell meiosis is initiated in the medulla and spreads radially into the cortex.

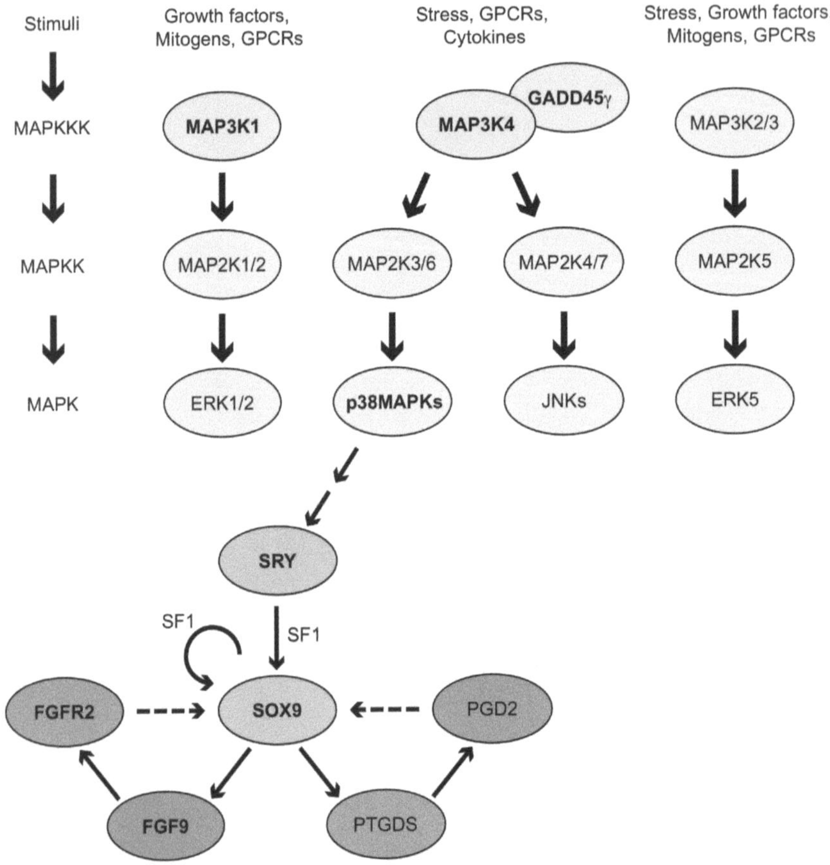

Dagmar Wilhelm *et al.*, Figure 3.3 Overview of the testis differentiation pathway. Recent data in mice indicate that activation of the mitogen-activated protein kinase (MAPK) pathway is required for induction of the testis-determining gene *Sry*. SRY, in concert with SF1, directly activates *Sox9* expression, which is maintained via an autoregulatory loop requiring SF1, and two positive feedback circuits involving prostaglandin D_2 (PGD_2) and FGF9/FGFR2. See text for additional details. Color code: kinases (yellow), transcription factors (green), signaling factors (blue), enzymes (pink), and receptors (purple). Genes in bold exhibit gonadal phenotypes when mutated in mouse and/or human.

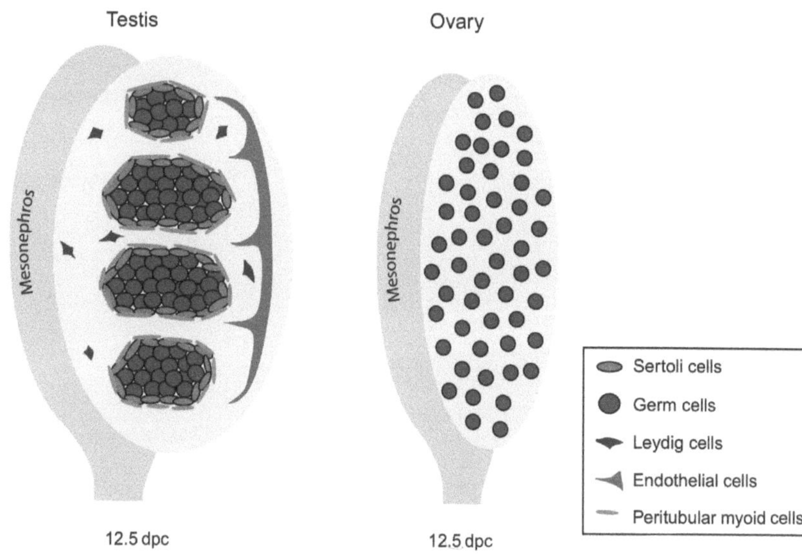

Dagmar Wilhelm *et al.*, Figure 3.4 Comparison of the developing murine testis (left) and ovary (right). By 12.5 dpc, approximately 48 h after the onset of *Sry* expression, the testis has undergone significant morphogenesis. Cords composed of clusters of primordial germ cells enclosed by a layer of Sertoli and peritubular myoid cells. Testis-specific vasculature such as the coelomic vessel has also begun to form and a population of fetal Leydig cells is present. In contrast, the ovary at this stage appears morphologically undifferentiated, containing germ cells and somatic precursor cell lineages.

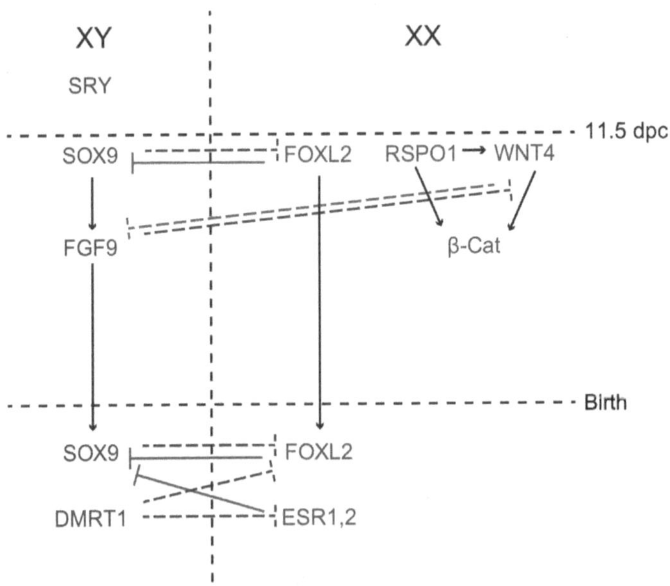

Dagmar Wilhelm *et al.*, Figure 3.5 Summary of key male-promoting (blue) and female-promoting (pink) factors and their regulatory relationship during embryogenesis and after birth. Mutual cross-repression of the transcription factors SOX9 and FOXL2 serves to canalize testis and ovarian differentiation during embryogenesis and maintain these phenotypes after birth. DMRT1 is also important for postnatal maintenance of the testes through repression of FOXL2 and ESR1/2. Antagonistic interactions between pro-ovarian canonical WNT signaling pathway components (RSPO1, WNT4, and β-catenin) and protestis FGF9 and SOX9 during embryogenesis also reinforce sex-specific gonadogenesis. See text for additional details.

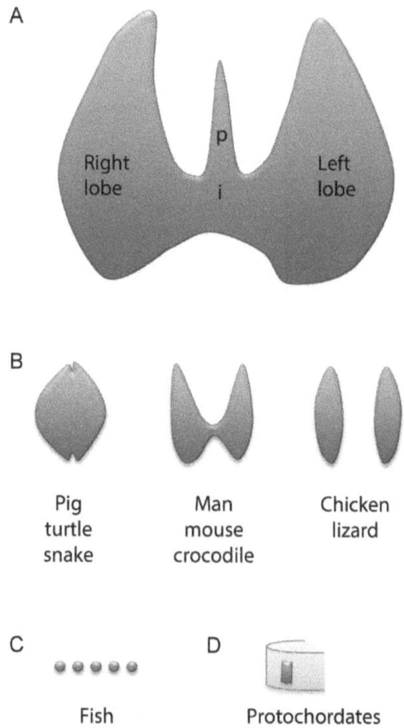

Mikael Nilsson and Henrik Fagman, Figure 4.1 Thyroid anatomy. (A) The bilobed human thyroid gland. p, pyramidal lobe; i, isthmus. (B) Shape variation of thyroid gland in (exemplified) different species. (C) Dispersed follicular thyroid in, for example, zebrafish. (D) Thyroid homolog in the wall of invertebrate endostyle.

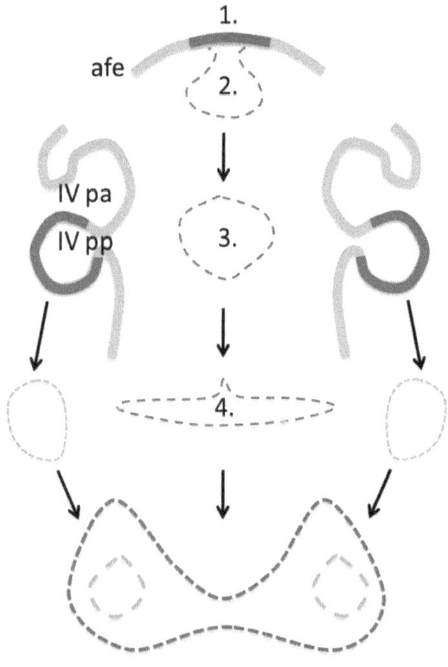

1. Placode formation
2. Budding
3. Migration
4. Bilobation

Mikael Nilsson and Henrik Fagman, Figure 4.2 Developmental stages of thyroid morphogenesis and organogenesis in mouse. Red (solid or encircled)—midline thyroid anlage and primordium. Green (solid or encircled)—ultimobranchial epithelium/body. afe, anterior foregut endoderm; pa, pharyngeal arch; pp, pharyngeal pouch.

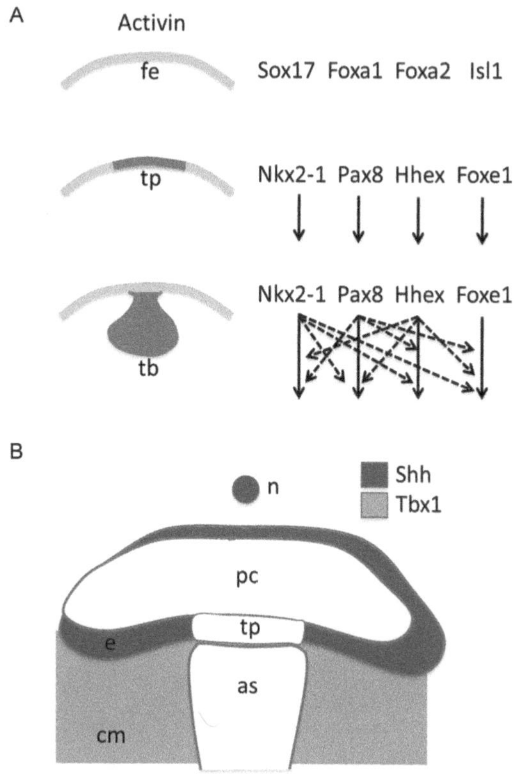

Mikael Nilsson and Henrik Fagman, Figure 4.3 Early thyroid development. (A) Transcriptional control of definitive endoderm formation as induced by Activin and regulatory role of Nkx2-1, Pax8, Hhex, and Foxe1 in thyroid progenitor cells. fe, foregut endoderm; tp, thyroid placode; tb, thyroid bud. (B) Expression pattern of sonic hedgehog (Shh) and Tbx1 implicated as non-cell-autonomous factors in early thyroid development. tp, thyroid placode; e, endoderm; pc, pharyngeal cavity; cm, cardiac mesoderm; as, aortic sac; n, notochord.

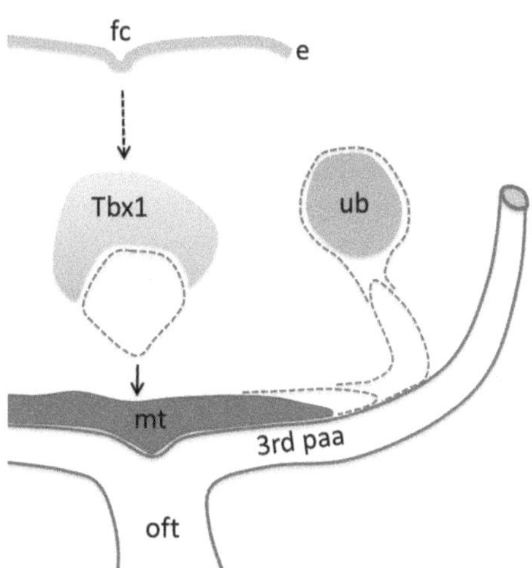

Mikael Nilsson and Henrik Fagman, Figure 4.4 Bilobation process. mt, midline thyroid; ub, ultimobranchial body; paa, pharyngeal arch artery; oft, cardiac outflow tract; e, endoderm; fc, foramen cecum.

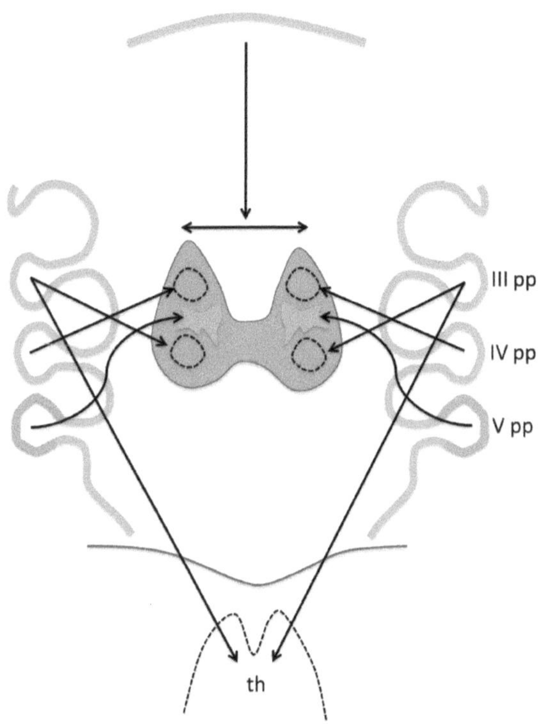

Mikael Nilsson and Henrik Fagman, Figure 4.5 Migration pathways of embryonic thyroid, ultimobranchial body, and pharyngeal pouch gland primordia. Parathyroid glands indicated by dashed circles. th, thymus, retrosternal.

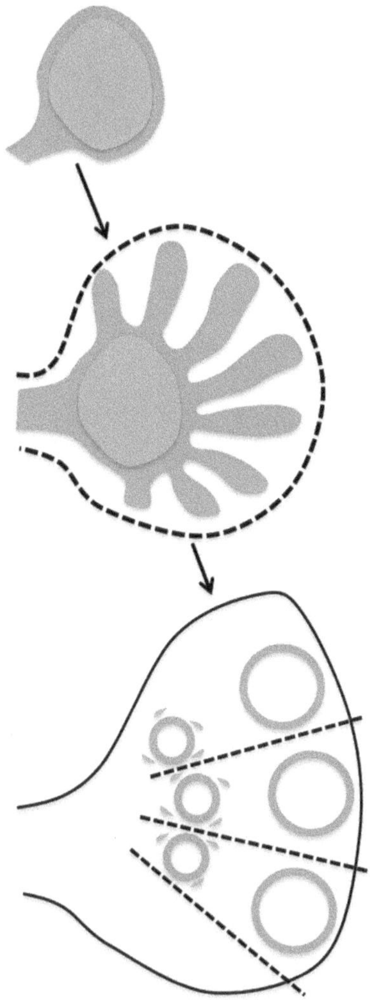

Mikael Nilsson and Henrik Fagman, Figure 4.6 Folliculogenesis based on model of parenchymal growth from embryonic thyroid in mouse. The ultimobranchial body remnant och thyroid C cells are indicated with green color. See text for further explanation.

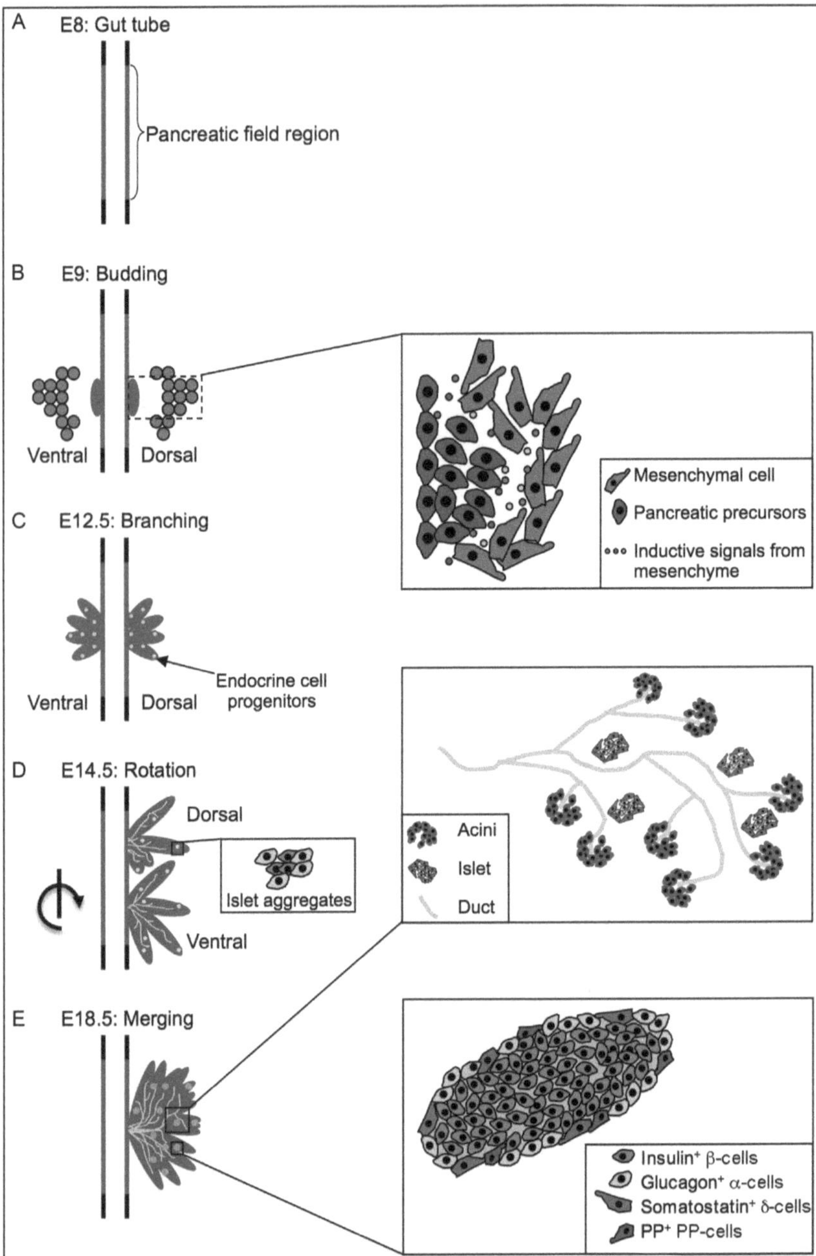

Fabio Avolio *et al.*, Figure 6.1 Schematic representation of the chronological progression of the embryonic mouse pancreatic morphogenesis. (A) Highlight of the pancreatic field of foregut tube. This region is composed of cells that will bulge from the foregut to give rise to the pancreas. (B) At ~E9, the dorsal and ventral foregut endoderm thickens and evaginates into the surrounding mesenchyme. (C) Active proliferation of pancreatic progenitors and formation of microlumens. The arrow outlines the emergence of endocrine precursor cells. (D) Rotation that starts at ~E14.5 and leads to the close contact of the ventral and dorsal counterparts prior to their fusion into a single organ. Subsequently, the first islet aggregates are detected. (E) Ventral and dorsal parts of the pancreas fuse in a single organ assuming the final organ structure. Concurrently, well-formed islets are detected within the exocrine population.

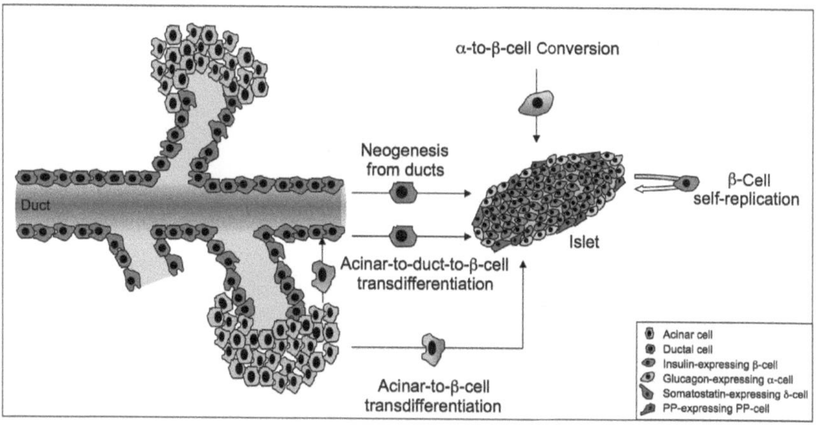

Fabio Avolio *et al.*, Figure 6.2 Different mechanisms leading to β-cell neogenesis. Several documented examples of reprogramming/regeneration induced by genetic- or injury-based approaches. Newly formed β-like cells could originate from the duct epithelium, endocrine cells, or acinar cells or be generated by self-replication.

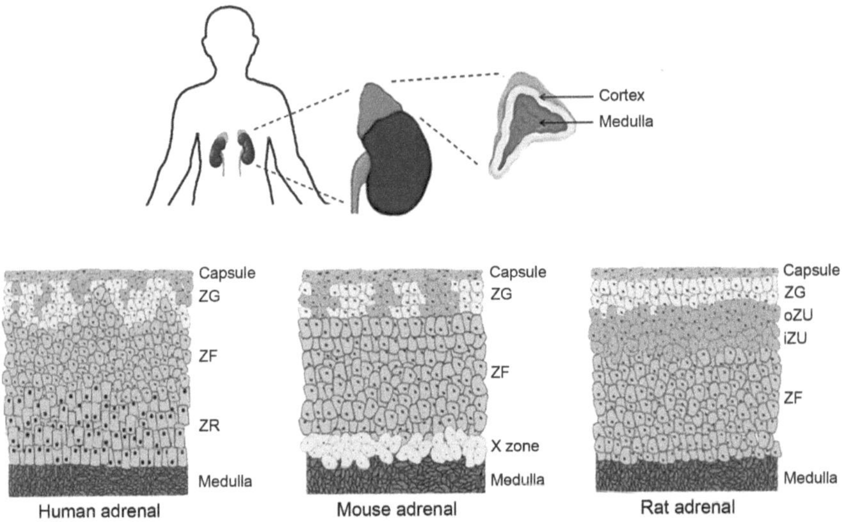

Rebecca Yates *et al.*, Figure 7.1 Anatomy and structure of the adrenal gland. (A) The adrenal glands are located at the upper poles of the kidneys. (B) Zonation differs between species. The adult human adrenal gland has three recognizable cortical zones, the ZG, ZF, and ZR, which surrounds the medulla. Clusters of CYP11B2-positive cells are observed in the ZG, separated from the CYP11B1-positive ZF cells, which can extend up to the capsule. SHH-expressing cells are also observed in clusters (pink), and by analogy with the mouse, these are depicted as separate from the CYP11B2 clusters, although this has not been directly proved. There is no sharp boundary between the ZF and the ZR. In the mouse, the CYP11B2-positive ZG and CYP11B1-positive ZF cells are in direct contact, with the CYP11B2-positive ZG cells arranged in clusters separated by clusters of SHH-positive, CYP11B2-negative cells (pink). The X zone (green) forms around the medulla following birth and persists until puberty in males or first pregnancy in females. In the rat, the zones are arranged as concentric shells with the undifferentiated zone (ZU) separating CYP11B2-positive cells from CYP11B1-positive cells. The ZU can be further divided into the Shh-positive, CYP11B2-negative outer ZU (oZU) and the inner ZU (iZU), which does not express Shh or CYP11B2. The DZ and TZ develop into the ZG and ZF, respectively, and the FZ regresses at birth to be replaced by the ZR in infancy. The islands of chromaffin cells coalesce to form the medulla after birth.

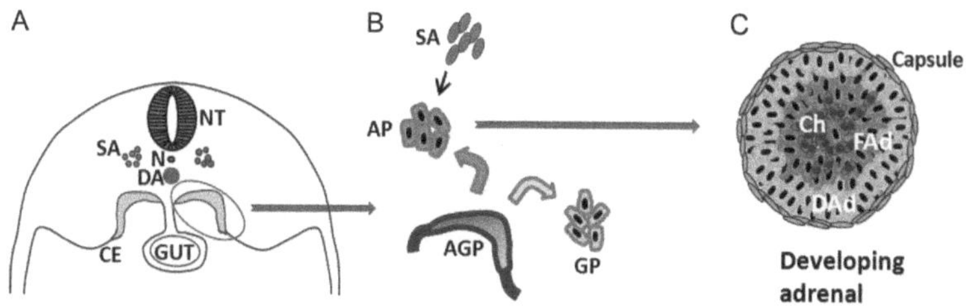

Rebecca Yates *et al.*, Figure 7.3 Development of the adrenal gland. (A) The adreno-gonadal primordium (AGP) develops as a thickening of the coelomic epithelium (CE) adjacent to the dorsal mesentery (DM). (B) These cells express SF-1 and the more medial population, which expresses higher levels of SF-1, delaminates into the mesenchyme to form the adrenal primordium (AP), while the cells adjacent to the mesonephros (M), expressing lower levels of SF-1, delaminate to form the gonadal primordium (GP). Migrating sympathoadrenal (SA) cells enter the AP. (C) The adrenal gland grows and surrounding mesenchymal cells form a capsule and the chromaffin (Ch) cells form the medulla. The AP cells form the fetal adrenal (FAd) gland and this is surrounded by the definitive adrenal gland (DAd). NT, neural tube; N, notochord; DA, dorsal aorta.

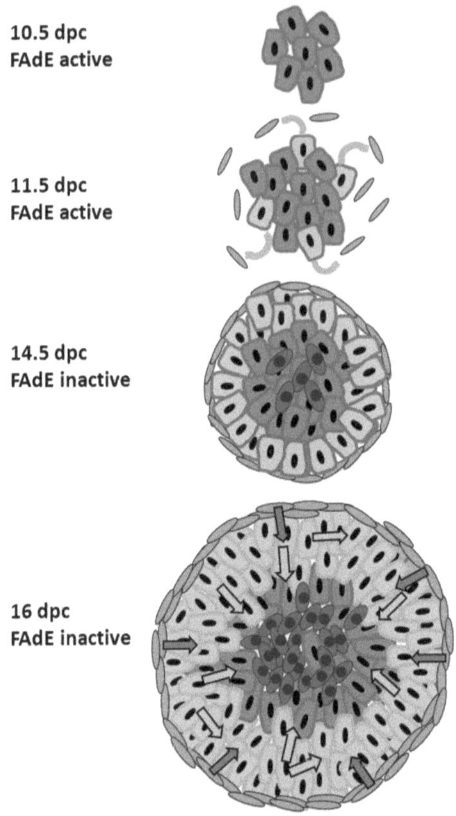

Rebecca Yates *et al.*, Figure 7.4 Lineages during adrenal development. SF-1 expression in the AP is driven by the FAdE from 10.5 dpc (dark cells). The AP grows and some cells (pink) express Shh from 11.5 dpc and signal to surrounding mesenchymal cells that express Gli1 and start to form the capsule (yellow). Some AP cells may enter the capsule and downregulate SF-1 at this point, although this is not proved. At around 14.5 dpc, the FAdE becomes inactive and SF-1 transcription is driven by the definitive enhancer from this point. Gli1-positive capsule cells enter the cortex and convert to an SF-1-positive identity in the steroidogenic lineage. At least some of these express Shh, and Shh-positive cells can differentiate into ZG and ZF cells of the definitive adrenal gland (light cells). Whether ZG and ZF cells can interconvert by transdifferentiation is not known at present.

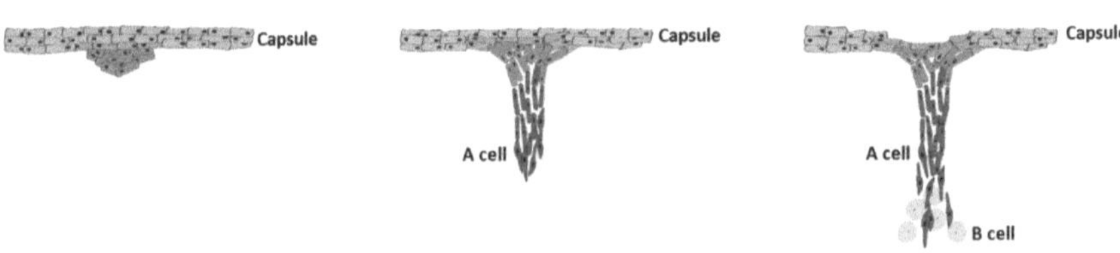

Rebecca Yates *et al.*, Figure 7.5 Spindle-cell hyperplasia. Spindle-cell hyperplasia develops in several mouse models. Spindle-shaped basophilic cells are observed near the capsule and extend into the cortex over time. These A cells are nonsteroidogenic but express Gata4, unlike the surrounding steroidogenic cells. They have also been shown to express Gli1 and Pod1, suggesting that they are capsular cells that have not differentiated down the correct adrenal lineage. A cells have also been shown to express Amhr2, which, like Gata4, is a marker of gonadal cells. In some models, B cells appear following gonadectomy as a consequence of the increase in LH. These are steroidogenic but resemble gonadal rather than adrenal cells, expressing CYP17, LHCGR, and gremlin.

FSC
www.fsc.org
MIX
Papier | Fördert gute Waldnutzung
FSC® C083411

Kolibri 360 GmbH
Ferdinand-Jühlke-Straße 7
99095 Erfurt, Deutschland
produktsicherheit@kolibri360.de